ATLAS OF HUMAN
ANATOMY

ATLAS OF HUMAN
ANATOMY

Thanks to Dr Giovanni Iazzetti, Department of Genetics, General and
Molecular Biology of the Università Federico II of Naples,
Dr. Enrico Rigutti, orthopedic and traumatology specialist
for their precious collaboration.

Ichonographic references
Page 23 on the to right:....
Where it is not expressly indicated, the images come from the Giunti
Ichonographic Archive.
As for the rights of reproduction, the Editor is fully available to settle
any possible fee due for those images whose source it was impossible
to retrieve.

This edition produced by
TAJ BOOKS LTD
27 Ferndown Gardens
Cobham
Surrey
KT11 2BH

info@tajbooks.com

SUMMARY

A CLOSER LOOK

MEDICAL PAGES

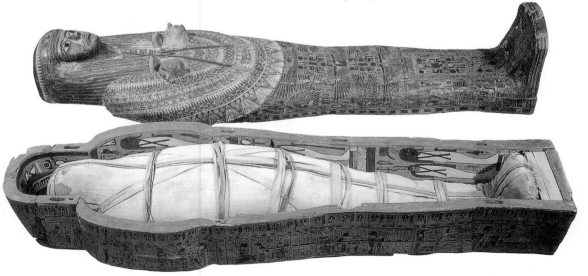

SINCE HUMANITY HAS STARTED TO INVESTIGATE THE WORLD, HUMAN ANATOMY HAS REMAINED CLOSELY LINKED TO MEDICINE AND SURGERY: THE TREATMENT OF DEEP WOUNDS, OF BROKEN BONES, OR OF A DIFFICULT BIRTH, HAVE ALL IMPROVED OUR KNOWLEDGE OF THE STRUCTURE OF OUR BODY.

DISCOVERING THE HUMAN BODY

THE ORIGINS

Even the most primitive cultures must have had some level of knowledge about anatomy: cannibalism and slaughtering of animals offered precise information regarding the main elements of the human body, long before more advanced civilizations would develop. It is certain that long before the Egyptians developed their complex embalming techniques, or the Chinese created their complex acupuncture methodologies, or Indians sharpened their surgical techniques (the first examples of nasal reconstructive surgery dates back to India, around 2000 years ago), humanity had already acquired a fair amount of knowledge about the human body. However, written testimonials are scarce; but this shouldn't surprise us: in ancient times knowledge was passed down from a shaman to his apprentice, from a scholar to his disciples, and medical notions were mixed in with magical rituals and religious beliefs, accessible only to few initiated people. It is improbable that this type of culture would leave any written evidence. Thanks to the Greeks, human anatomy established itself as a true science: in 700 BC the first medical school opened in Cnido. Here, religious traditions, linked to the cult of Asclepio, were completely abandoned and replaced by pure observation of sick patients. Alcmeone, the author of the first anatomical work, was one among those who worked at this school. Unfortunately only few fragments of his masterpiece remain intact. One hundred years later, Hippocrates, "the father of medicine", founded his medical school in Kos, a place destined to remain famous for thousands of years. However, anatomy developed most cogently at the medical school of Alessandria,

▶ Embalming: The embalming of a corpse involved the removal of the viscera (abdominal and thoracic) in one block, the removal of the brain, through a nostril, and the "stuffing" of the corpse with a rich compound of herbs and ointments. Finally, once reclosed, the body was rubbed with special ointments and wrapped into bandages. The mummies, closed and protected inside their precious sarcophaguses, have survived for thousands of years.

▲ Acupuncture:
Chinese acupuncture could have only developed as a result of an exceptional knowledge of human anatomy. In fact, it is assumed that the doctor would know with a high degree of accuracy where the main plexuses and nerve endings are located in the body.

in Egypt, under Macedonian rule (after 334 B.C). The discipline reached its apex during the III century B.C. This was the time of Europhiles of Calcedonia and Erasistrato of Keos: rumour has it that Europhiles was the first to conduct the public dissection of corpses. Thanks to his studies, he was able to give us an accurate description of the nervous system and of the way the liver and the small intestine are connected. He was the first one to discover that the brain is at the center of all mental functions, to distinguish between motor nerves and sensory nerves, to demonstrate that all nerves are linked to a single central system. He gave the name "duodenum" to the first part of the small intestine, and established that it is connected to the liver through large veins. Erasistratus discovered the great bile duct and it was he who analyzed blood flow in the liver, highlighting the parallel flow of hepatic veins and biliar capillaries. However, both Erasistrato and Europhiles believed that air ran inside arteries: this belief resulted from their work with corpses and the observation that arteries, unlike veins, could empty themselves.

IN ROME

While the Greeks were very interested in medicine, for the Romans it was unthinkable to "waste time" with such an unimportant discipline: in Rome, almost all the doctors came

▼ Miraculous surgery:
One of the most famous miracles of the saints Cosma and Damiano was a leg transplant affected by cancer; they substituted it with an healthy leg which belonged to a Moore who had just died (black)
Alonso de Sedano, about 1500, Wellcome Institute, London

THE ARABS, THE CHURCH AND THE MEDIEVAL ANATOMICAL STUDIES

Fortunately, the knowledge acquired in the school of medicine in Alessandria was not lost: in 642 the city fell in the hands of the Arabs who assured that its scientific knowledge would not fall into oblivion. In fact it is not by chance that the two saints Cosma and Damian, who had become patrons of surgery thanks to their miraculous interventions, were two Arab twins, who had converted to Christianity.
Thanks to the Arabs, past techniques and scientific knowledge emerged again in Europe during the middle Ages. At the end of the IX century, the Salernitan medi-

cal school became famous primarily because of the impulse given by Constantine the African. Originally from Carthage, he had learned Arabian, had studied the medical texts of Alexandria, had collected and translated into Latin all of the texts that he was able to find during his travels in the Orient. Consequently, Hippocrates ideas and the anatomical notions of the Alessandria's school filtered back into the Western culture. As a result, during the XII century, the first surgery textbook was drafted in Salerno. The book was entitled Chirurgia Maestri Rogeri and

Galen, natif de Pergame ville d'Asie, excellent Medecin vivoit du temps des Empereurs Antonin le Philosophe et de Commodus, on tient qu'il a vescu 140 ans.

▶ **Galen, in an etching of the XV century**
Undoubtedly a great surgeon, Galen had a strong personality and was very popular in Rome: he constantly advertised himself by holding lectures, participating in public debates and offering public demonstrations of his medical skills. He wrote this about himself: "I did for medicine what Trajan did for the Roman Empire, building roads and bridges in all Italy. I was the only one who revealed the true path of this science. However, I must admit that even if Hippocrates had already started to mark this path, he never dared to go where he could have gone: his writings are not without a flaw and they lack some basic instructions. Furthermore, his knowledge of some topics is insufficient, and often he is not very clear, just like older people generally tend to be. In summary, he has prepared the path, but it is I who has made it accessible

from abroad. However, a certain degree of anatomical knowledge must have been present among the people: in fact, in the pre-roman era, numerous clay "ex voto" representing different parts of the body (among which the uterus) were found inside temples, where people would ask for divine assistance. Such figurative skill could have only derived from a direct knowledge of anatomy, acquired through surgical procedures (such as cesarian delivery, quite common in those days) or through the dissection of corpses. Soon, though, the use of the human body for anatomical investigations was prohibited, mainly for religious reasons, and animals replaced human bodies. It is during this time of progressive medical-anatomical regression, that the great Galen lived. Born in Pergamo in 130 A.D, he lived in Rome right up to the year 200. He dedicated himself in the cure of those gladiators who would get wounded and injured after a fierce competition. In a short time he became so popular that the emperor Marcus Aurelius appointed him as his personal physician. Using his imagination, Galen applied his knowledge acquired with animals on human beings; that's why he believed that the heart had two cavities, that the brain would rhythmically pump the psychic essence into the body through the nerves, and that the intestine was long, so that one wouldn't have to eat continually. Galen left his descendants a lot of his written work. Starting from the knowledge acquired at the School of Alessandria, his writings were filled with extravagant "observations" and interpretations, which in rea-

soon it became well known in all Europe.
However the Church became one of the main reasons why anatomical knowledge couldn't develop further. In fact, in 1215, Pope Innocent III felt the need to officially discourage any activity on the human body, surgery included, issuing the encyclical *Ecclesia abhorret a sanguine* (the Church is horrified by blood). Only those schools and universities closely connected to ecclesiastic environments, such as those born in connection with cathedrals and rich monasteries (Cambridge, Monpellier, Padua, Bologna, Paris) could benefit from some privileges in this field, and could widen the medical science without risking to be excommunicated. AS a result, within the end of the XIII century, Montpellier had become the

most important European center for studying medicine: close to Italy and to the most "advanced' schools, but also to the Arabian Spain, this school was highly regarded by the Church. In 1350, Pope John XXII himself honored it with a silver banner", as a symbol of the glory, which radiates from the studies conducted there". A more significant step toward scientific knowledge was the permission granted by The Duke of Angio to the members of the school, to conduct each year the dissection of the corpse of a person condemned to death.
During this time of intellectual ferment and of growing interest in different aspects of the human body, Mondino de'Liuzzi wrote the book *Anatomia corporis humani*, a manual on how to dissect corpses. Although it was based on direct and practical

experience, it didn't overcome the errors made by Galen. For a long time this textbook remained one of the most influential works on human anatomy and was used by all medical students. During the XV century illustrations were added: they depicted only details of some dissecting techniques, since the interpretation of the structures done by Mondino (and therefore by Galen) did not correspond at all to reality.

◀ **A lecture at Montpellier:**
Henri de Mondeville was one of the leading French surgeons during the XIII century. He is depicted here while he is lecturing to his students. Even though he admitted "today we know many things which were unknown during Galen's times" even this great French surgeon accepted most of Galen's teachings.

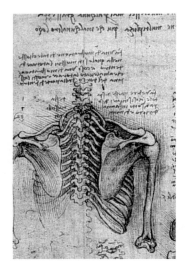

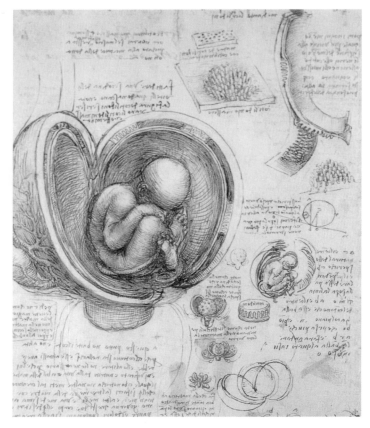

▲▶ Leonardo's autographs:
Leonardo Da Vinci was the first true scientific sketcher in history. In particular, his studies of the human body, hidden for many years for fear of a conviction by the ecclesiastic authorities, show his deep interest for the study of the "live" anatomical relationship between organs, muscles and the skeleton. In some cases, thanks to some "additional" drawings, we can sense his interpretation of how they functioned.

lity falsified the anatomical knowledge, based on the direct observation of the human body. In spite of that, for the next few centuries, his work was held in such high regard that any opinion, deviating from his ideas, was considered heretical.

THE RENAISSANCE

It is only during the Renaissance that the need to return to the classical sources of knowledge and inspiration enabled dogmatic prejudices, which had so far paralyzed the anatomical research, to be overcome. The artists of the 15th century became the representatives of a new way of interpreting reality, more faithful to objective observations than ideal standards. They played an important role in stimulating and developing a more modern way of studying anatomy. The same need for a direct, realistic and concrete contact with nature, which would later stimulate Galileo to formalize his revolutionary scientific method, would also inspire Pollaiuolo, Verrocchio and Leonardo Da Vinci, to embark in forbidden studies. They disinterred corpses, secretly dissected them, and reproduced, as accurately as possible, all of the features and details of the human body. Free from any prejudice, Leonardo Da Vinci was one of the first to become interested in the anatomical structures of the human body. His extensive research and systematic observations made him the first person in the world to elaborate an extremely detailed iconographic representation of the

The title page of Andrea Vesalio's master-piece, clearly depicts the innovation brought to education: around the dissecting table gathers not only Vesalio himself (on the left), but also his apprentices (older ones) and the students, who follow the anatomy lecture from everywhere, even the furthest steps.

male and the female body. Unfortunately, the potential "blasphemous" nature of his drawings caused them to remain hidden for a long time, and unknown to his contemporaries: almost a personal reflection of the artist himself on the truths of nature.

THE REVOLUTION OF ANDREA VESALIO

The time was ripe. Andrea Vesalio's splendid anatomy textbook, entitled De humani corporis fabrica, revolutioned the anatomical science and surgery.

Born in Bruxelles in 1514, he studied Hebrew, Arabian Greek and Latin, acquiring the opportunity to access the most remote and "pure" sources of scientific learning. After a period in Paris, he opened an Institute of anatomy in Loviano. Here he conducted the dissection of dead criminals. In 1537 he became professor of anatomy at the University of Padua. His students were enthusiastic and fascinated by his innovative and different teaching styles. He no longer lectured from his desk, simply analyzing Galen's writings. No more approximate demonstrations of various organs cut and displayed by apprentices. Vesalio did these dissections himself, and pushed his student to become directly involved in their personal quest for knowledge and to constantly challenge themselves. He was the first to place the anatomical table in the center of the classroom, and to show students the various organs. During his extensive stay in Padua, Andrea Vesalio collected the material for his masterpiece: for the first time in history, the most accurate anatomical descriptions were accompanied by extremely detailed illustrations. The creator of such illustrations was Jan Stephan Van Calcar, student of Titian and friend of Vesalio. Such illustrations were impressive and extremely beautiful. Published in Basil in 1543, Vesilio's work shook the academic environment unleashing a fierce dispute between the most famous doctors of those days, the majority of which supported Galen and not Vesalio. But time gave reason to the truth. Vesalio's work was so accurate and realistic that the corrections made later by the experts, were only marginal. Still today his work is very valid, and can be used in the study of human anatomy. Vesalio's work opened what some have described, the "century of anatomy", in which the Italian scientists had a prominent role. In few years, many eminent scholars such as Realdo Colombo, Bartolomeo Eustachio, Gabriele Falloppio, and Girolamo Fabrici d'Acquapendente, author, among other things, of a color anatomical illustration atlas (the Tabulae aatomicae), forerunner of some of the most advanced anatomical works, worked at the university of Padua, which had become the center of the anatomical knowledge.
The number of the most famous Italian (in Padua and

▼ **Portrait of Andrea Vesalio**
Andrea Vesalio when he taught at the University of Padua (1537-1544) Vesalio died in Zante, in 1554, on his way back from a pilgrimage in the Holy Land, which he conducted probably to "expiate" an homicide: having received permission to dissect the body of a Spanish nobleman who had died under his care, he removed his heart while it was still beating. He was brought in front of the Inquisition by the relatives of the deceased, but was saved by Philip II of Spain, since he was his personal physician

ANATOMICAL PLATES

Many of the biggest Plates illustrating Andrea Vesalio's work, do not simply reproduce the anatomical elements in a sterile way, as they would appear on a dissecting table, but they illustrate the body's muscular system, the skeletal system and all the organs, as if they were still alive and with facial expressions. So, as the Plates on the muscular system depict bodies in pain but serene, the Plates on the skeletal system appear somewhat ironic, resembling probably the medieval images of the Danses Macabre. The characters are framed in typically soft renaissance landscapes, becoming therefore active elements of the book, even if only made of muscles and bones. They are 'alive" almost as if they were trying to show themselves the progress of science, reminding us at the same time, about the imminence of death, and its usefulness for life

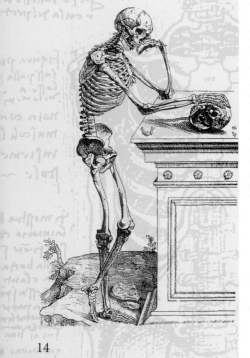

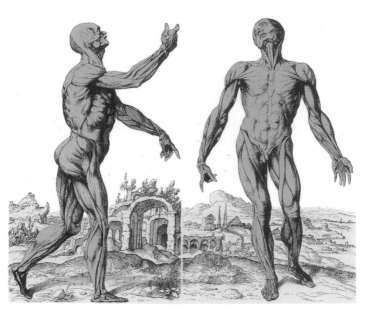

Rome), French (in Montpellier and in Paris) and German (in Basil) scholars, who specialized in anatomy, grew rapidly, allowing for the knowledge of the human body to become more and more detailed. But there was still a lot more to be discovered: how were the various tissues organized? How did the organs function? What was the physiology of different anatomical systems? These answers could not come solely from descriptions: as a result, the anatomical research discovered the microscope, and entered the XIX century.

TOWARD MODERN ANATOMY

The first people to study the human body through the lenses of the optical microscope were Antoni van Leewenhoek, its creator, and Marcello Malpighi. They described blood cells, the corpuscles present in the lining of the skin, the microstructures of the spleen, the kidney's glomeruli, the pulmonary alveoli, and the germinal cells…. In the X-VII century many more followed, and soon the technique of dyeing tissues allowed for the discovery of new structures, unknown until then. This widened the field of anatomical investigation. In the meantime, by linking the interpretation of symptoms and clinical interventions to the anatomical knowledge of healthy and sick organs, Giovan Battista Morgagni …..

SPECIALIZATION

Today's anatomy is divided into different "branches", speciali-

▲ Portrait of Giovan Battista Morgagni (1682-1771)
Professor of Anatomy at the University of Padua from 1711 until his death, he introduced anatomical concepts into pathology, discovering the connection between pathological anatomy and clinical problems

zing in very different aspects of the structure of the human body.

SYSTEMATIC NORMAL HUMAN ANATOMY (or descriptive) is the oldest and it analyzes the structure, the shape, the relationship and the development of different organs. It is divided into:

MACROSCOPIC: the elements are visible with bare eyes.

MICROSCOPIC: histological methodologies are used to describe the microstructures of various organs

TOPOGRAPHIC ANATOMY: studies the organs based on their location in the body. The body is divided into territories, regions and layers (from the most superficial to the deepest).

SURGICAL ANATOMY: studies the anatomical problems relative to those diseases engaged by surgery, their symptoms and the different surgical techniques.

PATHOLOGICAL ANATOMY: utilizing autopsy as the main investigative method, this discipline studies the macroscopic and microscopic changes in the organs due to diseases.

RADIOGRAPHIC ANATOMY: deals with the nomenclature and the healthy parts of the human body, as they appear in radiography. This discipline specifies the unique characteristics acquired by tissues consequently to overlapping, their projection on a screen and their different radiological density.

ARTISTICAL ANATOMY: it studies the external shape of the body, the proportions between the different parts, the organs directly visible and their external modifications, due to different postures and movement.

▶ Optical microscope
At the beginning of the Twentieth century, instruments similar to this one (latter part Nineteenth Century), enabled famous scholars such as Malpighi and Golgi to identify most of the cellular structures of our body, recognizing their primary functions

MODERN TECHNOLOGIES IN THE STUDY OF ANATOMY

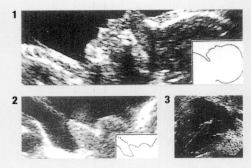

◀ **Ecography of a fetus:** The various details clearly show the contours of the head (1) and of the trunk (2). During the examination, by moving the sound wave generator, it is possible to highlight every time the different internal organs: in this case (3) the cardiac cavities.

As with all scientific disciplines, thanks to the development of new and more and more precise investigating techniques, even anatomy has taken some major steps forward. Let's take a look at the main ones, what they are, and when and how they are used.

AUTORADIOGRAPHY OR RADIOAUTOGRAPHY

What it is: it is a photographic technique that offers a detailed reproduction of an organ. Radioactive substances (radioisotopes) are injected into a vein: this enables us to look at the targeted organ using an instrument, which measures different radiation intensities.

BIOPSY

What it is: a small fragment of a live tissue is surgically collected. Depending on the type of tissue and on the type of collection, it can be removed with a surgical knife, a tenaculum or a syringe. After being treated with histological technique, it is then analyzed under a microscope. Even the analysis of a simple blood specimen can be considered to be a biopsy.
How it is used: it allows us to look at qualitative and quantitative characteristics of cells and tissues.
When it is used: It is primarily used in the early detection and diagnosis of pre-surgical tumors, but also to confirm or rule out the suspicion of the celiac disease in children, to detect various

organs' diseases (lungs, colon, spleen, pancreas, thyroid, mammary glands, prostate…) excluding the presence of any cancer formation, to clarify the status of the endometrium in case of female sterility, or to confirm, in more specific cases, the diagnosis of cirrhosis of the liver.

ECOGRAPHY

What it is: it is a technique that utilizes an ultrasound to visualize organs. A generator of high frequency waves is placed on the area of the body to be examined; the waves propagate deep into the organ from which they are then reflected: complex equipment is able to convert echoes from sound signals into images, which are then viewable on a monitor. The examiner can pho-

tograph the most important details (in order to register the clinical data), and make an initial diagnosis.
How it is used: to observe the structure of the organs and the presence on any unusual change (cysts, nodules, deformations, presence of fluids, etc…).
When it is used: mainly to verify the organs of the abdominal cavity (liver, pancreas, bladder, female reproductive apparatus, kidneys, intestine…), of the thoracic cavity (lungs, mammary glands, heart) and of the neck; it has also become a routine procedure in the diagnosis of extra-uterine and multiple pregnan-

cies, and for checking the development of the fetus.

ENDOSCOPY

What it is: it is an inspection of a non-directly accessible body cavity (for example the frontal and nasal sinuses, the esophagus, and the stomach, the trachea and the bronchi, the bladder, the intestine and the uterus….). Such inspection is possible thanks to the use of particular instruments equipped with a lighting system and different lenses. They can also enlarge images, allowing for a clearer view of even the smallest detail. Today, thanks to optical fibers and to the miracle of miniaturization, it is possible to examine and to videotape body cavities, reaching them through both a small surgical operation (cardiac cavities, blood vessels, peritoneum…), and directly (for example swallowing a micro-video camera).
When it is used: it is used to look at the internal structure of the hollow organs and the 'external" condition of those organs facing a cavity of the body.
How it is used: besides being an essential instrument for the detection of various diseases (inflammations, cysts, tumors….), this technique is currently used side by side with surgery and with bioptic investigations. Small surgical operations such as the removal of uterine fibroids, of nasal or intestinal

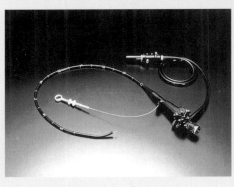

▲ **Endoscopy of the trachea:** This image was created by placing a micro camera inside the trachea: the inhaled smoke is entering the bronchi

◀ **Fibro scope:** This instrument allows the upper gastro intestinal tract to be examined: it is over a meter long and it contains over 40.000 optical fibers

polyps, of the medial or lateral meniscus of the knee, of foreign bodies in the airways, as well as the biopsies of internal organs and of abdominal cavities, are generally done with the help of specifically designed endoscopes. In some cases, endoscopy is also used to look at the fetus, and particularly to conduct prenatal operations.

RADIOGRAPHY

What it is: It is the "oldest" among the modern methods of anatomical investigation, which allows us to obtain information about the internal part of the body without the need of any surgical operation. This technique utilizes a beam of x-rays generated from high voltage electricity. Such beam goes through a vacuum tube and reaches molybdenum or a tungsten plate. This beam of x-rays is then directed on to the interested region: the body absorbs the radiation in a differentiated way, depending on

whether it goes through a "soft" organ, a bone tissue, or a cavity…Once it leaves the body, the beam ends up on a plate, a film, or another receptor able to highlight the intensity of the radiation. For example, if the x-ray beam is directed onto an arm, the rays going through only the muscle tissue, are absorbed less (they have less radiological density) and they darken the plate much more than those rays which have also passed through bone tissues. In fact, such tissues are much less "transparent" to X-rays (they have higher radiological density). The result is a "negative" image of the internal parts of the arm: dark for low radiological density (muscles), and light for high radiological density (bones). "Contrast mediums' must be used for tissues that do not absorb very many x-rays. These contrast mediums are substances that appear opaque or semi transparent to radiation. Once introduced into the tis-

sue or the targeted organ, they create an artificial contrast with the surrounding tissues, showing the interested structure. Adequate and specific contrast means are required to examine the heart, arteries and veins, the digestive tract, the respiratory tree, the lymphatic system, the gallbladder, the gallstones, the urinary tract, the kidney and bladder stones and of the spinal column. Since x-rays are very high energy electromagnetic waves, and can cause some permanent damage not only to tissues and organs, but also to the genetic makeup, today, the radiological equipment is built according to the highest radio-protective criteria, making the investigative process quite harmless and safe for the patient. Despite that, whomever is exposed to x-rays, either for therapeutic reasons or for work must be particularly careful not to receive high doses of

radiation. It is for this reason that the individual who manages the x-ray equipment does so from behind a protective screen; in fact the damage caused by x-rays are cumulative and continuous exposure even to the lowest dose can have long-term effects.

What we see: TX-rays allow us to see the deep structure and shape of tissues and organs, the presence (normal or abnormal) of air, of fluids or foreign objects, with a radiographic density different from that of bodily structures.

When it is used: It is used mainly in the orthopedic field, with traumas, to detect bone and joints problems, and in routine examinations for the prevention of breast tumors (mammography); it is also very helpful in the detection of foreign bodies, swellings or thickening of tissues, caused by inflammations or tumors, and also to check for the size and the functionality of the heart (angiocardiography) etc.

ANGIOCARDIOGRAPHY

What it is: It is a particular type of cardiac radiography. After injecting the contrast mediums, a

◀ **Marie Curie in her physics-chemistry laboratory**
Marie Curie created the first radiographic equipment used in medicine. After having mounted it on a car, she traveled along the eastern French side during WWI, allowing surgeons, for the first time, to rapidly and safely detect fragments of shells and bone fractures. For this type commitment during the war, she became the first person to receive two Nobel Prizes for her contribution to the knowledge of radioactive substances. She also received the French Legion of Honour.

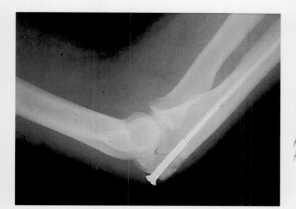

▲ **Radiography**
The contribution of radiography to the orthopedic field is essential. The image shows the result of a surgical procedure to strengthen the bone of an arm, by inserting a metal post

▶ **One of the first radiographies:**
It dates back to 1890 and was conducted by Wilhelm Conrad Röntgen, Nobel Prize for chemistry in 1901 for the discovery of X-rays

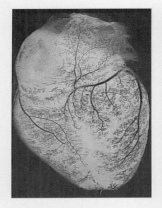

▲ An angiocardiography:
It highlights the outlines of the coronary arteries, the vessels that bring the oxygenated blood to the cardiac muscle

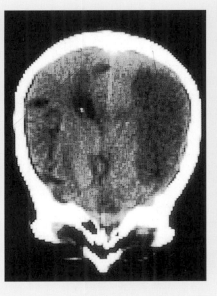

◄ NMR of a skull:
This technique allows us to highlight brain lesions, which are invisible through a radiographic examination. In this image, some dark spots are visible; they correspond to an ictus (central left area) and to damages areas (on the right side).

► Scintigraphy of a thyroid:
After being re-elaborated by a computer, the data is transformed into false dyes: the red areas correspond to a higher emission of radioactivity. In this case the thyroid is enlarged: the increased radiation is caused by the thickness of the gland

number of radiographies are shot in sequence. This, for example, allows us to see if the coronary arteries are obstructed or if the cardiac cavities are functioning normally. Usually, the angiocardiography is conducted before a surgical procedure.

RÖNTGENCINEMATOGRAPHY
What it is: Attaching a light amplifier (an instrument that allows to register an x-ray image by increasing its brightness) to a radiographic instrument, it is possible to obtain "x-rays" films. This allows us to collect important information on how certain organs function and on potential anomalies of the movement.

COMPUTERIZED TOMOGRAPHY (CT) better known as COMPUTERIZED AXIAL TOMOGRAPHY
What it is: This is a very expensive technique. Through a large number of steps utilizing x-rays, this technique mathematically re-elaborates the collected data, recreating the radiological image of transverse layers of the body. The patient is placed on a special table and in few minutes is bombarded by hundreds of radio-

graphies. After being elaborated, they will offer a complete anatomical picture.
What we can see: It allows us to see the smallest structures that have different radiological opacity> the se structures are normally not visible during a traditional radiological examination.
When it is used: Due to the high cost and the intensity of radiations absorbed by the patient, this technique is used only in very special cases, mainly to diagnose cancer in organs that are generally difficult to examine (pancreas, liver, brain).

NUCLEAR MAGNETIC RESONANCE (NMR) or simply MAGNETIC RESONANCE

What it is: It is an investigative technique that allows us to obtain exceptionally precise and incredibly detailed images, without the use of contrast mediums. The body is placed under a magnetic field; the atoms of each organ release extremely low wavelength radiations (radio frequencies). At this point very complex computerized equipment interconverts the released signals into images. Different areas of the body that have different emissions are colo-

red differently.
What we can see: It allows us to get a very detailed view of organs or tissues that are not easily accessible or that are very large (for example the circulatory or the lymphatic network).
When it is used: It is primarily useful in checking for the finest structure of organs such as the brain, the viscera, the circulatory and lymphatic systems, the muscles and the bones.

SCINTIGRAPHY

What it is: A radioactive substance (isotope) is injected into a vein. The radiation emitted by the radioisotope is then recorded. For example, the injection of a small dose of radioactive iodine rapidly accumulates into the thyroid. Through a computerized examination of the different degrees of registered radioactivity, an image of the thyroid is obtained. Such image shows the different concentration s of radioisotopes inside the organs, through false dyes.
What we can see: It allows us to see abnormalities (cysts, nodules, tumors) of

targeted organs (thyroid, heart, breast, kidneys).

THERMOSCOPY

What it is: This is a technique, which allows us to highlight in different colors areas of the body that have different temperatures. This is done through a computerized analysis of data collected by a very sophisticated thermometric instrument.
When it is used: It is helpful in the investigation of some pathology, and it is useful in diagnosing tumors.

◄ Thermography of a child's profile:
The white color indicates the warmer regions of the body, while dark blue indicates the coldest ones

When we talk about the "human body" we are actually talking about an abstract object: in fact, we can just think how different a newborn baby is from an adult, an old man from a woman, to immediately realize that, to be correct, we should always specify their sex and their developmental stage.

SIMILAR AND DIFFERENT

BABY BOY AND BABY GIRL….

At the beginning of our life sexual differences are almost insignificant: often a boy differs from a girl only in the color of his pajamas! But the differences relative to the development of the body are obvious: there are "long" and "short" newborns, some with lots of hair and some bold, some fat some big, skinny or tiny…. in the months to come the changes become very common: during the first year the brain doubles its weight, and the neuro-muscular system develops rapidly, while the skeleton strengthens, thanks to the fusion and the calcification of many cartilaginous bones. In a few months the equilibrium, develops as does the ability to see and to digest develop rapidly. The first teeth start to grow…and the baby starts to communicate and walk, to "grab" everything that surrounds him, to observe and taste things… Even the body proportions start to change rapidly: in a matter of few months height and weight double. Until about the third year, different body parts develop at different rates. From that point on they continue to grow regularly until the age of 12. This is the time of puberty, a new and rapid phase of development.

Therefore the "body" of a boy or a girl is only "similar" to the type of body studied by general anatomy: Almost all of the organs are in the 'right" position, but their development, their shape and function are often still growing. The same happens for their psychological development: between the ages of 6 and 11, the stimuli that need to be elaborated are endless, and each experience is something new, something that must be analyzed, interpreted and memorized. A more individualistic and social attitude replaces emulating behaviors: school, friends, television, they all stimulate the mind, bringing boys and girls rapidly toward adolescence.

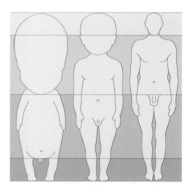

▲ **Proportions of the body**
At different ages, different body parts develop at a different rate.

YOUNG BOY AND YOUNG GIRL…

Around the ages of 10 for girls and 12 for boys, physiological changes linked to puberty begin to occur: suddenly the body starts to grow again (on average about 5cm each year!) due to the activity of the pituitary gland __°, which causes an increasing endocrine production of the adrenal gland, of the thyroid and of the gonads (ovaries and testicles): the production of muscles in the body increases, while the production of fat decreases…pubic and axillary hair start to grow, the sexual organs start to enlarge and the skin begins to change, showing an increasing activity of the sebaceous and sweat glands. Furthermore, the secondary sex characteristics start to develop: in females, subcutaneous fat begins to deposit around hips, thighs, buttocks, forearms, and under the nipples; mammary glands begin to develop as well. In males the osseous-muscular structure of the shoulders, the arms and the legs begin to strengthen. Hair starts to grow on the face, the arms, and the chest and sometimes even on the back. Furthermore, changes in the larynx cause the typical "change of the voice",

▼ **From childhood to adulthood:**
The graph highlights the two phases of accelerated metamorphosis, which characterize the first two years of life and puberty.

Height	Weight
☐ female	☐ female
☐ male	☐ male

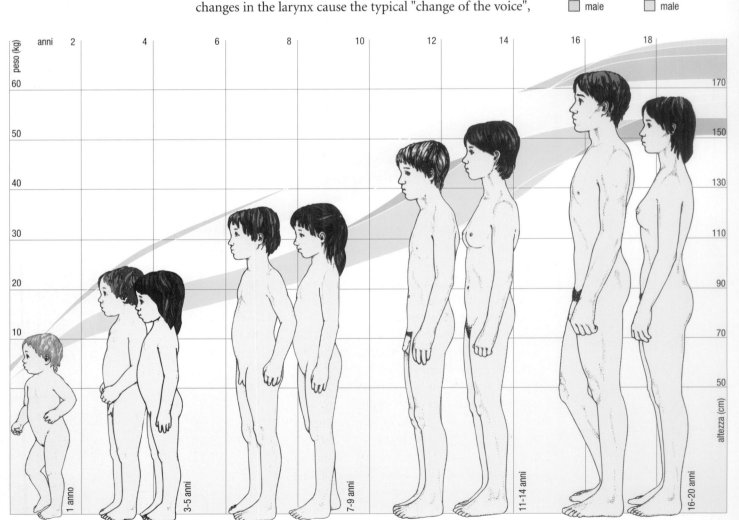

which becomes deeper.

This second metamorphosis ends around 18 years of age, when the human body reaches a size that is 20 times bigger than at birth, and acquires deeply diversified characteristics. From a psychological standpoint, adolescence is more devastating than childhood: often, from the beginning of puberty until the independent 'adult' life, there is a lapse of more than 10 years during which the individual is not yet considered an adult, nor a child anymore. Often, the confusing role of adolescents in this industrialized age, bring about many psychological conflicts.

MAN AND WOMAN

Passed puberty and physical development, the body of an adult remains unchanged for about 40 years, until old age starts to set in. This "adult human body" is the true object of study in anatomy. The adult body remains the same for a long time: same height, same distribution of the organs, same muscular development, same intellectual capabilities, and same physical abilities…. Obviously, with old age, aches and pains start to appear: metabolism slows down; neurons begin to die every day by the hundreds of thousands, causing the brain to loose about 3-4 gr. of weight each year; fat panels start to increase; if a lot of alcohol is consumed, the liver starts to enlarge, and the lungs begin to "pollute" themselves with smog and smoke…However, from an anatomical stand point, things remain more or less the same. This is exactly why every time we talk about the" human body" we refer to the drawing of an adult skeleton, of an adult circulatory apparatus, and of an adult nervous system…

But the difference between male and female are quite visible. For example, a male skeleton and a female skeleton are still distinguishable even after million of years: The skeleton of a male has a narrower pelvis and "straighter" legs compared to the skeleton of a female body. Furthermore, from a general standpoint, a male is bigger than a female: even his organs are bigger, bones are longer (or wider), and the brain is heavier…. The primary and secondary sex organs are also different, and consequently even the functions of the endocrine glands are different. However these glands control all the other activities of the body: It is not by chance that the average life span of women is longer than that of men.

In spite of this, the period of time defined as adulthood, is anatomically speaking "the most uniform" of our life; it is also the easiest to schematize, and to use, in order to reconstruct an ideal human body; by simply changing the reproductive system we can easily and rapidly turn the body into a male or a female.

OLD AGE

The "ideal body structure" remains unchanged even during old age. In industrialized countries old age starts later, lasts longer and it guarantees, on average, increasingly better physical conditions. Psychophysical changes, which have slowly started during adulthood, become more rapid: the skin begins to loose its elasticity and starts to wrinkle. Bones start to loose calcium, and the spinal column tends to shrink and to weigh down, with a noticeable loss of height. Muscles become weaker, fat layers begin to "dry up", and the senses (hearing, smell, sight…) loose sharpness. Blood circulation, respiration and the resistance to diseases become less effective.

However, even if the metabolism and the movements slow down, the general anatomical structure remains unchanged.

IF ANATOMY, BORN AS A PURELY DESCRIPTIVE AND MACROSCOPIC SCIENCE, HAD ONLY STUDIED THE ARRANGEMENT, THE STRUCTURE AND THE SPACIAL RELATIONSHIPS OF THE VARIOUS ORGANS OF THE BODY, IT WOULD HAVE BEEN OF LITTLE USE TO MEDICINE AND TO GENERAL SCIENCE.

THE CELLULAR BASIS OF ANATOMY

Initially, the human body was divided into systems or apparatuses, which are a group of many organs, (organs =complex structures with specific functions). Their functions are closely linked together in a general task. Later on, the attention fell on the study of tissues. These are clusters of cells that have very similar characteristics and are distributed inside the body in a "transverse" fashion: for example, the muscle tissue is part of the motor apparatus, the digestive system, the circulatory and the respiratory system…just as the nerve tissue, or connective tissue are. Today, anatomy investigates the human body from all these points of view: it studies cells, to understand their metabolism and the way they function; it studies tissues and their interactions, in order to understand the activity of various organs; it studies organs as if they were almost self sufficient structures, and apparatuses or systems, in order to understand the function and the relationship between the organs they are composed of.

Even though it is more convenient to study one system at the time, it is important not to forget that our body is a group of thousands of cells with an identical genetic make up. Each one has a specific characteristic because it has differentiated itself; each one "collaborates" with others, and all of them contribute in a perfectly synchronized and articulated way, to the "perfect functioning" of the "human machine"; each subdivision created to simplify the anatomical investigation is artificial.

OBSERVATIONAL INSTRUMENTS

The microscope played the same role in anatomy as the binocular did in astronomy: It allowed us to see things that nobody believed even existed and unimaginable details of what was already known.

This instrument opened new horizons for scientific research

▶ Images under an optical and an electron microscope

The two images show how the same cell (in this case a lymphocyte) appears under an optical (1) and an electron (2) microscope. The electron microscope allows us to see extremely small cellular details: in this case, HIV virus has affected the lymphocyte, and the viral particles are easily visible on its surface (dyed red by false dyes). On the bottom (3), a new electronic image shows a detailed image of the surface of the cell, and even the structure of some developing viral particles

▼ Phase contrast images (1), and ultraviolet light images (2) of the same blood specimen

These two images show how a single blood cell, dyed with an antibody conjugated to a fluorescent substance, becomes visible under ultraviolet rays. For this type of observations, the microscope must be specially equipped with a source of suitable ultraviolet and ocular rays.

and for innovative theories; the microscope has allowed us clarifying the nature of organs and tissues, their specific role, their structure and their metabolism, along with the nature and the origins of many diseases. It has also allowed us to learn about the embryonic origin of organs, facilitating the understanding of the chemical and functional relationships that link them together. The science that studies anatomical structures at a microscopic level is called histology. In this chapter, besides offering information on some of the instruments and procedures utilized, we will also summarize the main characteristics of tissues, and in some cases, we will even discuss how they "work".

A histological specimen can be looked under an optical or electron microscope. Both these instruments offer different information about differently prepared specimens. In the optical microscope, a system of lenses enlarges the image of certain objects up to about 1500 times; this allows us to distinguish details in the ranges of 1-2 mm. On the other hand, in the electron microscope, a beam of electrons passes through the specimen, and ends up on a photographic slide or into a computerized apparatus: this allows us to obtain an image which can be enlarged up 500.000 times, and the details can reach the size of 10_. As a result, the optical microscope is generally used to look at organs and tissues. Instead, with the electron microscope we are able to discover the internal structure of cells, up to the smallest details.

For specific anatomical purposes, most observations are done using an optical microscope modified according to the different staining techniques used in preparing the specimen: for example, if the specimen is treated with a fluorescent preparation, the microscope will be equipped in a way that will allow to see under ultraviolet light; it will have protective screens for the eyes, and a system which can alternate between normal and ultraviolet light.

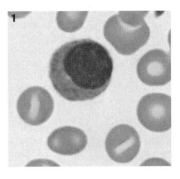

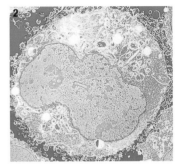

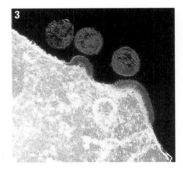

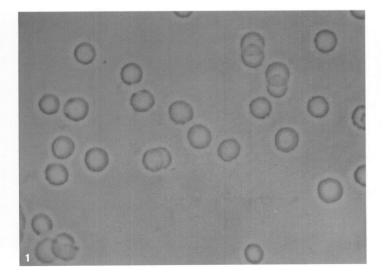

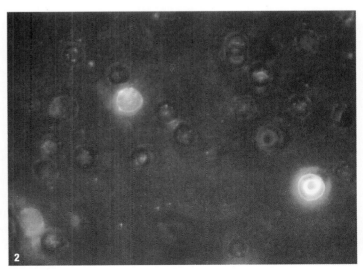

THE MICROSCOPES

The optical microscopes used in microbiology and histology have different characteristics, suitable for different types of observations.

CLEAR FIELD MICROSCOPE
It is the most popular optical microscope with dry or immersion objectives: the light reflecting from the slide shows dark images on a bright background.

DARK FIELD MICROSCOPE
The light reflects from the specimen in such a way that the image to be analyzed appears bright on a dark background. Phase contrast microscope Thanks to the use of different optical filters, this microscope diversifies the illumination of the specimen. In this way three-dimensional images are obtained.

ELECTRON MICROSCOPE
Its structure is similar to that of the compound optical microscope: an electron beam substitutes light, electromagnetic coils substitute lenses, the eyepiece is substituted by a screen or photographic slide. Specimens are sliced ultra thin with an ultramicrotome with diamond or glass blade, are collected from the surface of the water where they fall into, with the screen. The images obtained are always of cellular details of dead organisms, without color.

Eyepiece: Lens where the eye goes, or where a camera or a computer is connected.

Rotating plate or Revolver: a disk on which the objectives are screwed; it allows to switch from an enlargement to the next, without loosing the focus on the image.

Objective: Group of lenses placed close to the specimen in order to enlarge the image.

Object holder tray: it supports the specimen's slide. A little clamp holds it and moves it vertically and to the side. The hole in the middle allows for the beam of light to go through the slide.

Diaphragm: Device that allows us to adjust the intensity of the light on the specimen.

Condenser: A group of lenses that concentrate the light beam on the object.

Macrometer screw: allows for much vertical movement of the tube, focusing quickly but coarsely on the object.

Micrometric screw: allows for small vertical movements of the tube, in order to obtain a more precise focus.

Positioning screw: allows for slight movement of the slide, in order to change slowly the observation field.

Stand: it is the "backbone" of the microscope and it takes much of the weight from the foot.

Foot: usually is shaped as a horseshoe. It is very heavy in order to be able to support the weight of the microscope.

Condenser: electromagnetic coils that concentrate the electron beam on the specimen: they work simultaneously or separately (=lens of the condenser).

Screen slot: it allows us to shift the observation field, making micrometric movements along orthogonal axes (object holder tray).

Objective: enlarges the image, and by changing the tension, the focus changes. This differs from the optical microscope, where various objectives are used to shift to a different enlargement, here, there is only one objective made out of an electromagnetic coil.

Binocular: it allows us to look at detailed images reproduced on a fluorescent screen.

Observation screen: allows us to look at the image reproduced on the fluorescent screen.

Cathode: tungsten filament that produces electrons (=light source)

Projective: projects an image on a fluorescent screen. Often there are two projectives: one multiplies the selected enlargement (=eyepiece); the other one, with different degrees of power, allows us to change the size of the image.

Electrical shutter: when a picture is shot, it interrupts the flow of electrons.

Micrometric knob: it moves the screen

Fluorescent screen: visualizes the image

Shooting commands: they allow us to shoot and/or record images.

◀ **Scheme of an Optical Microscope:**

▲ **Scheme of an electron microscope:**

▶ The microtome:
Allows us to obtain extremely thin slices of tissues, which are then looked under a microscope

HOW TO PREPARE AND LOOK AT TISSUES

In order to be looked under the microscope (regardless of the type of microscope), a tissue must undergo a series of complex procedures. First of all it must be fixed: the process of fixation protects the tissue from any potential decomposition. The specimen is dipped into the ideal mixture of substances such as alcohol, formalin, or acetic acid, in order not to reduce any difficulty caused by the treatment. After that, the tissue undergoes dehydration. This process eliminates water and makes the tissue compatible with a water repellent substance that will be used in the inclusion. Usually dehydration occurs in stages, by dipping the specimen into an increasingly more concentrated alcohol solution. At this point the specimen is included: inclusion is a procedure that gives the tissue a homogeneous consistency, perfect for slicing. Usually the specimen is soaked in substances (such as paraffin, the most common, or epoxy resins) which tent to harden it considerably. At this point it is placed on the microtome's support _9, a type of slicing machine specifically designed to obtain extremely thin slices of histological specimens. If the specimen is being looked under an electron microscope, an ultra microtome is used. This slicing machine allows researchers to obtain slices that are only few hundreds of angstroms thick. Each of the slices is then placed on a slide (or on a screen, if using an electron microscope).

The tissue is then soaked in heavy metals (if looked under an electron microscope) or dyed: the process of dyeing with natural (such as carmine and haematoxylin) or synthetic dyes (such as the aniline dye) highlights the specific structures under investigation. In order to make the slides easier to handle, and to be able to also look at them with an immersion objective, that allows for bigger enlargements, the specimen is covered with a protective slide, sealed with natural resins. This process is called mounting. In order to improve the focus, a drop of an optically active substance is placed between the specimen and the protective slide.

▼ Outline of the normal procedures for an histological preparation:
usually the procedure is not as complicated and time consuming as the one outlined here, especially in the case of a biopsy where it is more important to obtain a quick response

Calibrating screw: it sets the thickness of the slices, by bringing the support of the specimen closer to the blade.

Support of the specimen: it moves from top to bottom and vice versa; it also moves forward and closer to the blade, in order to obtain slices of a specimen with the same thickness.

Crank: its rotary motion causes the support of the specimen to move vertically and horizontally

Support of the blade: it maintains the blade in a fixed position, at a constant angle, allowing it to be replaced

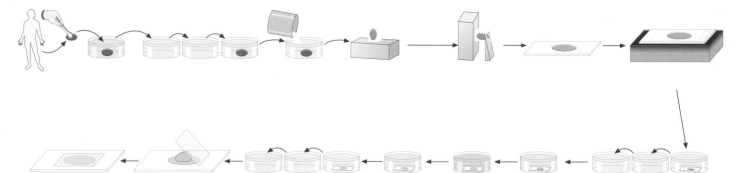

EPITHELIAL TISSUE

Characteristics: *cells that are tightly connected to each other, with very little free intercellular spaces.*

Where it is found: *it covers the entire body, externally (epidermis, dermis), and internally (mucous membranes); it also makes up all the glands, namely all those structures that release substances inside (endocrine glands) and outside (exocrine glands) the body. Furthermore it produces some particular structures (body hair, teeth, nails, eye's crystalline…). Depending on its functions and its characteristics, it is divided into different types: let's look at the main ones.*

LINING EPITHELIUM

Its function is to play an "interface' role between the inside and the outside of the body .its particular characteristics depend on the area or the organs that it coats.

SIMPLE PAVEMENT EPITHELIAL TISSUE

It outlines, for example, the air vesicles; it is mainly formed by laminar cells and by rare cuboidal or roundish cells, whose shape changes depending on the degree of dilation of the lung.

SIMPLE CUBOIDAL EPITHELIAL TISSUE

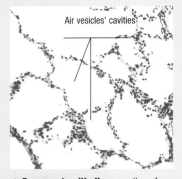

Air vesicles' cavities

▲ **Pavement epithelium:** section of an adult human lung. The cell nuclei are colored in fuchsia.

It covers, for example, all the tubular cavities, which divide the lung of a fetus. The cells, tightly aligned next to each other, have a cuboidal shape. At birth, with the first respiratory movements, the cavities begin to dilate, the connective tissue that separates the epithelial layers shrinks, and the cuboidal epithelium transforms into a simple pavement epithelial tissue ➤ 169.

SIMPLE PRISMATIC EPITHELIAL TISSUE

It constitutes, for example, the gastric mucous; it is made of prismatic cells. The cell nuclei are all aligned at the base of the cells, not too far from the underlying connective tissue.

PSEUDO-STRATIFIED

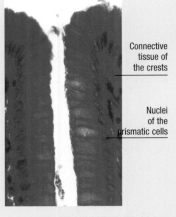

Connective tissue of the crests

Nuclei of the prismatic cells

▲ **Simple prismatic epithelium:** Section of a crest of the gastric mucous. The cytoplasm is in the color fuchsia, and the nuclei are colored blue

EPITHELIAL TISSUE

The mucous that coats the human trachea is an example of this type of tissue. It is made up by cells that have a different size and height, all found on the basal connective tissue ➤ 28 in a way that they seem to lies on different layers.

STRATIFIED PAVEMENT EPITHELIAL TISSUE

It coats for example the cornea ➤ 96: the epithelial cells grow in layers, starting from the basal connective tissue ➤ 28 . The deeper layer is made up by roundish, actively reproducing cells. A second layer of intermediate cells, which are more crushed together, follows this. The last one is a layer of completely flattened surface cells. Even the surface of the esophagus ➤ 148 is covered with this tissue, but the layers of the epithelial cells are much more numerous. On the foot sole, the

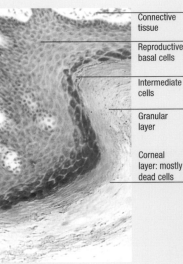

Connective tissue

Reproductive basal cells

Intermediate cells

Granular layer

Corneal layer: mostly dead cells

layers are even more differentiated: a germinative layer follows an intermediate layer of progressively flatter cells, then a granular layer with bigger cells, and finally a clear layer of cells much more elongated and flat. If the area of the body is mechanically stressed the superficial corneal layer becomes much more developed.

TRANSITIONAL EPITHELIUM

It is a characteristic of the urinary tract ➤ 210 and it represents a middle course between a stratified epithelium and a differentiated

◄ **Stratified pavement** epithelium: foot sole: the alive epithelial cells which are dyed violet.

▼ **Stratified pavement** epithelium: section of human cornea. The epithelial cells are in fuchsia; the connective tissue is in blue.

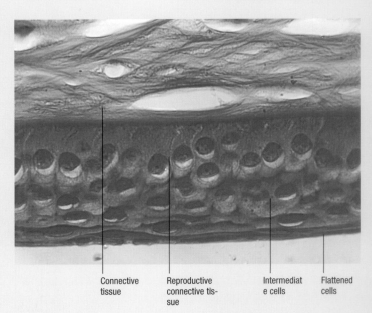

Connective tissue

Reproductive connective tissue

Intermediate cells

Flattened cells

epithelium. The lumen of the human ureter is covered with basal cells, club-shaped intermediate cells and with flattened superficial cells (to cover). Therefore, since the progressive flattening of the cells is missing. It is not considered a stratified pave-

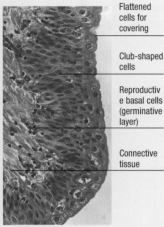

▲ **Transitional epithelium:** lumen of a human ureter.

ment epithelium.

DIFFERENTIATED EPITHELIUM

It takes on very different shapes, and it makes up specific body structures, such as hair, nails, teeth, eye crystalline....

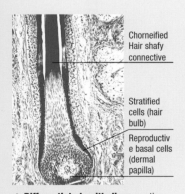

▲ **Differentiated epithelium:** section of human skin with hair.

SENSORY EPITHELIUM

There is an enormous number sensorial structures displaced in the epithelial tissues (in the skin ➤ [104], in the tongue and in the nose ➤ [102], inside the ear ➤ [98] ....). They

have very different characteristics depending on where they are located.

EXOCRINE GLANDULAR EPITHELIUM

It makes up those glands that secrete their products outside the body or inside a cavity connected to the outside (exocrine glands). There are different types of exocrine glands.

SIMPLE ALVEOLAR GLANDS

For example, they make up the Galeati's glands (or of Lieberkuhn), which secrete a lubricating mucous. In the small intestine ➤ [156] they open up primarily at the base of the villi ➤ [159]; inside the large intestine, where the villi are no longer present, they open up directly on the intestinal surface. Differently from the muciparous cells, which are distributed inside the mucous, different types of cells make up these glands. The Meibomio's glands (clustered around a general duct), and the sebaceous glands of the scalp, are also simple alveolar glands. The cells of the sebaceous glands become enlarged due to an accumulation of chole-

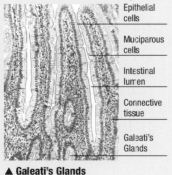

▲ **Galeati's Glands**

sterol, glycerides and fatty acids, until they "explode": in such case the cell transforms itself into a secretion, in a process called holocrine secretion.

BRANCHED TUBULAR GLANDS

Brunner's Gland (duodenal

gland) inside the intestine ➤ [156], is an example of a branched tubular gland: the structure of this type of gland is very similar of that of simple tubular glands.

SIMPLE TUBULAR GLANDS

The glands found at the bottom of the stomach ➤ [149] are an example

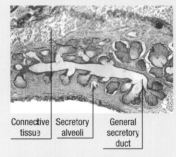

▲ **Simple alveolar gland:** inside the Meibomio's glands, the clustered alveoli open into a general duct of the eyelid

of a simple tubular gland: the main cells secrete the product of the gland inside a general duct, surrounded by enormous and round parietal cells.

SIMPLE BALLSHAPED TUBULAR GLANDS

An example of it is the sudoriferous gland ➤ [208]: a number of functional units (adenomere) which produce sweat are made up

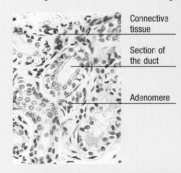

▲ **Simple ball-shaped tubular gland:** section of a human sudoriferous gland. The cell nuclei are in fuchsia

by an glandular epithelium, and they release their product into a sole excretory duct, delimited by a double layer of cubic cells.

ACINOSE (OR ALVEOLAR)

COMPOUND GLANDS

The parotid gland is an example of it: the adenomeres are extensive, and some simple ducts, which are bound by a double layer of cubic cells, run through the secretory mass. The submandibular glands, the sublingual, the mammary and the lacrimal glands are all acinose compound glands.

ENDOCRINE GLANDULAR EPITHELIUM

It makes up those glands that release their product into the blood stream (endocrine glands). Some endocrine glands (such as pancreas and the liver) possess both endocrine and exocrine characteristics: therefore, in the sectional view of some of these tissues, we can see some "endocrine" areas next to some "exocrine" ones.

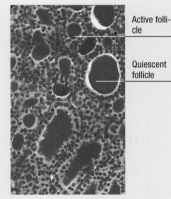

▲ **Thyroid (endocrine):** typical follicle structure full of colloid: the resting follicles are colored red, while the others are colored blue. The epithelial cell nuclei are pink.

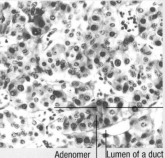

▲ **Anterior pituitary (endocrine):** the cell nuclei are in fuchsia

MUSCLE TISSUE

Characteristics: *it is made up by "elastic" cells, and plays a major role in movement and support.*
Where it is found: *Everywhere. Bundles of muscle tissue surround the digestive cavities and the blood vessels; they connect to bones; they intertwine inside the dermis, they attach to hair and so forth. There are three different types of muscle tissues. Not only they have different morphologies (structure of their cells) but also they have different functions (the type of role they have). Each type of muscle tissue is responsible for a very specific movement. Let's have a detailed look at this.*

STRIATED MUSCLE TISSUE

It is in charge of voluntary movements, which are movements made intentionally thanks to the brain activity (for example the movement of the hand and of the face). Inside the striated muscle, the myofibril of each cell is compact, crowded together, and the connective tissue ➤ 28, which separates them, is minimal. This tissue is called striated, because under the microscope the specimen shows bundles of fibers (fasciae) tightly together, with a characteristics striated coloration.

SMOOTH MUSCLE TISSUE

It is in charge of involuntary movements, which are automatic and spontaneous movements made by our body, without the active participation of the brain (for example the movements of the stomach and of the intestine). This tissue has this name because under microscope, the specimens show that the muscle fibers are immersed into a lot of connective material.

CARDIAC MUSCLE TISSUE

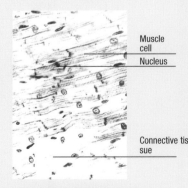

▲ **Smooth muscle fibers:** section of the muscle walls of a human artery. The fasciae of fibers have a concentric ring like characteristic compared to the lumen of the vessel

It makes up the heart (185). This tissue is very similar to the striated one, but is characterized by a higher quantity of connective tissue separating the fibers. This is probably due to its role of a striated but involuntary muscle.

Muscle cell (or fiber)

Nucleus

▲ **Striated muscle fibers:**
the long muscle cells (one for each nucleus, recognizable by a dark spot) are bundled together into fasciae: the transverse lines, next to each other, give a striated look to each fiber

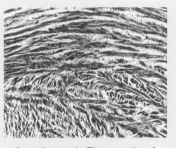

Myofibril

Nucleus

Membrane

▶ **Striated muscle fibers:**
transverse section of a muscle. Each fiber is made up by myofibrils (pink dots) and delimited by the membrane (pink line). The nucleus is laterally squeezed (dark pink spot).

Muscle cell

Nucleus

Connective tissue

▲ **Cardiac muscle fiber:** section of a human myocardium: the contractile cells of the cardiac muscle tissue are visibly separated one from the other, and the nuclei can also be found in the middle of each cell.

THE MUSCLE CONTRACTION

We have said that the cells of the muscle tissue are "elastic": in fact, they can shorten, and go back to their original length or they can stretch, in response to nerve stimulus ➤ 26-27 that they have received. But how does the movement occur?
Various muscle bundles, separated by connective tissue called motor units, make up a muscle. Each motor unit is made up by numerous muscle cells (or fibers): their number varies from 5 to 2000, and the lowest the number, the more precise the movement is. Each cell is made up by a bundle of many myofibrils, filiform structures enclosed in the plasmatic membrane. Each myofibril is made up by a series of contractile units called sarcomeres, and delimited by the Z lines: these are the ones that give the skeletal muscles ➤ 54 their striated look.
Two types of filamentous proteins, actin and myosin make up each sarcomere. These proteins are arranged parallel to each other. The actin molecules are filaments wrapped two by two in a spiral. Each spiral is connected on one end to a z line. The myosine molecules are also filamentous, but they have a "head", or a terminal protuberance. They are gathered into bundles, from which the "heads" of each molecule come out.
Actin molecules surround the myosine bundles, and the "heads" can bond with the actin creating transverse bridges. The formation of these bonds depends on the presence of calcium ions (Ca_+) inside the muscle cell, which not only bond to the double helix of the actin, but also to the 'heads" of the myosine. Depending then on the concen-

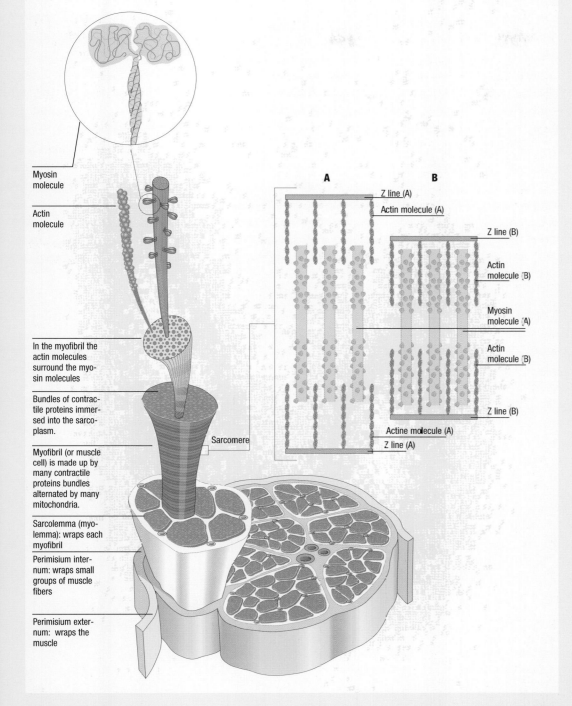

Myosin
molecule

Actin
molecule

In the myofibril the
actin molecules
surround the myo-
sin molecules

Bundles of contrac-
tile proteins immer-
sed into the sarco-
plasm.

Myofibril (or muscle
cell) is made up by
many contractile
proteins bundles
alternated by many
mitochondria.

Sarcolemma (myo-
lemma): wraps each
myofibril

Perimisium inter-
num: wraps small
groups of muscle
fibers

Perimisium exter-
num: wraps the
muscle

Sarcomere

A   B

Z line (A)

Actin molecule (A)

Z line (B)

Actin
molecule (B)

Myosin
molecule (A)

Actin
molecule (B)

Z line (B)

Actine molecule (A)

Z line (A)

◀ **Structural diagram of a striated
muscle, and of the contraction of
a sarcomere.**
A. Relaxed sarcomere: few calcium ions
are available, few transverse bridges
connect the "heads" of the myosin to the
double helixes of the actin molecules; the
Z lines are far
B. Contracted sarcomere; many calcium
ions are available, many transverse bridges connect the "heads" of the myosin
molecules to the double helixes of the
actin molecules; the Z lines are close

*bril. However, at the same time,
the presence of ATP (the molecule rich in energy, which is produced by the mitochondria, numerous inside the muscle cells) causes a rapid change in the shape of
the myosin's "heads", and the
rupture of the transverse bridges.
As a result, the "heads" attach
themselves to the nearby actin
helixes and then detach themselves in a rapid sequence: this causes the Z lines to come closer and
a progressive shortening of the
sarcomere. Because this happens
simultaneously to all sarcomeres,
the entire myofibril shortens.
Since these reactions occur simultaneously in all the myofibrils of
all muscle cells, the entire muscle
mass contracts: its length can
decrease up to 65% of its length
at rest.
Without any nerve stimuli, the
calcium ions are then "pumped"
back up into the cisterns by the
cell: as their concentration
decreases, the sarcomere stretches
out and the myofibril relaxes
itself.*

*tration of the Calcium ions, the
bridges are more or less numerous.
The muscle cells are able to regulate the internal concentration of
calcium ions: they are stored
inside a network of tubes and*

*cisterns, by special "pumps",
found inside the cell membranes.
Near the Z lines, this network is
connected to the outside membrane: when stimulated by the
nerve impulse ➤ 26, it changes,
propagating up to the cisterns*

*and the tubules. In a short time
they will release the ions.
Therefore, whilst the nerve stimulus lasts, the calcium ions
continue to increase in number
inside the cell, creating the transverse bridges inside the myofi-*

NERVE TISSUE

Characteristics: It is made up by "excitable" cells, which are specialized in transmitting stimuli or nerve impulses thanks to a series of very sophisticated and complex chemical-physical activities of their membrane. The nerve tissue makes up the encephalon, the bone marrow and all the nerves' and nerve endings networks, which run through the body. Particularly, it connects to muscles, regulating their movements, and also to the glandular tissues, regulating their secretory activity.

Specific Characteristics: The cells that make up this tissue can have very different shapes, characteristics, lengths and functions, depending on their role.

THE TRANSMISSION OF NERVE IMPULSES

Neurons, along with al the living cells, have an internal concentration of ions that differ from the external ones. Therefore, each neuron has a membrane electric potential that, at rest, is about 70mV: usually, compared to the outside, the inside charge is negative. Also the distribution of positive ions is not homogeneous: usually, on the outside, there are more sodium ions (Na+); while on the inside there are more potassium ions (K+). This is not a "spontaneous" condition: the cell burns energy to be able to "expel" the sodium ions, keeping the membrane's potential constant.

In a normal cell, things remain the same for the entire duration of its life. Neurons, on the other hand, are very unique: they can "excite", which means that they can respond to a stimulus with a rapid polarity change inside the membrane. The stimulus, regardless of its nature (sonorous, luminous, chemical, pain-producing…) causes a change in the structure of the nerve membrane: for a few thousands of a second the sodium ions enter freely, then everything goes back to normal. If the stimulus is strong enough (higher that a certain "threshold" value, below which nothing happens), such change spreads out extremely fast (even 120Km/second!) trough the cell membrane; the stimulus changes into an electrochemical impulse, and is transmitted to the neuron. The reversing charge and its immediate reinstatement (a process named action potential, which only lasts few milliseconds in each localized area of the neuron) spreads like a wave along the entire nerve cell, increasing its intensity as it moves along. After each action potential, each area of the membrane remains "refractory" to any stimulus, for a very short period of time: this ensures that the action potentials only spread in one direction, without "coming back". The nerve fibers which make up the nerves, are a particular type of neuron which connect body parts located very far from each other: their dendrite is much longer than the others' (axon or cylindraxis); this dendrite is wrapped into a myelin sheath, made up by membranes of the Schwann's cells. The sheath prevents the ions to go through the membrane, acting as an "electrical insulator" of the axon.

Since the Schwann's cells grow away from each other, the myelin sheath is also interrupted at regular intervals. The areas in which the membrane of the nerve fibers remains free, are called "Renvier's Rings": particularly rich of sodium pumps, it is only in this area that the membrane is able to develop an action potential. "Skipping" from one ring to the next, the nerve impulse spreads along the axon in a much faster way. When the nerve impulse reaches the end of the neuron, it can get to the next cell the same way it is transmitted from one Ring of Renvier to the next (electrical synapses). The neuron that transmits the impulse is called pre-synaptic neuron, while the cell is called postsynaptic. However, more often, specific substances called chemical mediators or neurotransmitters mediate the connection between the neuron and the next cell.

The arrival of the action potential at the end of the pre-synaptic neuron, causes the release of the

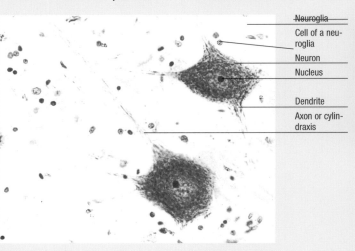

▲ Nerve cells and neurons:
They are immersed in the characteristic connective tissue of the central nervous system (neuroglia), where some cells, other than nerve cells, are visible. They play a supportive role. The nucleus is clear; the dendrites, which are filaments that connect the neuron to other neurons or with the target organ, are almost transparent

Labels for image 1:
Neuroglia
Cell of a neuroglia
Neuron
Nucleus
Dendrite
Axon or cylindraxis

▶ Human nerve:
Transverse section. The nerve is made up by many nerve fibers, each made up of a central axon (purple spots), coated with a myelin sheath (surrounding white space). The fibers are separated by a special connective tissue called neurolemma (purple fibers).

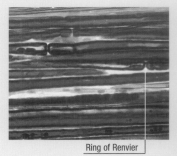

Ring of Renvier

▲ Myelinic nerve fibers:
Longitudinal section of a nerve. The myelin sheath coating each nerve cell is purple. The interruptions are called Renvier's Rings.

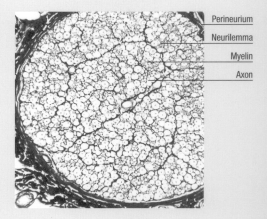

Perineurium
Neurilemma
Myelin
Axon

► Chemical synapses between two neurons

The vesicles inside the neuron (top) get close to the membrane releasing the neurotransmitters into the inter synaptic space

▼ Structure of a nerve and process of a nerve impulse mediated by a neurotransmitter

1. The action potential brings the vesicles of neurotransmitters closer to the neural membrane, fusing them together: the release of a chemical mediator takes place.
2. The molecules of the mediator attach to specific molecules (receptors) of the membrane of the postsynaptic cell, modifying them, causing therefore the inflow of sodium ions (Na+).
3. When the modified receptors are enough, an action potential is released, spreading in the opposite direction of the cell nucleus.
4. While the inflow of Na+ spreads, the membrane goes back to pump out Na+

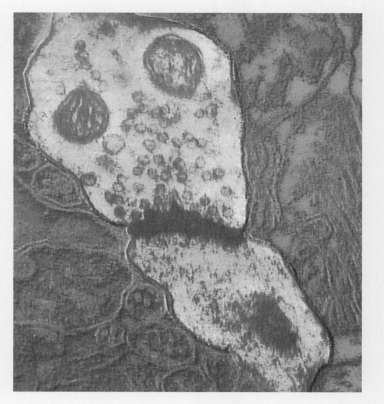

chemical mediators: these molecules go through the extremely short synaptic gap between the two cells, they attach themselves to the postsynaptic membrane, modifying it and causing the "excitation" or the "depression" of the cell. If the stimulus causes an "excitation", and if the postsynaptic cell is a neuron, the latter will be stimulated in order to produce an action potential similar to the first one, which will spread in a centrifugal direction (opposite to that of the cell nucleus). If the postsynaptic cell is a muscle fiber, we will have a contraction; if it is a glandular cell, we will have a secretion, and so on. Vice versa, if the stimulus is repressive, the activity of the post-synaptic cell will be inhibited.

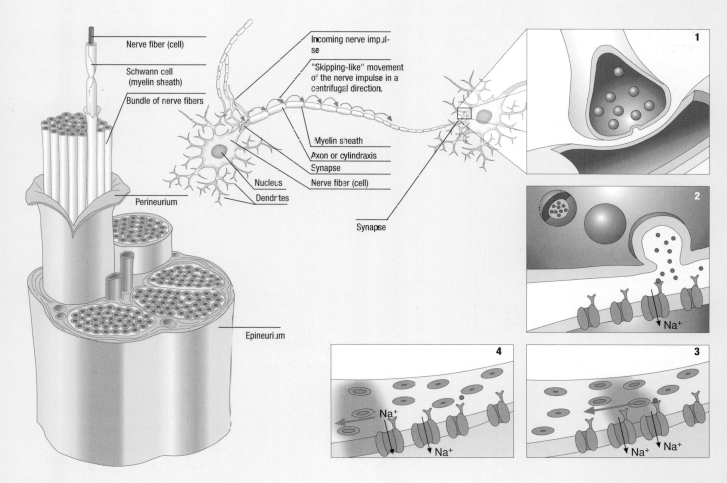

CONNECTIVE TISSUE

Characteristics: *It is a tissue where cells are "immersed" into an abundant " amorphous" intercellular substance, primarily made up by water and proteins. Its role is to support and connect different tissues.*

Where it is found: *It occupies almost all of the spaces left free by other tissues; therefore it is found in all region of the body.*

There are also very well characterized tissues that many consider to be particular types of connective tissue: let's take a brief look at them.

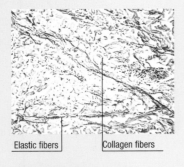

▲ Section of human dermis:
The elastic fibers are in fuchsia; the collagen fibers, one of the substances that make up the amorphous substance of the connective, are in violet.

Elastic fibers   Collagen fibers

CARTILAGINEOUS TISSUE

According to some authors it must be considered as a particular type of connective tissue, since its cells (chondrocytes) are immersed in

an abundant intercellular substance, which, in turns, is wrapped into a more or less elastic or solid amorphous substance.

This type of tissue represents the embryonal state of bone tissue: in fact, during growth, most of the cartilage found in the body becomes rich in mineral salts and transforms into bone tissue.

In an adult, cartilage is found in highly localized areas of the body, such as in the outer ear, in the nose, the trachea, in the bronchi, on the front part of ribs and on the surface of joints. Its consistency changes depending on its role.

BLOOD AND LYMPHATIC TISSUE

Many cells with different structures and roles make up these two types of tissue. Such cells circulate inside specific vessels of the body, immersed in a liquid or semi liquid amorphous substance. Therefore they also considered to be a particular type of connective tissue.

The typical cellular elements of these two tissues are produced primarily by the bone marrow, and from here they enter the bloodstream through different "maturation" stages ➤ 178-179.

They reach part of the body, thanks to a thick network of vessels.

BONE TISSUE
It derives directly from the carti-

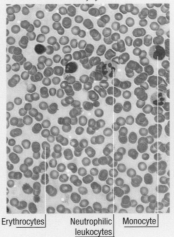

Erythrocytes   Neutrophilic leukocytes   Monocyte

▲ Human blood cells:
Here, we can see some components of the blood and lymphatic tissue. The most numerous ones (pink) are red blood cells (erythrocytes); the bigger ones are white blood cells: neutrophilic leukocytes, monocytes and lymphocytes (almost completely purple).

laginous tissue and, according to some authors, it should also be considered as a particular type of connective tissue, in which the cells (osteocytes) are immersed into an abundant solid and

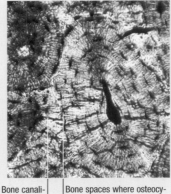

Bone canali-culus   Bone spaces where osteocytes are found

▲ Bone tissue:
The coloration highlights the system of Volkman's canals (transverse) and Haversian canals (osteone).

amorphous substance (bone). Its role is primarily to support, but also to protect some internal organs. It makes up all the skeletal bones ➤ 36.

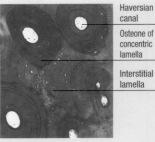

Haversian canal
Osteone of concentric lamella
Interstitial lamella

▲ Compact bone tissue: transverse section of the bone, with the typical concentric lamellar structure (osteones)

ADIPOSE TISSUE

It is made up by fat collecting cells (adipocytes). Usually they are globule-like, very large, and the cellular structure (including the nucleus) is laterally squeezed by the fat.

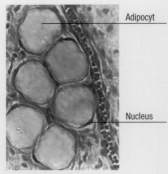

Adipocyt
Nucleus

G Adipose tissue:
Group of adipocytes (fat cells). In pink, the cellular cytoplasm; the nuclei are darker.

The role of this tissue is to store energy, and to insulate. Often, the adipocytes collect themselves in subcutaneous layers, but they are also frequently found scattered inside other types of tissues.

▶ Hyaline cartilage of the trachea:
In the same cartilage space (purple), we can see cells that have been generated by the same osteocyte (chondrocytes).

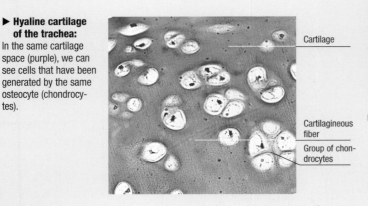

Cartilage
Cartilaginous fiber
Group of chondrocytes

TERMS IN ANATOMY

Because the details of our body's change drastically depending on the angle from which we look at them, often, in order to indicate, "what we are talking about", a brief and concise terminology is adopted. It is important to understand if we are looking at an organ "from the top", "from the side", "from the front" or "from the back"…

As a result, for example, when we look at some histological specimens we talk about "transverse section" and "longitudinal section": this terminology allows us to understand right away that what we are looking at has an elongated structure, and that the image is available both "longitudinally" and "widely". Both of them are very different from each other.

The same cunning devices are used when we need to describe the position of various organs of the body or their parts.

In order to standardize the language, a specific terminology has been adopted. Such terminology refers to a set of planes. The human body is portrait in an erect position, arms straight alongside the hips; it is divided into different **regions** by imaginary planes, called "**sectional planes**", perpendicular to each other.

The sagittal planes vertically divi-

◄ **Section:**
Imagine cutting the human body according to various planes parallel to each other (in this case the sagittal planes) in order to accurately describe each anatomical element. This image also shows a cross section separating the bust from the rest of the body

▢ A sagittal plane
▢ A sagittal plane
▢ A sagittal plane

de the body into two parts: left and right; only one of these planes divides the body into two parts approximately symmetrical. Also, the frontal planes vertically divide the body, but they are perpendicular to the sagittal planes; they divide the body into two parts: an anterior or ventral and a posterior or dorsal.

The transverse planes are horizontal, and perpendicular to the other two types of planes; they divide the body into two parts: superior or cranial, and inferior or caudal.

THE MOST COMMONLY USED TERMS

In order to describe the position of anatomical structure, we generally refer to their sectional planes; to

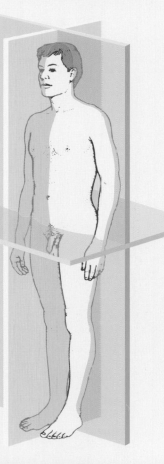

understand where a particular anatomical structure is, we will need to "visualize" the plain it refers to.

Anterior or ventral: its meaning is the opposite than posterior or dorsal, but a reference always needs to be specified; for example, the eyes are "ventral" in regard to the brain, but they are "posterior" in relation to the tip of the nose.

The term **lateral** is always referred to a sagittal plane; the targeted organs can be either on the 'right side" or on the "left side" of the plane, but often it is sufficient to say "lateral" to understand right away their location: the left arm is lateral to the left shoulder, but also the neck is lateral to the left shoulder. When needed, "left" or "right" are also used to describe the position in more detail: the left lateral vessel, the right lateral lobe…

Superior (or **cranial**) is the opposite of **inferior** (or **caudal**): we always refer to a horizontal plane, above or below which, the organ under exam is located (the liver is above the kidneys, and the pelvis is below the viscera).

The term **Proximal** is opposite than distal: these are terms which indicate, within an elongated structure, respectively, the part which is the closest or the farthest

from the body or the body part which we are referring to (for example: in the arm, the hand is at the distal extremity, while the armpit is located at the proximal extremity; or, inside a motor neuron, the neuromuscular plaques are found in its distal part).

Superficial is the opposite of **deep**: the plane, which it refers to, is that of the surface of the body or the organ under investigation (the heart is a deep organ; the deep layer of the cerebellum is made up by neuronal fibers, the villi are found on the surface of the intestine….).

Palmar is the opposite of **frontal**: the first term refers to the face of the hand, which closes (palm).

Plantar is the opposite of **frontal** (or anterior): the first term refers to the part of the foot, which closes (foot sole).

Obviously all these terms can be used in various combinations in order to give the best description of what is being observed. Therefore we will say that the kidneys are poster-inferior to the liver, poster-superior to the bladder, ventral –lateral to the spinal column (which is then posterior in the back), posterior to the intestine, superior to the pelvis…

SKELETON AND MUSCLES:
SUPPORT, PROTECTION AND MOVEMENT

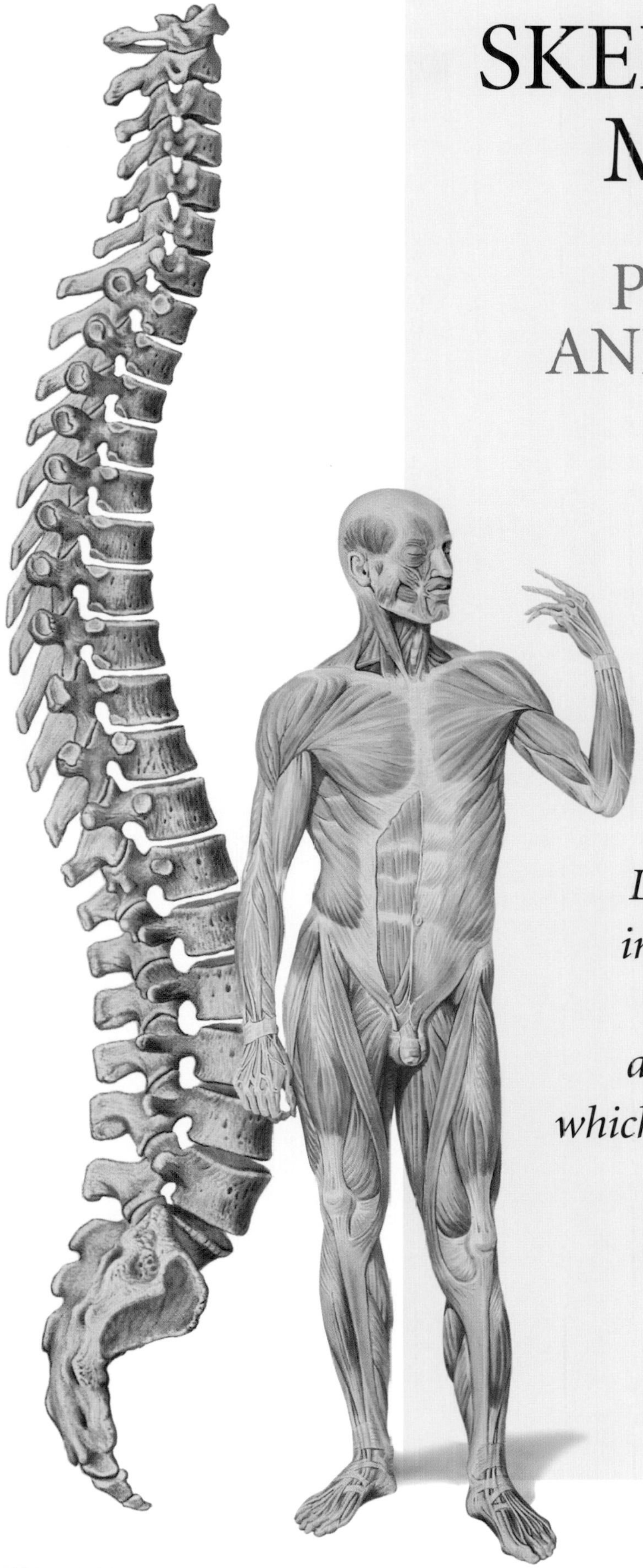

*Let's start our journey
inside the human body
by analyzing those
anatomical structures,
which characterize our species:
the erect posture
and the biped walk.*

In an adult, the skeleton is made up of an average of 206 bones and by different types of joints, which connect them to each other and give them a wide range of movements.

THE SKELETON
OR SKELETAL SYSTEM

The skeleton of an adult is made up on average of 206 bones; on the other hand, the skeleton of an embryo is made up by about 350 completely cartilaginous bones ➤38: as they grow, many of them fuse together, while the development of the circulatory and nervous system transforms the cartilage ➤24 into bone tissue, richer in mineral salts.

In an adult skeleton, the cartilage almost completely disappears: it remains only in certain parts of the ear, of the nose, the mouth, of the trachea and the bronchi, the anterior part of the ribs, and on the surface of the joints.

Our skeleton reaches full maturity at about 25 years of age, and during its entire lifetime it will not only enable movement, but also protect and support the internal organs.

HOW THE SKELETON IS DIVIDED

Out of the 206 bones in our body, 29 make up the cranium ➤42, 26 make up the spinal column ➤44, 25 make up the thorax ➤46, 64 make up both upper limbs (including the hands)➤48, and 62 make up both lower limbs ➤50.

Also, the skeleton is divided into two parts with clearly different functions:

-The **axial skeleton**, or **skeleton of the trunk**, or **central skeleton**,

Includes the cranium, the spinal column, and the thorax; its primary function is to support and protect the internal organs;

The appendicular skeleton includes the limbs (upper and lower), and the girdle; its primary function is enable movement and to support.

▲ **Fetal skeleton:**
The parts already ossified are colored gray-brown

◄ **How the skeleton is divided:**
☐ Axial skeleton
☐ Appendicular skeleton
☐ Girdle
☐ Cartilages

▼ Adult human skeleton:
Frontal and posterior view

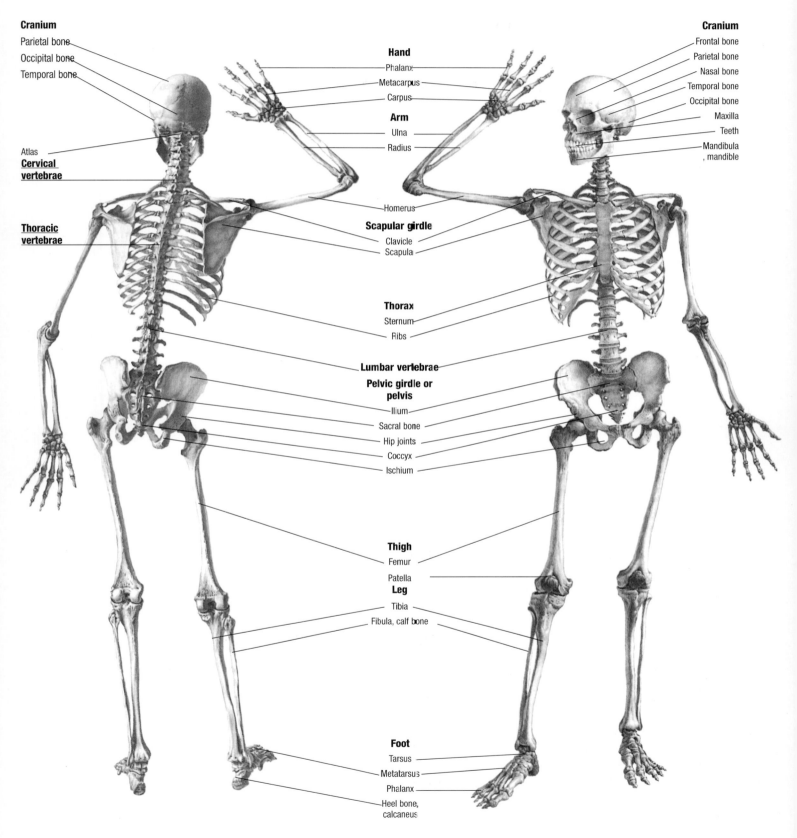

Cranium
Parietal bone
Occipital bone
Temporal bone

Atlas
Cervical vertebrae

Thoracic vertebrae

Hand
Phalanx
Metacarpus
Carpus

Arm
Ulna
Radius

Homerus

Scapular girdle
Clavicle
Scapula

Thorax
Sternum
Ribs

Lumbar vertebrae

Pelvic girdle or pelvis
Ilium
Sacral bone
Hip joints
Coccyx
Ischium

Thigh
Femur

Patella
Leg
Tibia
Fibula, calf bone

Foot
Tarsus
Metatarsus
Phalanx
Heel bone, calcaneus

Cranium
Frontal bone
Parietal bone
Nasal bone
Temporal bone
Occipital bone
Maxilla
Teeth
Mandibula, mandible

THE BONES

Bones are very strong, but at the same time very elastic structures. Depending on their shape, they are identified as long, short or flat. The can also come in very different sizes: from the femur ➤ 50, which, at times, can reach more than half a meter, to the tiny bones of the middle ear, only few millimeters long. The surface of the bone is full of protrusions (apophysis, processes, or tubers), localized relieves (tubercles), sharp spines or thin grooves, round fossae, and elongated canals that often play very important articular roles, or enable tendons and ligaments to be inserted on the bone. At times the movement of the muscles themselves, shaping the bone that they are in contact with, creates this roughness. Each bone is coated with a fibrous membrane called periosteum. Such membrane is rich with blood vessels and nerve endings and it ends at the edge of the joint area or where the ligaments and the tendons insert themselves. Inside the periosteum, a microscopic membrane (endosteum) coats the outer layer of the compact bone tissue; this membrane is very strong and has a lamellar structure. On the inside we can find the spongy and more elastic bone tissue, which is permeated with the bone marrow.

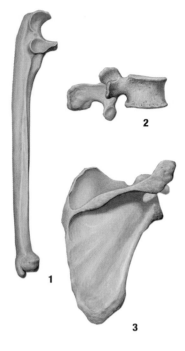

▲ **Examples of bones:**
Long bone (1. femur), short bone (2. vertebra) flat bone (3. Scapula)

▲ **Compact bone tissue under electron microscope**

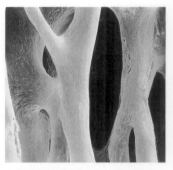

▲ **Spongy bone tissue under electron microscope**

BONE GROWTH AND SALINE EQUILIBRIUM OF THE BODY

Bone growth, along with the development of all the organs depends first on genetic factors: for example, among ethnic groups such as the African Pygmies, the reduced skeletal growth is regulated by genetic factors and does not depend on the activity of the body.

Within the same population, and under the same conditions, bone growth depends a lot on hormonal factors ➤ 126, which constantly regulate the production and the distribution of the bone tissue.

While the parathormone and the calcitonine promote respectively the breakdown and the building up of bone tissue, thanks to the osteocytes, the hypophyseal (pituitary) growth hormone (somatotropin or human growth

hormone) and the sexual hormones stimulate both the production of cartilage and its ossification.

If an endocrine gland, which produces one of these hormones doesn't work well, and if the hormone is not produced in a calibrated fashion, the development of bone will be abnormal. Therefore, for example, when the hypophysis doesn't work well and produces an insufficient amount of growth hormone (STH), the person doesn't develop normally (pituitary dwarfism). However, if during growth he receives the right amount of somatotropic hormone, the skeletal development can almost normalized.

Even diet is very important for bone growth: the supply of nutrients such as phosphorous,

calcium, vitamin A, C and D is essential for a correct growth of the bone tissue. In fact, the development of a bone is the result of a dynamic equilibrium between bone tissue cells, which promote the formation of calcium deposits (osteoblasts and osteocytes), and cells from the same tissue, which promote their dissolution (osteo-

clasts), by producing enzymes which are able to "melt" salt crystals and "digest" the collagen fibers.

In other words, the bone tissue doesn't reach a final growth, but it continues to grow and to get destroyed at the same time. This "reshuffling" of the bone continues throughout one's life: in

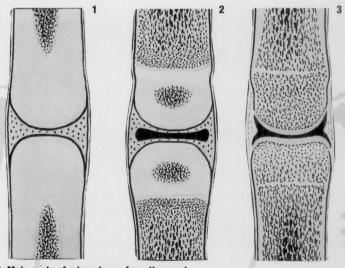

▲ **Main parts of a long bone, from the most superficial layer to the deepest ones**

▼ **Main parts of a long bone, from the most superficial layer to the deepest ones.**

The bone marrow is divided into yellow bone marrow, rich in fat (adipose tissue ➤[24]) and red bone marrow, is where cells are constantly splitting. Besides producing osteoblasts, osteocytes and osteoclasts (cells which make up the bone), the cells of the red bone marrow produce also the majority of the cellular elements of the blood and of the lymph ➤[24,173]. A thick network of interconnecting Haversian and Volkmann's canals, which bring nourishment to all its cells, runs through the bone.

In the long bones, the central cylindrical part (diaphysis) marks the medullary cavity, which is rich with red bone marrow: the spongy part is only found in the enlarged extremities (epiphysis).

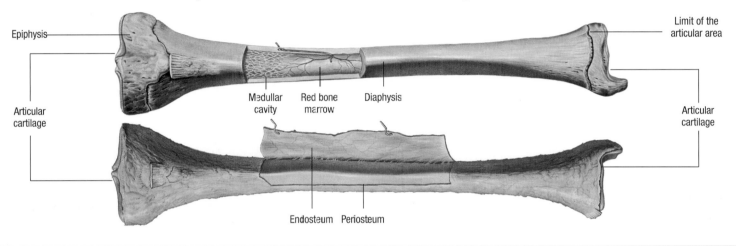

Epiphysis

Articular cartilage

Medullar cavity

Red bone marrow

Diaphysis

Endosteum   Periosteum

Limit of the articular area

Articular cartilage

response to the salt balance in the blood, there is a constant change of mineral salts between bones and blood, while mechanical needs (gravity, muscles…) determine a continuous structural bone change. As a result, while osteoblasts and osteocytes "fix" mineral salts, the osteoclasts, with their destructive action, free them again into the bloodstream, maintaining the phosphorus/calcium ration, constant.

So, while the osteocytes build the bones, the osteoblasts shape them, "guiding" the tissue's growth, and modifying its shape and the mass, according to different structural needs.

On the other hand, starting from the fetal cartilaginous buds, the bone formation follows different rules: it can be direct, as in the case of the mandible, or indirect (or substitutive), as in the case of long bones. With direct ossification, the cartilage precursor doe-sn't get involved in the formation of the definitive bone. Its role is simply of "a guide"; instead, with indirect ossification, the cartilaginous tissue progressively transforms into bone tissue. This process occurs according to different processes, and at different times. Around the 7th week of a fetus' life, the central cartilage begins to calcify (1). At that same time, a stratified cuff of twisted fibers' bone tissue begins to develop. This constitutes a primitive diaphysial case. The growth of the blood vessels through the cuff and their ramification inside the cartilage, promotes the formation of the medullary cavity. The extremities remain "obstructed" by the cartilaginous epiphysis.

The ossification (2) proceeds and increases the external diameter of the cuff. The fibers are still concentric to the axis of the bone, while on the inside, the twisted layers are eroded and the diame-ter and the length of the medullary cavity expand. At the extremities, the osteocytes and the osteoclasts begin to "shape" the bone. Subsequently, even after few years from birth, ossification nuclei begin to appear inside the epiphyses. During a very complex process, a spongy bone tissue begins to form and to expand toward the walls of the epiphysis. A section of connecting cartilage, which contributes to the growth in length of the diaphysis, remains for a long time between the spongy tissue of the epiphysis and the compact one of the diaphysis (3). When this layer of cartilage disappears, the bone stops lengthening.

▶ **Architectural structure of long and short bones:**
The different distribution of the spongy tissue corresponds to precise mechanical needs

THE JOINTS

The term joint or articulation refers to an anatomical structure in which two or more bones are touching. Each joint can be classified differently, depending on the degree of reciprocal mobility of the bones involved, on the type of movement produced by the bones, or the type of tissue, which connects them to one another.

Therefore, depending on the degree of reciprocal mobility of the bones, there are three different types of joints: diarthrosis or mobile joints (knee, shoulder, fingers…), amphiarthrosis or semi-mobile (spinal column, foot bones.) and synarthrosis or fixed joints (skull cap).

On the other hand, depending on the type of movement performed by the bones, the following types of joints can be distinguished: hinge joint (elbow), condyle joint (knee), ellipsoidal joint (wrist), spheroidal joint (shoulder), plane joint (bones of the foot), pin joint (neck) and saddle joint (ankle).

Doctors prefer to distinguish joints into sutures, that connect two or more flat bones with dense connective tissue (in the skull cap); symphyses, characterized by more or less compact fibrous cartilage placed between two bones (in the spinal column); arthrodiae, characterized by plane joint surfaces (in the foot); enarthroses, with spheroid or partially spheroid joint surfaces (in the shoulder); condylarthrosis, where the joint surfaces are ellipsoidal (in the wrist and the ankle); ginglymus joints, characterized by cylindrical or partially cylindrical joint surfaces (in the elbow).

The surface of the bone that is part of a joint, made up by individual parts called condyles, is covered with hyaline cartilage

▼ **Type of joints distinguished by the type of bones' movement**
1. Hinge joint (elbow)
2. Condyle j. (knee)
3. Plane j. (bones of the foot)
4. Saddle j. (ankle)
5. Ellipsoidal j. (wrist)
6. Spheroid j. (shoulder)
7. Pin j. (neck)

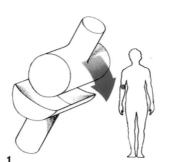

1

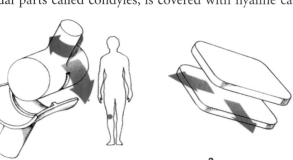

2 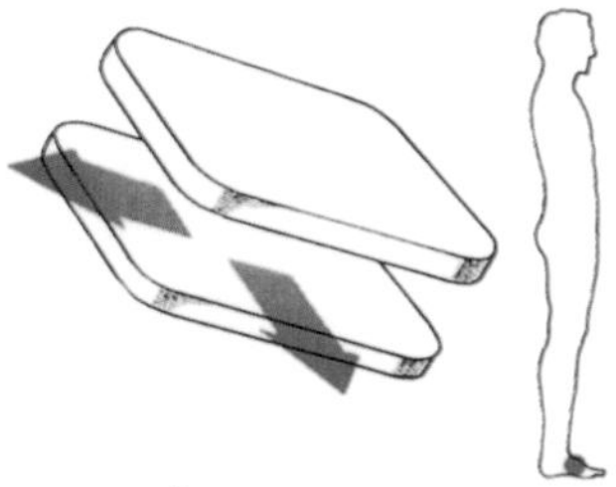   3 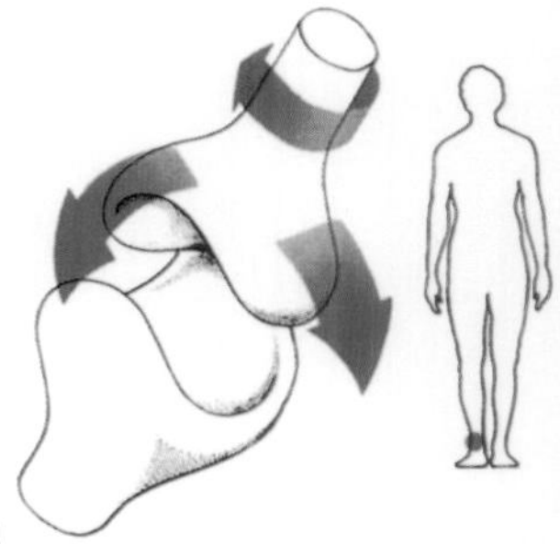   4

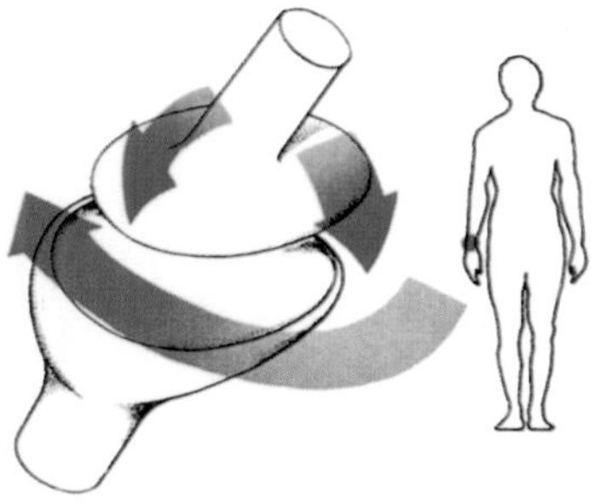

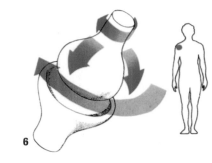

5   6

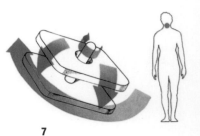

7

(24). In amphiarthrosis or diarthrosis, the synovial membrane wraps the cartilage. Such membrane contains the synovial fluid or synovia. Such fluid reduces friction between the touching surfaces, acting as a lubricant and facilitating movement. Also, a fibrous cuff, called joint capsule, which inserts from both sides, on the edges of the cartilage, and continues on inside the two periostea, covers the joint heads of the connected bones. The joint capsule is made up by bundles of dense twisted connective tissue, often infiltrated with fat. The joint capsule shows in depth a layer of a well-characterized morphology, called synovial layer. Two layers are distinguishable: a simple synovial layer, limited to those areas more subjected to movement related traumas, and a complex synovial layer, richly innervated and vascularized.

Joint ligaments in this function also assist the function of the joint capsule, as a connector between bones. Ligaments are bundles of non-elastic connective tissue, which prevent bones from moving away from each other (prevent dislocations). They are distinguished into:

Internal ligaments, which appear to be inside the joint, but are actually separated by the synovial membrane;

The **peripheral ligaments** insert into the joint or the parajoint.

Distal ligaments, insert into the bone even at a remarkable distance from the joint.

While the bones of the amphiarthroses and diarthroses are kept together by the joint ligaments, the bones of the synarthrosis are "stuck" together, "cemented" together by the dense connective tissue.

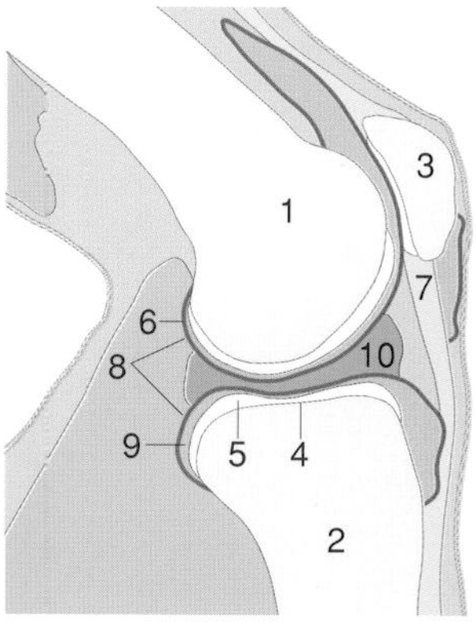

▲ **Structure of a diarthrosis: the knee**
1. Femur
2. Tibia
3. Patella
4. Condyles of the head and the tibia
5. Hyaline cartilage
6. Joint capsule
7. Tendon of the thigh muscle which keeps the patella in position
8. Synovial membrane
9. Synovial fluid or synovia
10. Meniscus (crescent cartilage)

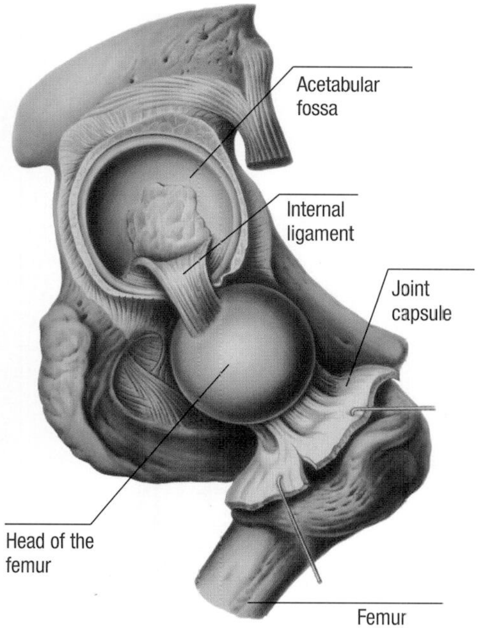

▼ **Ligaments of the elbow joint**
● peripheral ligaments
■ distal ligaments

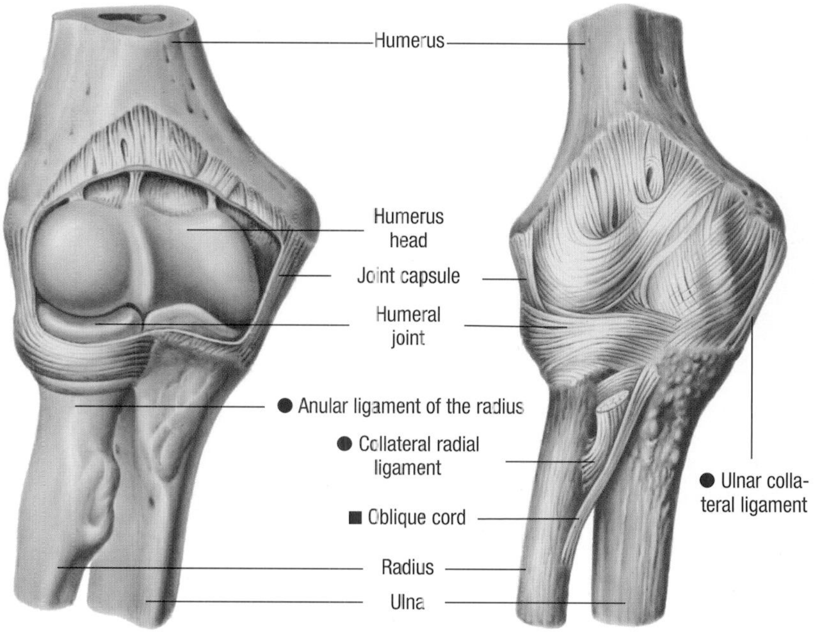

▼ **Example of an internal ligament: joint of the femur on the hip**

Acetabular fossa

Internal ligament

Joint capsule

Head of the femur

Femur

Humerus

Humerus head

Joint capsule

Humeral joint

● Anular ligament of the radius

● Collateral radial ligament

■ Oblique cord

Radius

Ulna

● Ulnar collateral ligament

THE SKULL

The term skull refers to a complex group of bones, which make up the head: these bones are primarily flat bones, connected to each other by synarthrosis (sutures). The joint found between the mandible and the spleno-cranial bone, which makes mastication (chewing) possible, and the junction between the two occipital's condyles and the atlas (the first cervical vertreba), which allows for all the movements of the head on the neck are the only two exceptions.

Of the 29 bones of the head (22 of the skull, 3 little bones inside each ear, and the … at the base of the tongue), 8 make up only one resistant cap (the skull cap), which contains and protect all the brain ➤ 82, the most important and delicate organ of the body. 14 bones make up the face, which protects the delicate sensory organs (eyes, ears, nose and mouth); while the organs of sight and hearing fit into bone cavities of the skull, leaving only a little opening, the olfactory organ is completely protected by the nasal bone, and the taste organ (tongue) remains protected between the bones at the basis of the skull and the mandible. The bones of the skullcap

have a very characteristic structure: an internal lamina of compact bone tissue; the diploe (1), a spongy bone tissue full of cavities, and an external lamina (2) of compact bone tissue, usually thicker than the internal one. This particular structure makes the skullcap sturdy, shock resistant and light at the same time.

The diploe is run through by diploic canals (3), where we can find the blood

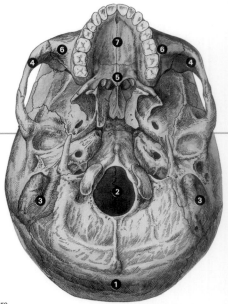

Lateral, ventral view of the skull ▼ ▶ View without the mandible

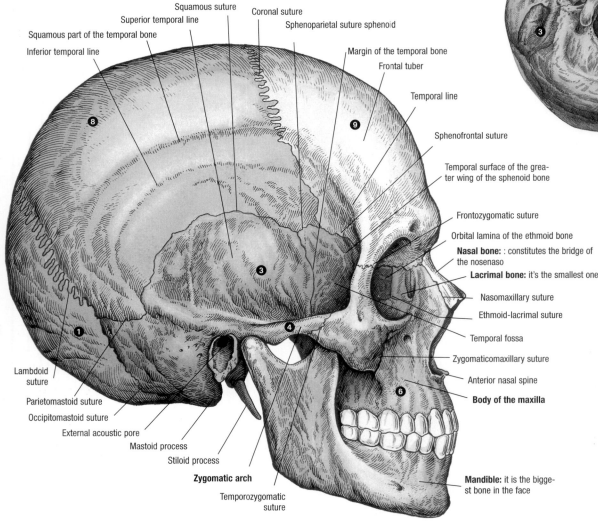

Squamous suture
Superior temporal line
Squamous part of the temporal bone
Inferior temporal line
Coronal suture
Sphenoparietal suture sphenoid
Margin of the temporal bone
Frontal tuber
Temporal line
Sphenofrontal suture
Temporal surface of the greater wing of the sphenoid bone
Frontozygomatic suture
Orbital lamina of the ethmoid bone
Nasal bone: : constitutes the bridge of the nosenaso
Lacrimal bone: it's the smallest one
Nasomaxillary suture
Ethmoid-lacrimal suture
Temporal fossa
Zygomaticomaxillary suture
Anterior nasal spine
Body of the maxilla
Mandible: it is the biggest bone in the face
Lambdoid suture
Parietomastoid suture
Occipitomastoid suture
External acoustic pore
Mastoid process
Stiloid process
Zygomatic arch
Temporozygomatic suture

❶ **occipital bone**
It is the biggest; it closes the skullcap on the back. It has two excrescences (condyles) that pivot on the first cervical vertebra (atlas)

❷ **occipital foramen**
The bone marrow goes through it

❸ **temporal bones**
form the inferior and lateral part of the skullcap, and are part of the ear.

❹ **Zygomatic arch**
They are Zygomatic bones that make up the walls and the base of the eye sockets

❺ **Sphenoid bone**
Has a very uneven shape; it makes up the base of the skullcap and the bottom of the eye sockets. It protects and hosts the pituitary gland (hypophysis) inside a bony niche (sella turcica)

❻ **Maxillary bone**
Are fused, and make up the maxilla

❼ **Bony palate**

❽ **Parietal bones**
Are fused, make up the central part of the cranial vault

❾ **Frontal bones**
Are fused, delimit the skullcap in the front, and the eye sockets in the back

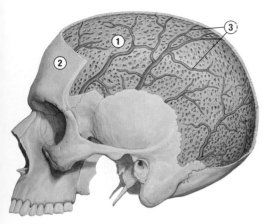

cranial vault. Often vessels give the name to the foramens, the canals and the grooves, in which they reside. Along with the vessels even the nerves of the facial muscles and of the sensory organs pass through the foramens, grooves and canals of the cranial bones; similarly, traces of blood vessels and of muscles attachments and their activities, are often found on the external surface of the facial bones.

Teeth are also part of the skull: even if not made of bone tissue, the 32 definitive teeth, remain permanently fixed into the dental alveoli, bony niches along the border of the mandible and of the maxilla. Here, the teeth grow starting from the specialized epithelial tissue ➤ 22.

vessels that keep the bone tissue alive. Starting from the neck, veins and arteries reach the brain and the facial organs, passing through a series of foramen in the inferior bones of the skull. The path of the arteries remains "imprinted" on the internal surface of

THE SKULL OF NEWBORNS

The bones of the skull have a certain degree of reciprocal movement only during infancy. At this time the sutures of connective tissue, which permanently ties them together, have not yet been developed. The six membranous spaces (fontanels), that separate the adjacent bones of the skull in newborns, facilitate childbirth and the subsequent and rapid expansion of the skull.

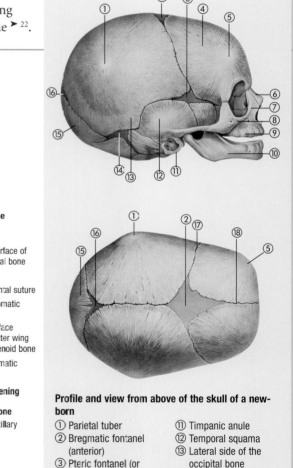

Profile and view from above of the skull of a newborn

① Parietal tuber
② Bregmatic fontanel (anterior)
③ Pteric fontanel (or sphenoid)
④ Greater wing
⑤ Frontal tuber
⑥ Nasal bone
⑦ Lacrimal bone
⑧ Zygomatic bone
⑨ Maxilla
⑩ Mandible
⑪ Timpanic anule
⑫ Temporal squama
⑬ Lateral side of the occipital bone
⑭ Mastoid fontanel
⑮ Occipital squama
⑯ Lambdoid fontanel (or posterior)
⑰ Coronal suture
⑱ Me topic suture

Frontal view of the bones of the skull

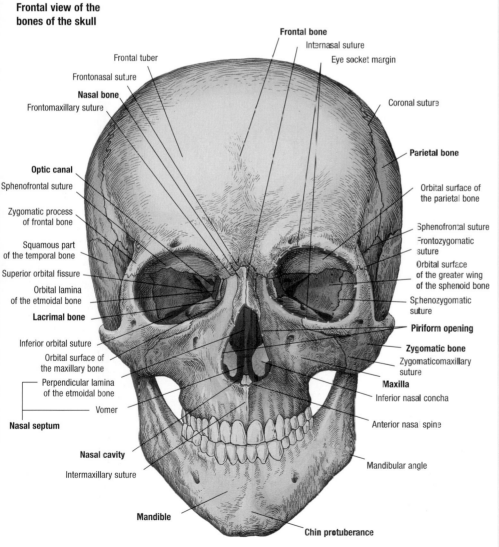

Frontal tuber
Frontonasal suture
Nasal bone
Frontomaxillary suture

Frontal bone
Internasal suture
Eye socket margin
Coronal suture

Optic canal
Sphenofrontal suture
Zygomatic process of frontal bone
Squamous part of the temporal bone
Superior orbital fissure
Orbital lamina of the etmoidal bone
Lacrimal bone
Inferior orbital suture
Orbital surface of the maxillary bone
Perpendicular lamina of the etmoidal bone
Vomer
Nasal septum
Nasal cavity
Intermaxillary suture
Mandible

Parietal bone
Orbital surface of the parietal bone
Sphenofrontal suture
Frontozygomatic suture
Orbital surface of the greater wing of the sphenoid bone
Sphenozygomatic suture
Piriform opening
Zygomatic bone
Zygomaticomaxillary suture
Maxilla
Inferior nasal concha
Anterior nasal spine
Mandibular angle
Chin protuberance

43

THE SPINAL COLUMN

It is the central support of the body. It is made up by 33-34 bone elements, called vertebrae, placed on top of each other. Depending on their function, they are characterized by different shapes. The spinal column is divided into 5 regions, in which the vertebrae have similar characteristics:

Cervical region: in mammals, it comprises of 7 vertebrae, which facilitate the rotation of the head.

Thoracic region: 12 vertebrae, from which the ribs articulate. The ribs make up the thorax ➤ 46

Lumbar region: 5 vertebrae. Their size is bigger than that of the other vertebrae; in fact, they are designed to support most of the body weight and to sustain the great efforts made to keep the an erect posture;

Sacral region: 5 vertebrae fused with each other. They form the sacral bone ➤ 50, on which pelvic bones articulate.

The Coccyx: is made up by 4 or 5 extremely small sealed vertebrae: in most vertebrata, they support the tail, are separated, and their number can vary. Vertebras have a central foramen; because they are piled on top of each other, the foramens make up a cylindrical canal, which hosts the bone marrow ➤ 106. Each vertebra has also some arched protrusions (spinal processes), on which the trunk muscles

and ligaments attach. The first cervical vertebra (atlas) and the occipital bone of the skull make up a condylar joint which allows the head to move forward and backward. Furthermore the joint of the atlas pivots on the second cervical vertebra (axis): a cylindrical protrusion of the superior surface of the axis, inserts into a ring of the inferior surface of the atlas.

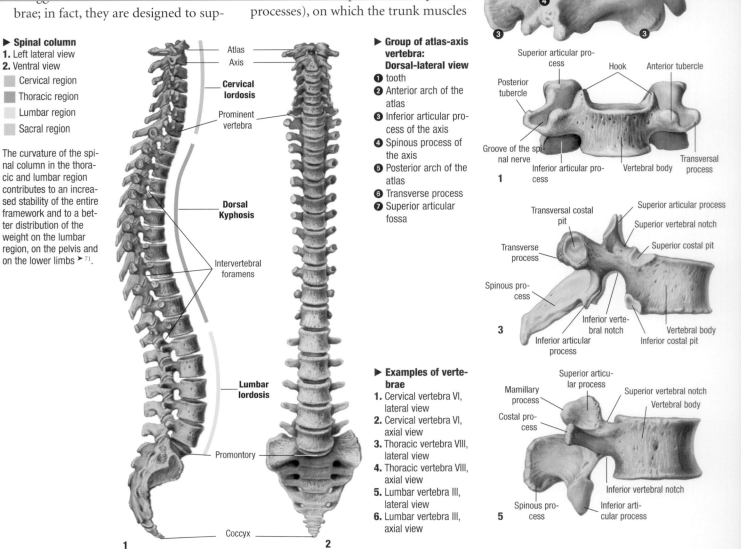

▶ **Spinal column**
1. Left lateral view
2. Ventral view

⬜ Cervical region
⬜ Thoracic region
⬜ Lumbar region
⬜ Sacral region

The curvature of the spinal column in the thoracic and lumbar region contributes to an increased stability of the entire framework and to a better distribution of the weight on the lumbar region, on the pelvis and on the lower limbs ➤ 71.

Atlas
Axis
Cervical lordosis
Prominent vertebra
Dorsal Kyphosis
Intervertebral foramens
Lumbar lordosis
Promontory
Coccyx

1

2

▶ **Group of atlas-axis vertebra: Dorsal-lateral view**
❶ tooth
❷ Anterior arch of the atlas
❸ Inferior articular process of the axis
❹ Spinous process of the axis
❺ Posterior arch of the atlas
❻ Transverse process
❼ Superior articular fossa

Superior articular process
Hook
Anterior tubercle
Posterior tubercle
Groove of the spinal nerve
Inferior articular process
Vertebral body
Transversal process
1

Transversal costal pit
Superior articular process
Superior vertebral notch
Transverse process
Superior costal pit
Spinous process
Inferior vertebral notch
Vertebral body
Inferior articular process
Inferior costal pit
3

▶ **Examples of vertebrae**
1. Cervical vertebra VI, lateral view
2. Cervical vertebra VI, axial view
3. Thoracic vertebra VIII, lateral view
4. Thoracic vertebra VIII, axial view
5. Lumbar vertebra III, lateral view
6. Lumbar vertebra III, axial view

Superior articular process
Mamillary process
Superior vertebral notch
Vertebral body
Costal process
Inferior vertebral notch
Spinous process
Inferior articular process
5

This joint is kept in place by a specific ligament and enables to rotate and tilt the head. The movements allowed by these two joints, guarantee a wide mobility of the head, which is the main stimuli receptor in our body.

While bones give power to the spinal column, the intervertebral or spinal disks are made up by cartilage and by alternating with the vertebrae, they give it flexibility. They are pressure resistant but being more elastic than bones, they are shock resistant.

Furthermore, thanks to these flat joints, the spinal column can rotate and curve.

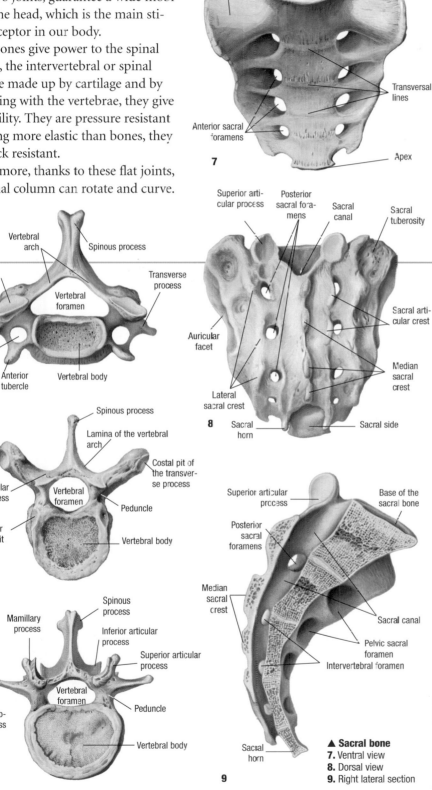

7

Sacral wing
Base of the sacral bone
Superior articular process
Transversal lines
Anterior sacral foramens
Apex

2

Vertebral arch
Spinous process
Superior articular process
Transverse process
Posterior tubercle
Vertebral foramen
Transversal foramen
Anterior tubercle
Vertebral body

4

Spinous process
Lamina of the vertebral arch
Costal pit of the transverse process
Superior articular process
Vertebral foramen
Peduncle
Superior costal pit
Vertebral body

6

Mamillary process
Spinous process
Inferior articular process
Superior articular process
Vertebral foramen
Costal process
Peduncle
Vertebral body

8

Superior articular process
Posterior sacral foramens
Sacral canal
Sacral tuberosity
Auricular facet
Sacral articular crest
Lateral sacral crest
Median sacral crest
Sacral horn
Sacral side

9

Superior articular process
Base of the sacral bone
Posterior sacral foramens
Median sacral crest
Sacral canal
Pelvic sacral foramen
Intervertebral foramen
Sacral horn

▲ Sacral bone
7. Ventral view
8. Dorsal view
9. Right lateral section

✚ PROBLEMS ASSOCIATED WITH GROWTH

The Deviations of the spinal column area are a quite popular problem, even though the number of very serious cases, for which surgery is needed, is very low (1 ‰). In our country on average 1 out of 5 children suffer from a spinal column problem. Usually girls are more prone to these problems than boys. Particularly during puberty, the skeleton undergoes incredible efforts and this can cause bones' deformities: between the ages of 10 and 14, bones grow a lot and quite rapidly. Since they still haven't reached the level of appropriate and final calcification characteristic of a mature bone, they can become permanently deformed.

There are three types of spinal column deviations, which can affect posture: kyphosis, lordosis and scoliosis. With kyphosis the dorsal curvature is more accentuated than normal. Usually remaining curved for a long time (e.g. Studying) can cause the development of such curvature. Often the dorsal curvature is compensated by a more accentuated lumbar curvature. Often the use of high hill shoes causes lordosis to worsen. Scoliosis is a lateral pathological deviation of the column. It is often cause by an uneven length of the lower limbs, and it is accentuated by standing for long periods of time and by keeping the wrong posture while sitting down.

Up to a certain degree of curvature, Lordosis and Kyphosis are considered to be physiological, which means that they are linked to a static posture.

Deviations of the spinal column
In people affected by lordosis ① the natural curvature of the pelvis is altered.
② Example of scoliosis

THORAX

The thorax is a framework consisting of 12 pairs of bones called ribs, which articulate on the back of the thoracic vertebrae. Such structure protects the heart, the lungs, and the primary blood vessels. It also supports the muscles that support the other abdominal organs. The thorax can expand and contract thanks to the rib muscles. Starting from the top, each pair of ribs is marked by a progressively bigger roman number. Ribs are flat arched bones: although they are long, they are not classified as such, since they lack the marrow canal. All ribs, except for pair I, have a costal groove inside which, a

bundle of blood vessels and of intercostals nerves, run. The length increases from rib I to rib VIII and then starts to gradually decrease; furthermore, starting from the first pair, the obliquity with which they are joined on the vertebrae, increases. A cartilaginous tract (costal cartilage) completes each rib, and its shape is similar to that of the rib it extends from.

The first 7 pairs are also articulated in the front with the sternum, a flat uneven median bone, which closes the thorax. The sternum is made up by three segments (manubrium, body and xiphoid process) often fused to each

other. It possesses an anterior facet (the sternal plane), which is convex and rugose, from which many muscles of the neck, of the thorax and of the abdomen originate. The posterior facet is longitudinally concave and smooth. From this side, muscles originate only from the top (upper limbs muscles) and from the bottom (abdominal muscles). Each rib of pair VIII, IX and X is connected to the tip of the previous one, with costal cartilage. Finally, the ribs of the 11th and 12th pair, remain floating (floating ribs): these bones are articulated only to the spinal column and they do not "close" in the front. They are quite different.

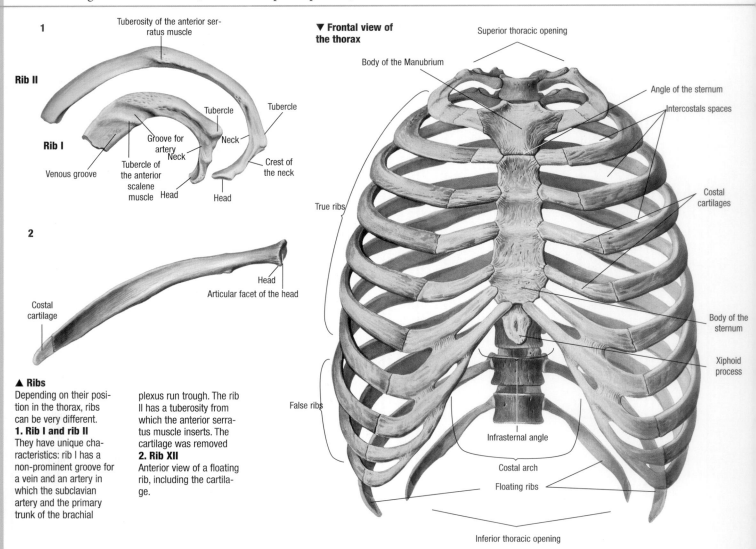

1

Tuberosity of the anterior serratus muscle

Rib II

Rib I

Tubercle

Tubercle

Groove for artery

Neck

Neck

Venous groove

Tubercle of the anterior scalene muscle

Head

Head

Crest of the neck

Head

2

Head

Articular facet of the head

Costal cartilage

▼ **Frontal view of the thorax**

Superior thoracic opening

Body of the Manubrium

Angle of the sternum

Intercostals spaces

True ribs

Costal cartilages

False ribs

Body of the sternum

Xiphoid process

Infrasternal angle

Costal arch

Floating ribs

Inferior thoracic opening

▲ **Ribs**
Depending on their position in the thorax, ribs can be very different.
1. Rib I and rib II
They have unique characteristics: rib I has a non-prominent groove for a vein and an artery in which the subclavian artery and the primary trunk of the brachial

plexus run trough. The rib II has a tuberosity from which the anterior serratus muscle inserts. The cartilage was removed
2. Rib XII
Anterior view of a floating rib, including the cartilage.

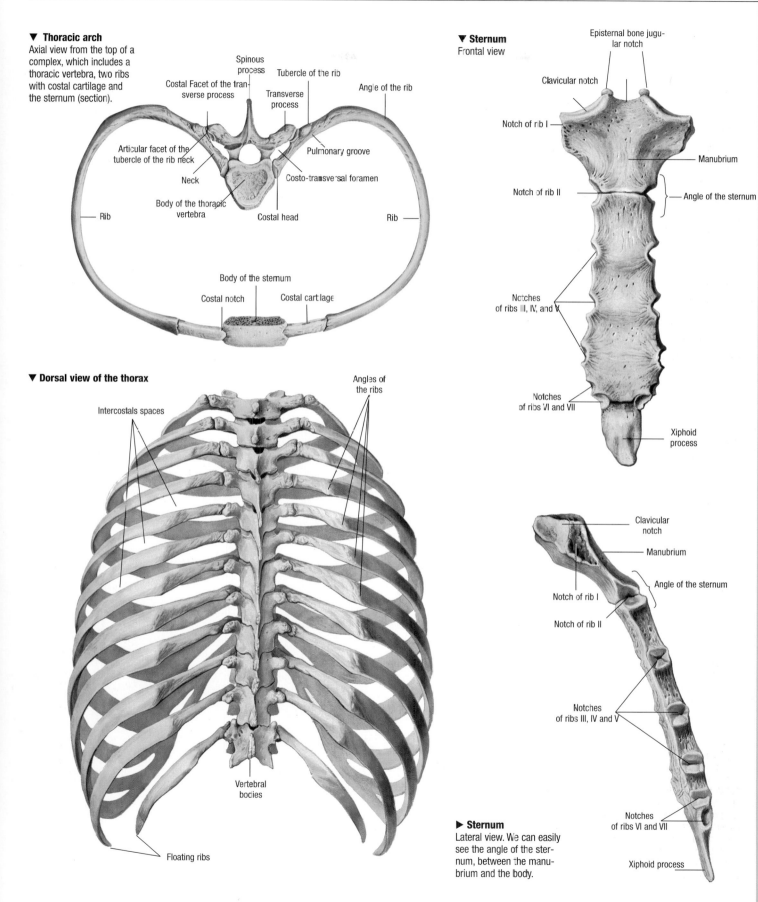

▼ Thoracic arch
Axial view from the top of a complex, which includes a thoracic vertebra, two ribs with costal cartilage and the sternum (section).

Spinous process
Tubercle of the rib
Costal Facet of the transverse process
Angle of the rib
Transverse process
Articular facet of the tubercle of the rib neck
Pulmonary groove
Neck
Costo-transversal foramen
Body of the thoracic vertebra
Costal head
Rib
Rib
Body of the sternum
Costal notch
Costal cartilage

▼ Dorsal view of the thorax

Intercostals spaces
Angles of the ribs
Vertebral bodies
Floating ribs

▼ Sternum
Frontal view

Episternal bone jugular notch
Clavicular notch
Notch of rib I
Manubrium
Notch of rib II
Angle of the sternum
Notches of ribs III, IV, and V
Notches of ribs VI and VII
Xiphoid process

▶ Sternum
Lateral view. We can easily see the angle of the sternum, between the manubrium and the body.

Clavicular notch
Manubrium
Angle of the sternum
Notch of rib I
Notch of rib II
Notches of ribs III, IV and V
Notches of ribs VI and VII
Xiphoid process

SCAPULAR GIRDLE AND THE BONES OF THE UPPER LIMBS

The upper limbs have the same structure as the lower limbs. They include a girdle, or belt, which is made up of a series of bones connected to each other so as to surround the body. The girdle connects a series of long bones to the spinal column. At the end of each long bone there is a group of specialized smaller bones. As for the upper limbs, the girdle is actually the scapula. The scapula is located behind (posterior) the thorax, and along with the clavicle, which is located in the front. It makes up the shoulder joint. This joint connects the humerus to the body (the humerus is the first long bone of the upper limb). The bones of the scapular girdle do not articulate on the vertebrae, but are tightly connected to the trunk through muscles and ligaments.

The humerus represents the bone structure of the arm, and it articulates into the two long bones of the forearm: the radius and the ulna. These two, in turn, articulate with the carpus, a group of small bones that, along with the little bones of the metacarpus, make up the hand. The shoulder, the elbow (between the arm and the forearm) and the wrist (between the forearm and the hand) are extremely mobile joints, but are much weaker than their equivalent joints of the lower limb. In fact, they have to guarantee a high degree of movement, a strong grip and refined manipulation, rather than the ability to lift and sustain physical effort. For example, the humerus rests on a much shallower groove than the hip (52): this allows the arm to make a wide range of movements, but exposes the joint to potential dislocations. On the other hand, because the two bones of the forearm are more mobile than the corresponding bones of the leg, the elbow not only allows for "hinge" movements, but also for ample torsions.

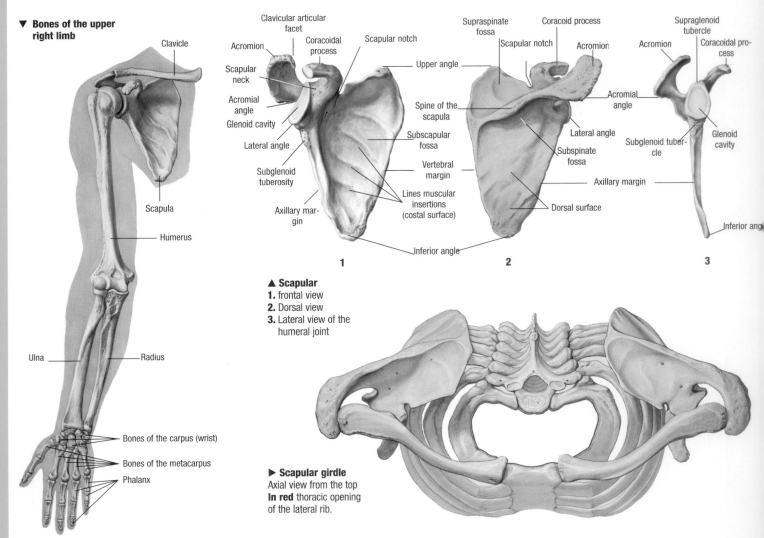

▼ **Bones of the upper right limb**

Clavicle

Scapula

Humerus

Ulna

Radius

Bones of the carpus (wrist)

Bones of the metacarpus

Phalanx

Clavicular articular facet

Acromion

Scapular neck

Acromial angle

Glenoid cavity

Lateral angle

Subglenoid tuberosity

Axillary margin

Coracoidal process

Scapular notch

Upper angle

Spine of the scapula

Subscapular fossa

Vertebral margin

Lines muscular insertions (costal surface)

Inferior angle

1

Supraspinate fossa

Scapular notch

Coracoid process

Acromion

Acromial angle

Lateral angle

Subspinate fossa

Axillary margin

Dorsal surface

2

Supraglenoid tubercle

Acromion

Coracoidal process

Subglenoid tubercle

Glenoid cavity

Axillary margin

Inferior ang

3

▲ **Scapular**
1. frontal view
2. Dorsal view
3. Lateral view of the humeral joint

▶ **Scapular girdle**
Axial view from the top
In red thoracic opening of the lateral rib.

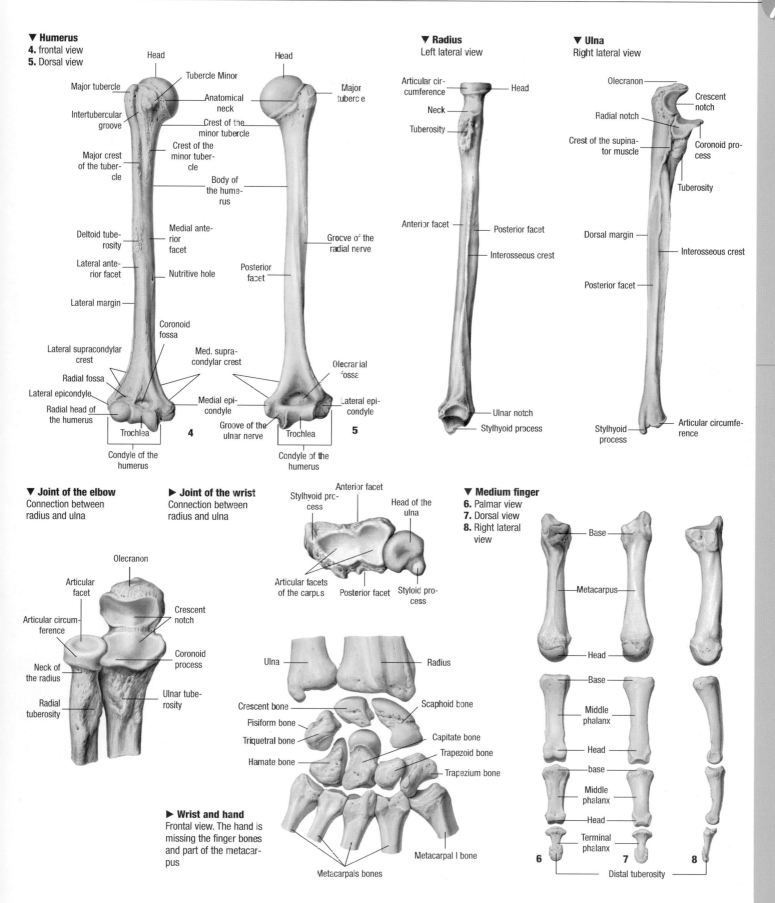

▼ Humerus
4. frontal view
5. Dorsal view

Head
Tubercle Minor
Major tubercle
Anatomical neck
Intertubercular groove
Crest of the minor tubercle
Crest of the minor tubercle
Major crest of the tubercle
Body of the humerus
Head
Major tubercle
Deltoid tuberosity
Medial anterior facet
Lateral anterior facet
Nutritive hole
Groove of the radial nerve
Posterior facet
Lateral margin
Coronoid fossa
Lateral supracondylar crest
Med. supracondylar crest
Olecranial fossa
Radial fossa
Lateral epicondyle
Radial head of the humerus
Medial epicondyle
Lateral epicondyle
Trochlea
4
Groove of the ulnar nerve
Trochlea
5
Condyle of the humerus
Condyle of the humerus

▼ Radius
Left lateral view

Articular circumference
Head
Neck
Tuberosity
Anterior facet
Posterior facet
Interosseous crest
Ulnar notch
Stylhyoid process

▼ Ulna
Right lateral view

Olecranon
Crescent notch
Radial notch
Coronoid process
Crest of the supinator muscle
Tuberosity
Dorsal margin
Interosseous crest
Posterior facet
Stylhyoid process
Articular circumference

▼ Joint of the elbow
Connection between radius and ulna

Olecranon
Articular facet
Articular circumference
Crescent notch
Neck of the radius
Coronoid process
Radial tuberosity
Ulnar tuberosity

► Joint of the wrist
Connection between radius and ulna

Anterior facet
Stylhyoid process
Head of the ulna
Articular facets of the carpus
Posterior facet
Styloid process

► Wrist and hand
Frontal view. The hand is missing the finger bones and part of the metacarpus

Ulna
Radius
Crescent bone
Scaphoid bone
Pisiform bone
Triquetral bone
Capitate bone
Hamate bone
Trapezoid bone
Trapezium bone
Metacarpal I bone
Metacarpals bones

▼ Medium finger
6. Palmar view
7. Dorsal view
8. Right lateral view

Base
Metacarpus
Head
Base
Middle phalanx
Head
base
Middle phalanx
Head
Terminal phalanx
6   **7**   **8**
Distal tuberosity

PELVIS AND LOWER LIMBS

Like the upper limbs, a girdle, called pelvic girdle, pelvis or hip, also connects the lower limbs to the spinal column. As distinct from the scapular girdle, the pelvic girdle is made up of three hipbones, which firmly articulate with each other, in the backside, on the sacral bone and on the coccyx, and in the front. Unlike the scapular girdle, the pelvis changes into a relatively rigid, massive and well anchored osseous complex. This complex is cha-

racterized by a concave shape, which supports the inferior abdominal organs, and offers a solid articulation of the inferior joints.

In fact, thanks to the strong articulations of the hip, the pelvis connects the free part of the limbs to the body, allowing for a biped walk and an erect sitting posture ➤ 70.

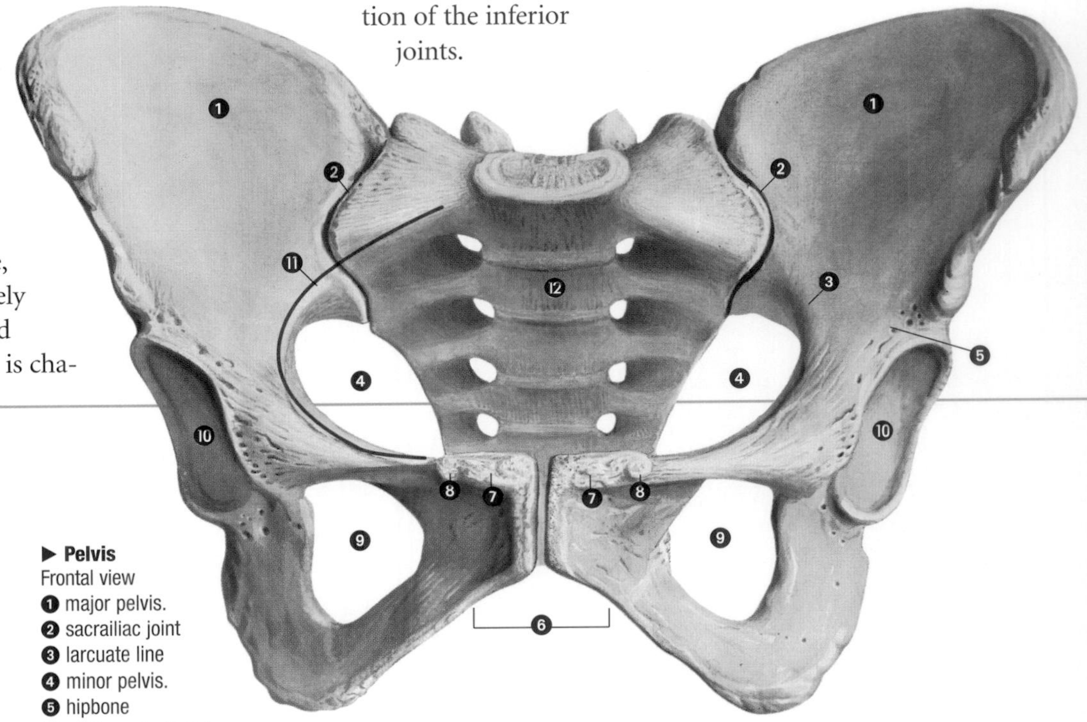

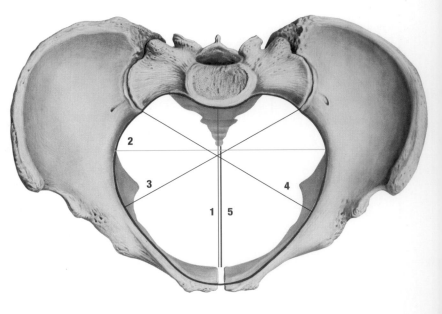

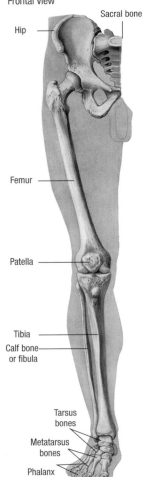

▼ Bones of the right inferior limb
Frontal view

Sacral bone
Hip
Femur
Patella
Tibia
Calf bone or fibula
Tarsus bones
Metatarsus bones
Phalanx

▶ Pelvis
Frontal view
❶ major pelvis.
❷ sacrailiac joint
❸ larcuate line
❹ minor pelvis.
❺ hipbone
❻ pubic angle (arch)
❼ pubic crest
❽ tpubic tubercle
❾ Obturator foramen
❿ acetabulum
⓫ terminal line
⓬ Sacral bone

▶ Top view of a female pelvis
The anatomical, anthropological, and practical dimensions are visible (in obstetrics).
We can see **1**. an antero-posterior diameter or anatomic conjugate (ca. 11-9,5 cm);
a maximum transverse diameter; **3**. a left oblique diameter and a right oblique diameter (**4**) (ca. 12 cm); **5**. a true or obstetrical conjugate

▼ Pelvis
Dorsal view that highlights the articular ligaments

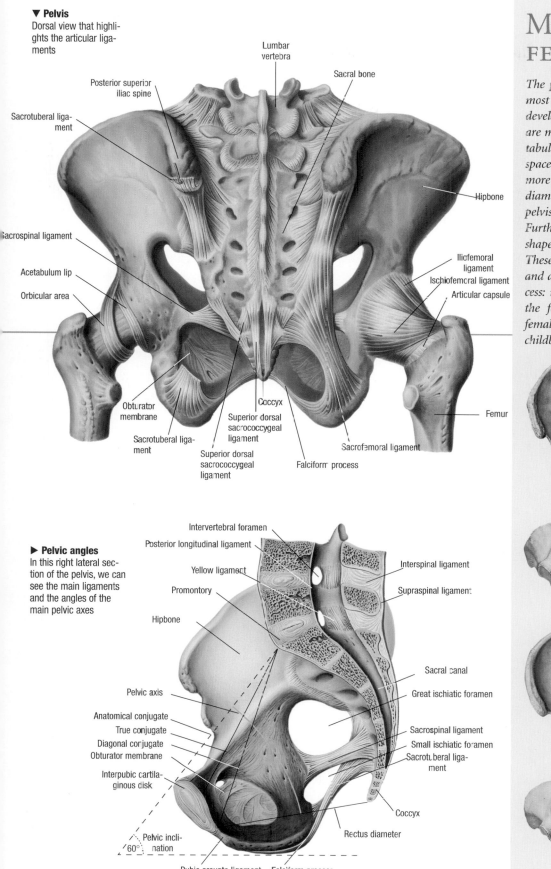

Lumbar vertebra

Sacral bone

Posterior superior iliac spine

Sacrotuberal ligament

Sacrospinal ligament

Acetabulum lip

Orbicular area

Hipbone

Iliofemoral ligament

Ischiofemoral ligament

Articular capsule

Obturator membrane

Coccyx

Superior dorsal sacrococcygeal ligament

Femur

Sacrotuberal ligament

Superior dorsal sacrococcygeal ligament

Falciform process

Sacrofemoral ligament

► Pelvic angles
In this right lateral section of the pelvis, we can see the main ligaments and the angles of the main pelvic axes

Intervertebral foramen

Posterior longitudinal ligament

Yellow ligament

Promontory

Hipbone

Interspinal ligament

Supraspinal ligament

Sacral canal

Pelvic axis

Anatomical conjugate

True conjugate

Diagonal conjugate

Obturator membrane

Interpubic cartilaginous disk

Great ischiatic foramen

Sacrospinal ligament

Small ischiatic foramen

Sacrotuberal ligament

Coccyx

Pelvic inclination 60°

Rectus diameter

Pubic arcuate ligament   Falciform process

MALE AND FEMALE

The pelvis is the anatomical part that is the most distinct between the sexes: in female, it develops primarily in width. The iliac wings are more flared and tilted outwards. The acetabuli and the ischiatic tuberosities are more spaced, and the walls of the little pelvis are more vertical. In males, the pelvis is higher; the diameter of the large pelvis of the small inferior pelvis and the pubic angle are more acute. Furthermore, the pelvic inlet (In color) is oval-shaped in females, and heart-shaped in males. These differences first appear during puberty, and are firmly linked to the reproductive process: in fact, besides supporting the weight of the fetus during the entire pregnancy, the female pelvis must hold it and release it during childbirth

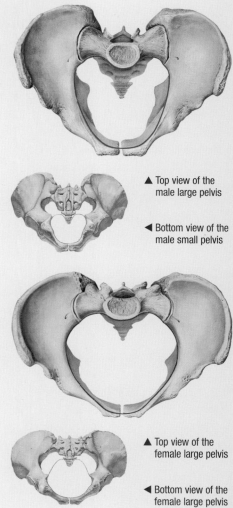

▲ Top view of the male large pelvis

◄ Bottom view of the male small pelvis

▲ Top view of the female large pelvis

◄ Bottom view of the female large pelvis

51

Each hipbone is a flat bone and it is made up of three bones (ileum, ischium, and pubis) that fuse together during the body's development. Each hipbone articulates with the femur, the thigh bone.

In turn, the femur distally articulates with three bones, which make up the skeleton in the leg: 2 long ones (the tibia and the calf bone or fibula) and a short flat bone (the patella) in the knee joint ➤ 50.

This is one of the most complex joints in our body: while the articular surfaces of the bones would allow for a wide range of movements, the bonds of the ligaments only allow the knee to flex and extend.

The tibia, just like the femur, is a long, voluminous and slightly curved bone: it constitutes most of

the distal part of the articulation of the knee, because the fibula, although equally long, is much thinner. The patella offers a point of insertion for the quadriceps muscle of the femur.

From a distal standpoint, the long bones articulate with the short bones of the foot: the tarsus bones, organized into 2 rows: a posterior one, which includes the astragalus, or talus, and the heel), and an anterior one, which includes scaphoid, cuboids and 3 cuneiform bones. These are followed by the bones of the metatarsus (5 small long bones) and the phalanxes, similar to the corresponding bones of the hand in number and shape.

Starting from the first finger, the phalanxes get smaller. Each finger, except

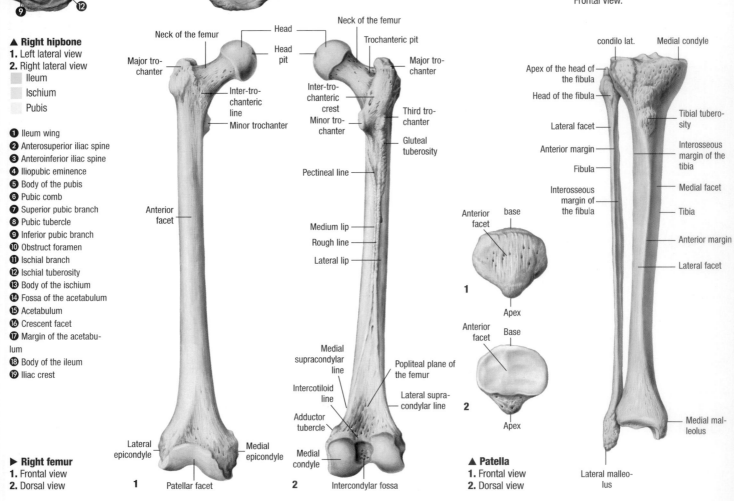

▲ **Right hipbone**
1. Left lateral view
2. Right lateral view
   Ileum
   Ischium
   Pubis

❶ Ileum wing
❷ Anterosuperior iliac spine
❸ Anteroinferior iliac spine
❹ Iliopubic eminence
❺ Body of the pubis
❻ Pubic comb
❼ Superior pubic branch
❽ Pubic tubercle
❾ Inferior pubic branch
❿ Obstruct foramen
⓫ Ischial branch
⓬ Ischial tuberosity
⓭ Body of the ischium
⓮ Fossa of the acetabulum
⓯ Acetabulum
⓰ Crescent facet
⓱ Margin of the acetabulum
⓲ Body of the ileum
⓳ Iliac crest

Right femur
1. Frontal view
2. Dorsal view

Femur 1 labels:
Neck of the femur
Head
Head pit
Major trochanter
Inter-trochanteric line
Minor trochanter
Anterior facet
Lateral epicondyle
Medial epicondyle
Patellar facet

1

Femur 2 labels:
Head
Neck of the femur
Trochanteric pit
Major trochanter
Inter-trochanteric crest
Minor trochanter
Third trochanter
Gluteal tuberosity
Pectineal line
Medium lip
Rough line
Lateral lip
Medial supracondylar line
Intercotiloid line
Adductor tubercle
Medial condyle
Popliteal plane of the femur
Lateral supracondylar line
Intercondylar fossa

2

▲ **Patella**
1. Frontal view
2. Dorsal view

Patella 1 labels:
Anterior facet
base
Apex

1

Patella 2 labels:
Anterior facet
Base
Apex

2

▼ **Right fibula and tibia**
Frontal view.

condilo lat.
Medial condyle
Apex of the head of the fibula
Head of the fibula
Lateral facet
Anterior margin
Fibula
Interosseous margin of the fibula
Tibial tuberosity
Interosseous margin of the tibia
Medial facet
Tibia
Anterior margin
Lateral facet
Medial malleolus
Lateral malleolus

for the hallux, is made up by 3 pha-lanxes which starting from the metatar-sus, are called proximal, medial and distal; or first, second and third). The bones of the foot are connected to each other and to the bones of the leg through fibrous articular capsules. These capsules are reinforced by ligaments that guarantee strength, resistance and at the same time, the necessary mobility for walking.

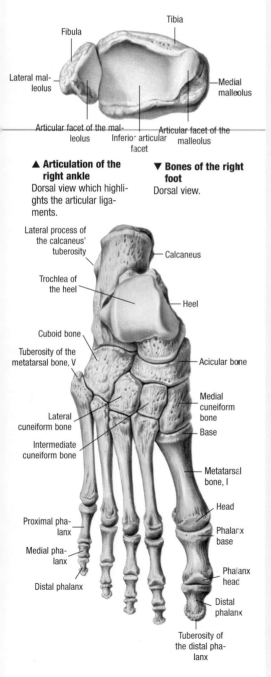

Fibula — **Tibia**

Lateral mal-leolus — **Medial malleolus**

Articular facet of the mal-leolus — **Inferior articular facet** — **Articular facet of the malleolus**

▲ Articulation of the right ankle
Dorsal view which highlights the articular ligaments.

▼ Bones of the right foot
Dorsal view.

Lateral process of the calcaneus' tuberosity — **Calcaneus**

Trochlea of the heel — **Heel**

Cuboid bone

Tuberosity of the metatarsal bone, V — **Acicular bone**

Lateral cuneiform bone — **Medial cuneiform bone**

Intermediate cuneiform bone — **Base**

Metatarsal bone, I

Proximal pha-lanx — **Head**

Medial pha-lanx — **Phalanx base**

Distal phalanx — **Phalanx head**

Distal phalanx

Tuberosity of the distal pha-lanx

✚ OSTEOPOROSIS

Osteoporosis is the name commonly given to primary involutional osteoporosis, and unlike secondary osteoporosis, it is not correlated to other diseases. Osteoporosis is a degenerative syndrome of the bones: the bone tissue decalcifies and atrophies, becoming more fragile.

In everyone, a progressive loss of bone density starts at the age of 40. While men loose about 0,3-0,5 % of their bone density each year (senile osteoporosis or Type II), women, can loose up to 3-5% of bone density, during menopause (postmenopausal osteoporosis or Type I). After reaching 65-70 years of age, this rapid rate of bone density loss decreases, reaching the same levels as men. This difference between men and women is caused by a profound hormonal change during menopause, and particularly a drop in estrogen levels, which control the equilibrium of the skeletal metabolism ➤38.

Osteoporosis strikes the entire body, but the most affected areas are those, which support most of the body's weight: the spinal column and the pelvic-leg system (mainly the femur). Osteoporosis causes spontaneous fractures and painful crushing of vertebrae and of the knee. Computerized instruments are used to diagnose osteoporosis, and x-rays are used to view the areas of the body affected by decalcification.

In both men and women, osteoporosis can be embarked or favored by circumstances independent from age, such a poor diet (lack of calcium), smoke, alcohol abuse, overweight, and a reduced physical activity.

Pharmaceutical drugs can reduce the bone resorption, and can stimulate new bone growth. However the best cure is prevention. A healthy lifestyle, a good diet, a constant and moderate outdoor physical activity, is all great ways to prevent osteoporosis. Multivitamins, salt supplements and hormonal therapies during menopause are also very efficient

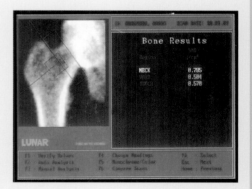

▲ Osteoporotic femur
The double photonic computerized bone mineralometry (DEXA, dual energy Xray Absonometry) shows the decalcified areas and quantifies the loss of bone density.

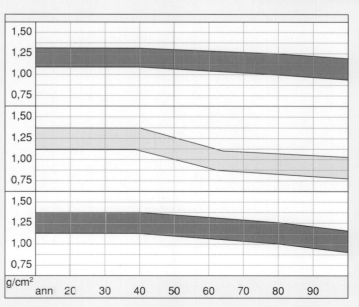

◄ Loss of bone density
Average in males (blue); average in females (pink); average in a group of women who have taken an additional daily dose of mineral salts, and who have been physically active, in order to strengthen the skeleton (walking, playing tennis…etc) and have been sunbathing, in order to stimulate the production of Vitamin D (Red).

THE MUSCLE TISSUE IS SENSITIVE TO NERVOUS STIMULI AND MAKES UP 35-40% OF THE
BODY WEIGHT. IT IS ORGANIZED IN MUSCLES AND IT IS RESPONSIBLE FOR ALL THE
MOVEMENTS AND FOR THE "TONICITY" OF OUR BODY

THE MUSCLES
OR THE MUSCULAR SYSTEM

The muscles are organs primarily made of muscular tissue.
They are able to move both the bones connected by diarthrosis
or amphiarthrosis, the skin, and the internal organs, such as
the stomach and the intestine; in fact the contractions
and the extensions of muscles can be transmitted to
other parts of the body.
Muscles are divided into three main types, according
to the type of muscle tissue, which characterizes
them:
-The **skeletal muscles** are made up by striated
muscle tissue ▸[24] and they insert into the bones.
They are further divided into two types:
voluntary and involuntary.
The voluntary muscles are muscles controlled
by the central nervous system. They are able
to rapidly contract, generating a brief but
powerful burst of energy.
The involuntary muscles are controlled
by the peripheral nervous system; they
are not controlled by will (posture
muscles) and are able to generate
medium power, for longer periods of time.
-The **cardiac muscle** is made up of cardiac fibers,
laid into spiral bundles; each cell can contract
rhythmically: the entire tissue contracts in a
coordinated fashion, due to a particular anatomical
element, from which, "waves" of contraction
emanate. These waves spread to the heart, regulating
its beat. The cardiac muscle is able to sustain
strong and continuous contractions, without
getting tired.

-The **smooth muscles** are made up by smooth muscular tissue
and they control involuntary movements of the internal
organs (blood vessels, bronchi, digestive tract, uterus, etc.).
They are controlled by the autonomous nervous system and
they react to impulses by generating long lasting slow and
regular contractions.

THE SKELETAL MUSCLES

The skeletal muscles (almost all voluntary) are made
up of almost 650 layers of muscles stratified around
bones. Their size varies greatly: from the gluteus,
(many layers of thousands of fibers), to the stapedial
muscle (few fibers in the medial ear). When a
muscle connects two bones, we say that it
"originates" from one bone and that it "inserts"
into the other: the origin is recognized by the fact
that muscle fibers start directly from the
periosteum; the insertion is recognized by the
shape of the muscle, which usually becomes
thinner and ends into the tendon. A tendon is an
extremity of hard and non-elastic connective tissue.
Tendons can either have an elongated shape, and be
connected to a precise point in the bone, or they can be
shaped as a flattened layer (aponeurosis). A muscle can
originate from more than one bone: an example of

◀ **Types of muscle tissue**
  Smooth musclesi
  Skeletal muscles
  Cardiac muscles

▼ Surface skeletal muscles in an adult body
Front and back views.

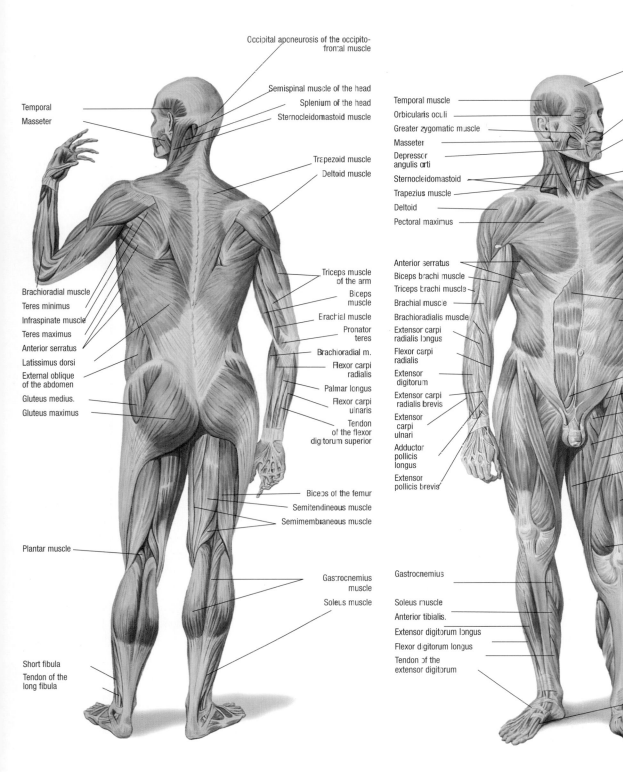

Occipital aponeurosis of the occipito-frontal muscle

Semispinal muscle of the head
Splenium of the head
Sternocleidomastoid muscle

Trapezoid muscle
Deltoid muscle

Temporal
Masseter

Triceps muscle of the arm
Biceps muscle
Brachial muscle
Pronator teres
Brachioradial m.
Flexor carpi radialis
Palmar longus
Flexor carpi ulnaris
Tendon of the flexor digitorum superior

Brachioradial muscle
Teres minimus
Infraspinate muscle
Teres maximus
Anterior serratus
Latissimus dorsi
External oblique of the abdomen
Gluteus medius.
Gluteus maximus

Biceps of the femur
Semitendineous muscle
Semimembraneous muscle

Plantar muscle

Gastrocnemius muscle
Soleus muscle

Short fibula
Tendon of the long fibula

Frontal aponeurosis of the occipitofrontal muscle
Orbicular muscle of the mouth
Mentalis muscle

Temporal muscle
Orbicularis oculi
Greater zygomatic muscle
Masseter
Depressor angulis orti
Sternocleidomastoid
Trapezius muscle
Deltoid
Pectoral maximus

Sternoid muscle

Extensor carpi ulnaris
Extensor digiti minimi
Extensor digitorum
Anconeus muscle
Flexor carpi ulnaris
Rectus abdominis
External oblique muscle of the abdomen

Anterior serratus
Biceps brachi muscle
Triceps brachi muscle
Brachial muscle
Brachioradialis muscle
Extensor carpi radialis longus
Flexor carpi radialis
Extensor digitorum
Extensor carpi radialis brevis
Extensor carpi ulnari
Adductor pollicis longus
Extensor pollicis brevis

Pyramidal muscle
Iliopsoas
Tensor fasciae latae muscle
Pectineal muscle
Adductor longus
Sartorius
Straight muscle of the thigh
Musculus rectus femoris

Vastus lateralis muscle
Vastus medialis muscle

Gastrocnemius
Soleus muscle
Anterior tibialis.
Extensor digitorum longus
Flexor digitorum longus
Tendon of the extensor digitorum

Adductor hallucis

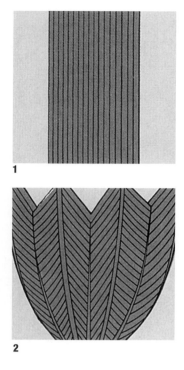

1

2

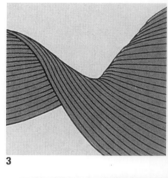

3

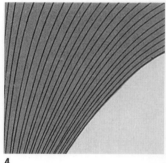

4

this is the biceps, which originates from two different bones; or the quadriceps, which originates from four bones.

A muscle can also insert into the fibrous sheath of another muscle, increasing its power. It can also connect to the subcutaneous tissue, contributing to the voluntary movement of the skin (mimic facial muscle).

Even the fibrous distribution inside the muscle is important for the effectiveness of these organs: let us take a look at the main muscular structures.

In the long muscles (1), the fibers are parallel, and their synchronized contractions are all moving in the same direction: this produces the maximum, most precise and powerful contraction, but not necessarily the most powerful. In the multipinnate muscles (2), the fibers are short and grouped into thousands of bundles, arranged obliquely to the traction axis: this allows them to be very powerful, even though the total shortening of the muscle is relatively limited. In those muscles that generally rotate around a joint (3), the fibrous fasciae are arranged in a spiral fashion. In fan-shaped muscles, which produce strong traction in a localized area, the bundles of fibers are triangular in shape and are arranged obliquely to the traction axis. A specific characteristic of skeletal muscles is to "work in pairs": if the bone can not constantly work against an equal strength, the single muscle will not be able to move the bone. In fact it can only "pull" a muscle when it contracts and it can not "push" it when it relaxes. Therefore, in order to contract and stretch a leg, there must be at least two muscles, which act in opposition; by contracting, one is able to bend the leg, while the other is able to stretch it. Because of this characteristic, those skeletal muscles (or groups of muscles), which "work in pairs", are called antagonists. When one of them contracts, the antagonist relaxes, and vice versa. This way, each movement is the result of a reciprocal and balanced muscular interaction, under the

METABOLISM OF THE MUSCULAR MOVEMENT

In order to be able to contract, muscle fibers need chemical energy. Energy is produced by the mitochondria, through respiration, a process that can take place with or without oxygen (anaerobic respiration). The muscular use of chemical energy is not very efficient: only 20%

transforms into movement (muscular work); the rest is dispersed as heat. Therefore, when a muscle contracts, it produces heat and the bi-products of cellular respiration: carbon dioxide and water. If the respiration is anaerobic the bi-product is lactic acid. If the

muscle undergoes a prolonged effort, or is not well wetted with blood (which guarantees a sufficient oxygen supply), the prolonged anaerobic respiration produces lactic acid. A water and mineral salts' deficiency inside the muscle, causes what we call a cramp. The accelerated breathing and cardiac rhythm increases the flow of oxygenated blood to the muscles, and rapidly compensates for this deficiency

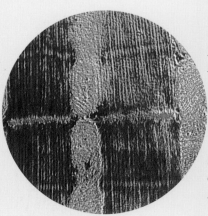

◀ Mitochondria (yellow) inside a muscle fiber.

cerebral control. However, this action is not necessarily voluntary: for example, when we travel on a bus we are consciously able to get closer to the door, but the thousands of necessary muscular movements needed to keep us balanced, are not controlled by the cerebral trunk ▸ [90] . The cerebral trunk is the encephalic center, which regulates a complex flow of sensorial information, sending the necessary signals to the muscles, in order to control the erect posture.

INVOLUNTARY MUSCLES

The number of involuntary muscles is considerable: there are those that enlarge and contract the iris ▸ [96], those that straighten the hair on our skin when it's cold, those that empty the bladder and those that allow us to swallow food even when we are upside down. They are regulated by nervous impulses of the peripheral nervous system: the sympathetic ▸ [111], activates when the body needs to fire up fast (for example, an increasing heart rate in case of danger); the parasympathetic activates when the body is at rest. For this reason the involuntary muscular system is so sensitive even to certain hormones: for example, the adrenaline produced by the adrenal glands ▸ [138] is also a chemical mediator of the sympathetic, and it directly influences the muscular activity. The endocrine controls of the involuntary muscles are parallel to those of the secondary nervous system.

Although we do not have to think about breathing, swallowing, adjusting the crystalline to focus an object or making the heart beat, there are times when we need to voluntarily intervene on "involuntary" muscles, and have the central nervous system modulate the involuntary messages: this is what happens, for example, when the bladder is full, and we are able to voluntarily go against the involuntary action which is responsible for its emptying.

▼ Involuntary muscles
The nervous impulses, which create movement, come from the cerebral trunk and the bone marrow

✚ TORN MUSCLES AND BURSITISE

These are the most popular and most frequent muscular problems. A torn muscle occurs when one or more muscles make an excessively brusque effort or sudden movement. In fact, just as any other mechanical structure, even muscles have a "breaking threshold", after which, the fibers "break". This fracture can be more or less extensive, depending on the effort that has caused it. A clear symptom of a torn muscle is a persistent muscle pain: the only cure is rest.

Bursitis is the inflammation of a mucous bursa; this is one of those small pockets, found in between muscles and joints, between tendons and bones, that allows for these elements to have the maximum mobility, and to reduce friction. Lined by cells, which under normal conditions secrete a small quantity of fluid, these pockets can get inflamed due to a non-articular rheumatic process triggered by a traumatic lesion, a dislocation or a bacterial infection. If a quantity of fluid fills the pockets beyond normal levels, violent pains can develop, especially at night, and can intensify with each movement.

As the swelling increases, an increase in temperature and pressure causes pain to intensify in the affected area.

The remedies are numerous: from cold packs to surgery. Usually though, the aspiration of the fluid, the injection of an anti-inflammatory liquid inside the bursa, along with cortisone, can decrease the symptoms.

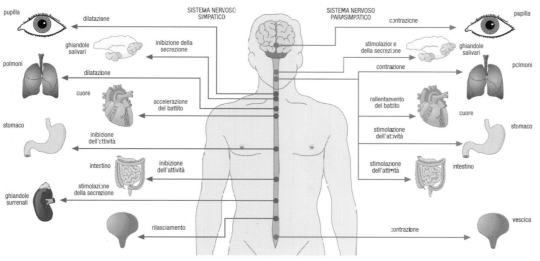

CRANIAL MUSCLES

With the exception of the small muscles of the auricle, of the middle ear, the ocular globe, the tongue, the soft palate, and of the pharynx, the cranial muscles are divided into two groups; one group is located completely inside the head, and the other connects the head to the body.

EXTRINSIC MUSCLES

They originate in different points of the axial skeleton (shoulders, neck, chest…) and they insert into the bones of the skull; these are skeletal muscles, which set the total mobility of the head on the trunk. They are divided into areas, which generally take the name of the bones they connect to, or of the main muscle.

INTRINSIC MUSCLES

They originate and insert inside the head. These are skeletal muscles needed for mastication, and mimetic muscles (or cutaneous), whose function is to mimic and protect. In animals, the voluntary mimic muscles are distributed around almost the entire body: for example, a horse is able to drive away a fly moving its skin, and a cat twitches when it is stung by a flea. In humans, on the other hand, the muscular tissue of the skin has lost this voluntary ability to move. The face, however, is an exception: the mimic muscles of the face are the only ones that originate from the bone tissue, and insert themselves directly into the deep fascia of the skin. They allow the skin to corrugate and to relax in thousands of ways, and to keep its tension, giving us facial expressions. This is the reason why they are also called mimic muscles, which means that they are designed to non-verbally communicate. Non-verbal communication dates back those days when verbal communication had not yet developed: all animals, but mainly

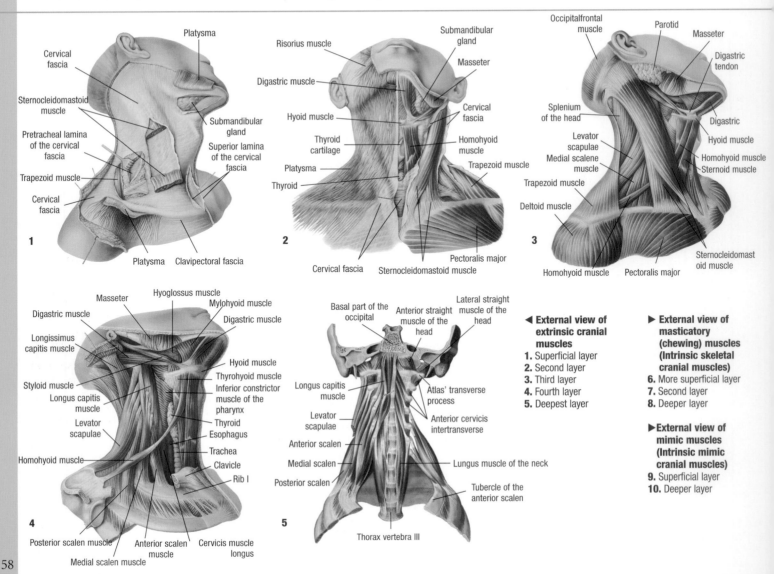

1. Cervical fascia
Platysma
Sternocleidomastoid muscle
Pretracheal lamina of the cervical fascia
Trapezoid muscle
Cervical fascia
Submandibular gland
Superior lamina of the cervical fascia
Platysma
Clavipectoral fascia

2. Risorius muscle
Digastric muscle
Hyoid muscle
Thyroid cartilage
Platysma
Thyroid
Submandibular gland
Masseter
Cervical fascia
Homohyoid muscle
Trapezoid muscle
Pectoralis major
Cervical fascia
Sternocleidomastoid muscle

3. Occipitalfrontal muscle
Parotid
Masseter
Digastric tendon
Splenium of the head
Levator scapulae
Medial scalene muscle
Trapezoid muscle
Deltoid muscle
Homohyoid muscle
Pectoralis major
Digastric
Hyoid muscle
Homohyoid muscle
Sternoid muscle
Sternocleidomastoid muscle

4. Masseter
Hyoglossus muscle
Mylohyoid muscle
Digastric muscle
Longissimus capitis muscle
Styloid muscle
Longus capitis muscle
Levator scapulae
Homohyoid muscle
Hyoid muscle
Thyrohyoid muscle
Inferior constrictor muscle of the pharynx
Thyroid
Esophagus
Trachea
Clavicle
Rib I
Posterior scalen muscle
Anterior scalen muscle
Cervicis muscle longus
Medial scalen muscle

5. Basal part of the occipital
Anterior straight muscle of the head
Lateral straight muscle of the head
Longus capitis muscle
Levator scapulae
Anterior scalen
Medial scalen
Posterior scalen
Atlas' transverse process
Anterior cervicis intertransverse
Lungus muscle of the neck
Tubercle of the anterior scalen
Thorax vertebra III

◄ **External view of extrinsic cranial muscles**
1. Superficial layer
2. Second layer
3. Third layer
4. Fourth layer
5. Deepest layer

► **External view of masticatory (chewing) muscles (Intrinsic skeletal cranial muscles)**
6. More superficial layer
7. Second layer
8. Deeper layer

► **External view of mimic muscles (Intrinsic mimic cranial muscles)**
9. Superficial layer
10. Deeper layer

the anthropoid apes, communicate with each other through "facial expressions". By changing their facial expression, people are able to communicate even the slightest emotional changes. The mobility of the face is not culturally bound, and it has been valid since the most ancient of times: each gesture that curves the lips and the eyebrows toward the top, has a positive connotation, and vice versa. Not all mimic muscles are the same: for example, while many of us can move their nose, only few can move the scalp or roll down the tongue. These are genetically determined capabilities.

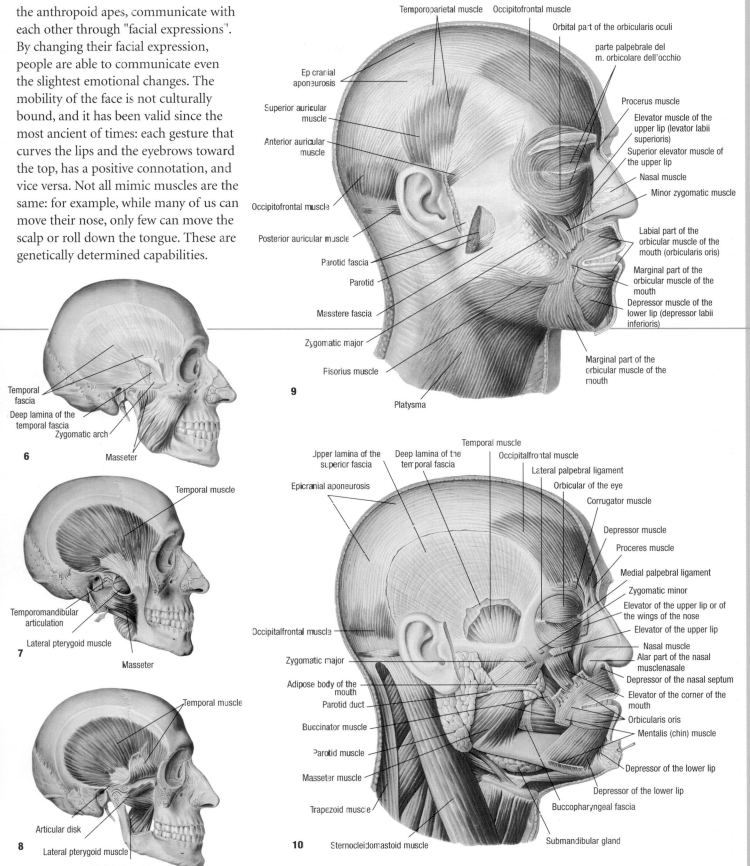

Temporoparietal muscle

Occipitofrontal muscle

Orbital part of the orbicularis oculi

parte palpebrale del m. orbiolare dell'occhio

Ep cranial aponeurosis

Superior auricular muscle

Anterior auricular muscle

Procerus muscle

Elevator muscle of the upper lip (levator labii superioris)

Superior elevator muscle of the upper lip

Nasal muscle

Minor zygomatic muscle

Occipitofrontal muscle

Posterior auricular muscle

Parotid fascia

Parotid

Labial part of the orbicular muscle of the mouth (orbicularis oris)

Marginal part of the orbicular muscle of the mouth

Depressor muscle of the lower lip (depressor labii inferioris)

Masstere fascia

Zygomatic major

Fisorius muscle

Marginal part of the orbicular muscle of the mouth

9

Platysma

Temporal fascia

Deep lamina of the temporal fascia

Zygomatic arch

Masseter

6

Temporal muscle

Temporomandibular articulation

Lateral pterygoid muscle

7

Masseter

Temporal muscle

Upper lamina of the superior fascia

Deep lamina of the temporal fascia

Temporal muscle

Occipitalfrontal muscle

Lateral palpebral ligament

Orbicular of the eye

Corrugator muscle

Epicranial aponeurosis

Depressor muscle

Proceres muscle

Medial palpebral ligament

Zygomatic minor

Occipitalfrontal muscle

Zygomatic major

Adipose body of the mouth

Parotid duct

Buccinator muscle

Parotid muscle

Masseter muscle

Trapezoid muscle

Elevator of the upper lip or of the wings of the nose

Elevator of the upper lip

Nasal muscle

Alar part of the nasal musclenasale

Depressor of the nasal septum

Elevator of the corner of the mouth

Orbicularis oris

Mentalis (chin) muscle

Depressor of the lower lip

Depressor of the lower lip

Buccopharyngeal fascia

10

Sternocleidomastoid muscle

Submandibular gland

Articular disk

Lateral pterygoid muscle

8

Medial pterygoid muscle

59

DORSAL MUSCLES

The spinal column is a very articulated, extremely resistant and flexible structure, due to a series of ligaments and skeletal muscles which hold the vertebrae tightly together. The muscles that support and move the spinal column are located mainly on the back of the vertebrae, right behind the skeleton: these are called spinodorsal muscles, and represent the deeper muscular layer, and the spinocostal muscles, located at more superficial levels. The spinodorsal muscles in particular, are for the most part made up of small bundles of fibers, which run parallel to the spinal column. The deeper ones are very short, and they join the vertebrae together. The intermediate ones are longer and connect bones 2-3 vertebrae away. The more superficial layers are much longer, and they connect vertebrae, which are far from each other. Other skeletal muscles assist these different muscle groups in supporting the body. Such muscles are located on even more superficial layers, and are directly responsible for movements. They are:

-The spinoappendicular: they structurally connect the limbs to the trunk;
-The suboccipital: they insert into the occipital bone and in the temporal bones of the skull, allowing the head to move;
-The prevertebral of the neck and the anterior straight muscle of the head: relative to the spinal column, they are ventrally located, and are responsible for the movement of the head and of the arms;
Finally, the rudimental sacralcoccygeal muscles: they are the only muscle group of the spinal column located in a ventral position.

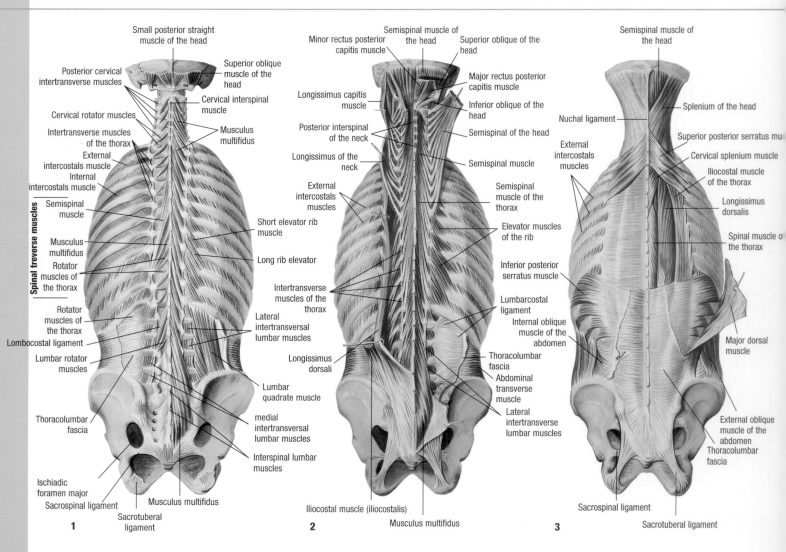

1

Small posterior straight muscle of the head
Posterior cervical intertransverse muscles
Superior oblique muscle of the head
Cervical interspinal muscle
Cervical rotator muscles
Intertransverse muscles of the thorax
Musculus multifidus
External intercostals muscle
Internal intercostals muscle
Semispinal muscle
Spinal treverse muscles
Musculus multifidus
Rotator muscles of the thorax
Rotator muscles of the thorax
Lombocostal ligament
Lumbar rotator muscles
Thoracolumbar fascia
Ischiadic foramen major
Sacrospinal ligament
Sacrotuberal ligament
Musculus multifidus
Short elevator rib muscle
Long rib elevator
Intertransverse muscles of the thorax
Lateral intertransversal lumbar muscles
Longissimus dorsali
Lumbar quadrate muscle
medial intertransversal lumbar muscles
Interspinal lumbar muscles

2

Semispinal muscle of the head
Minor rectus posterior capitis muscle
Superior oblique of the head
Longissimus capitis muscle
Major rectus posterior capitis muscle
Inferior oblique of the head
Posterior interspinal of the neck
Semispinal of the head
Longissimus of the neck
Semispinal muscle
External intercostals muscles
Semispinal muscle of the thorax
Elevator muscles of the rib
Inferior posterior serratus muscle
Lumbarcostal ligament
Internal oblique muscle of the abdomen
Thoracolumbar fascia
Abdominal transverse muscle
Lateral intertransverse lumbar muscles
Iliocostal muscle (iliocostalis)
Musculus multifidus

3

Semispinal muscle of the head
Nuchal ligament
External intercostals muscles
Splenium of the head
Superior posterior serratus mu
Cervical splenium muscle
Iliocostal muscle of the thorax
Longissimus dorsalis
Spinal muscle o the thorax
Major dorsal muscle
External oblique muscle of the abdomen
Thoracolumbar fascia
Sacrospinal ligament
Sacrotuberal ligament

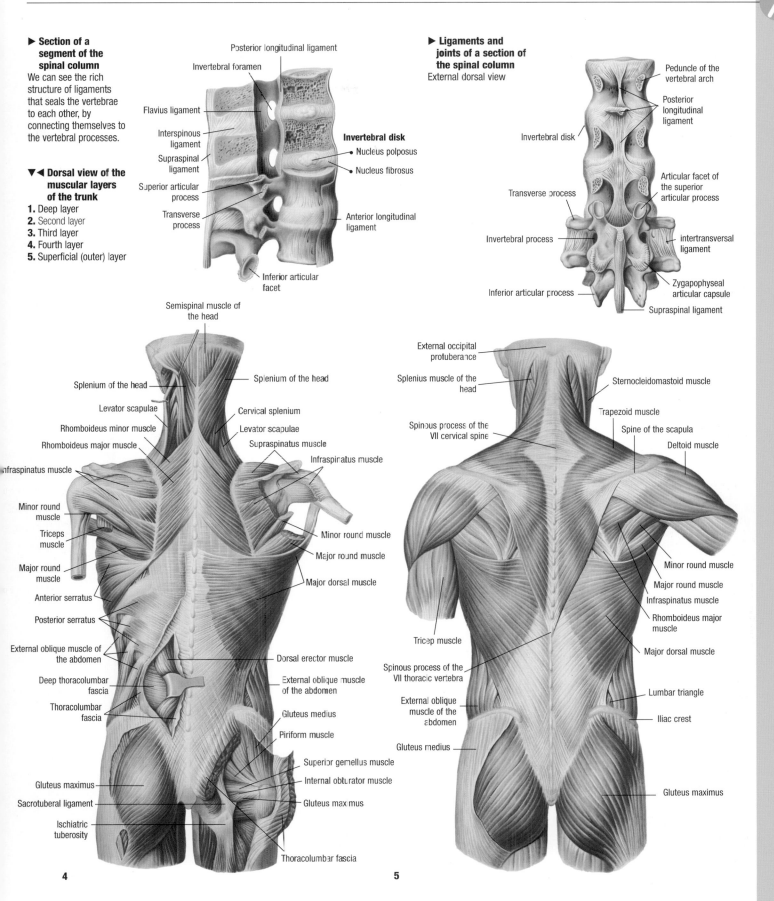

▶ **Section of a segment of the spinal column**

We can see the rich structure of ligaments that seals the vertebrae to each other, by connecting themselves to the vertebral processes.

▼◀ **Dorsal view of the muscular layers of the trunk**

1. Deep layer
2. Second layer
3. Third layer
4. Fourth layer
5. Superficial (outer) layer

Posterior longitudinal ligament

Invertebral foramen

Flavius ligament

Interspinous ligament

Supraspinal ligament

Superior articular process

Transverse process

Inferior articular facet

Invertebral disk
• Nucleus polposus
• Nucleus fibrosus

Anterior longitudinal ligament

▶ **Ligaments and joints of a section of the spinal column**
External dorsal view

Peduncle of the vertebral arch

Posterior longitudinal ligament

Invertebral disk

Transverse process

Invertebral process

Inferior articular process

Articular facet of the superior articular process

intertransversal ligament

Zygapophyseal articular capsule

Supraspinal ligament

Semispinal muscle of the head

Splenium of the head

Levator scapulae

Rhomboideus minor muscle

Rhomboideus major muscle

Infraspinatus muscle

Minor round muscle

Triceps muscle

Major round muscle

Anterior serratus

Posterior serratus

External oblique muscle of the abdomen

Deep thoracolumbar fascia

Thoracolumbar fascia

Gluteus maximus

Sacrotuberal ligament

Ischiatric tuberosity

Splenium of the head

Cervical splenium

Levator scapulae

Supraspinatus muscle

Infraspiratus muscle

Minor round muscle

Major round muscle

Major dorsal muscle

Dorsal erector muscle

External oblique muscle of the abdomen

Gluteus medius

Piriform muscle

Superior gemellus muscle

Internal obturator muscle

Gluteus maximus

Thoracolumbar fascia

External occipital protuberance

Splenius muscle of the head

Spinous process of the VII cervical spine

Tricep muscle

Spinous process of the VII thoracic vertebra

External oblique muscle of the abdomen

Gluteus medius

Sternocleidomastoid muscle

Trapezoid muscle

Spine of the scapula

Deltoid muscle

Minor round muscle

Major round muscle

Infraspinatus muscle

Rhomboideus major muscle

Major dorsal muscle

Lumbar triangle

Iliac crest

Gluteus maximus

4

5

61

MUSCLES OF THE TRUNK

They are divided into **muscles of the thorax** (**intrinsic** and **extrinsic**, depending on their insertion and their origin) and **abdominal muscles**.

INTRINSIC MUSCLES OF THE THORAX
Elevator muscles of the ribs: 12 pairs of muscles close to the spinal column.
Intercostal muscles: they occupy the spaces between the ribs; they are 11 per side and are divided into external, medial and internal.
Subcostal muscles: are found in the internal posterior of the thorax; they connect the vertebrae to the ribs.

The transverse muscle of the thorax: is located on the internal surface of the anterior thoracic wall; the endothoracic fascia protects its internal surface

EXTRINSIC MUSCLES OF THE THORAX
Thoracoappendicular muscles: they are the pectoralis major, the pectoralis minor, the subclavian muscle and the anterior serratus.
Spinoappendicular muscles: they originate from the spinal column and insert into the bones of the thoracic girdle and on the humeri; they are connected to the thorax. They

include the trapezoid muscle, the major dorsalis , the rhomboid muscle and the levator scapulae.
Spinocostal muscles: are large, thin and quadrilateral muscles and they extend from the intermediate dorsal layer. They are the posterior superior and inferior serratus muscles.
Diaphragm: it is a flat muscle, which separates the thoracic cavity from the abdominal one. It originates from the lumbar bones, the ribs and the sternum, a dome–shaped structure whose top is located inside the thoracic cavity. The diaphragm has various orifices through which the esophagus and the blood vessels run. Both surfaces are covered by a light diaphragm fascia, which connects,

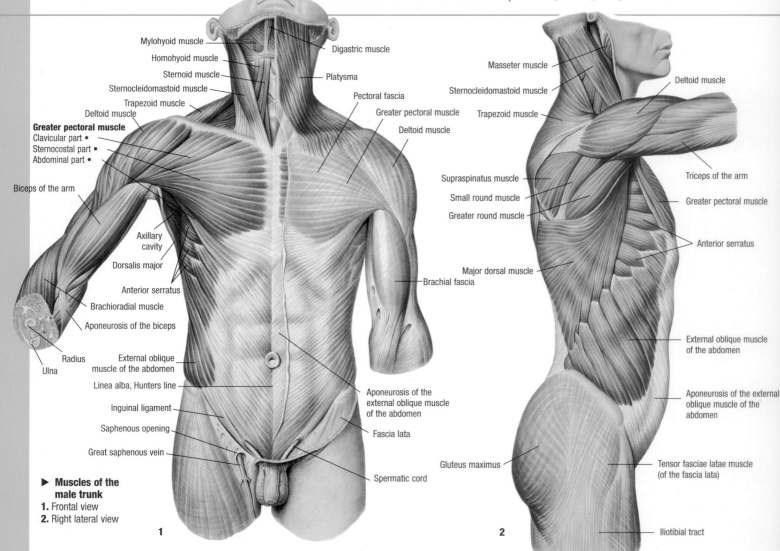

► **Muscles of the male trunk**
1. Frontal view
2. Right lateral view

at the top, to the pleura ▸[161], and at the bottom, to the peritoneum. Its movements contribute to respiration ▸[163].

ABDOMINAL MUSCLES

Rectus abdominis: it makes up the anterior abdominal wall.

Pyramidal muscle: it is a small part of the inferior and medial abdominal wall.

External and internal oblique muscle: they cover the lateral and anterior abdominal side, ascending also laterally on the thoracic wall.

Transverse muscle: runs on the inside of the internal oblique muscle.

Cremaster muscle: it extends into the genital region.

Lumbar quadrate muscle: it covers the posterior abdominal wall.

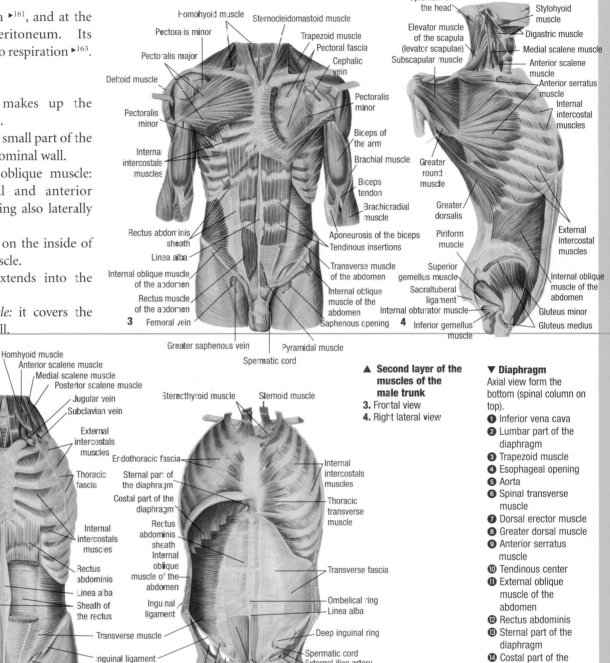

▲ Second layer of the muscles of the male trunk
3. Frontal view
4. Right lateral view

▼ Diaphragm
Axial view form the bottom (spinal column on top).
❶ Inferior vena cava
❷ Lumbar part of the diaphragm
❸ Trapezoid muscle
❹ Esophageal opening
❺ Aorta
❻ Spinal transverse muscle
❼ Dorsal erector muscle
❽ Greater dorsal muscle
❾ Anterior serratus muscle
❿ Tendinous center
⓫ External oblique muscle of the abdomen
⓬ Rectus abdominis
⓭ Sternal part of the diaphragm
⓮ Costal part of the diaphragm

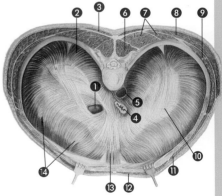

▲ Deep muscular structure of the male trunk
Frontal view.

▲ Muscles of the male abdominal cavity
Posterioranterior view (the spinal column was removed)

SCAPULAR GIRDLE AND UPPER LIMBS

Even these muscles are divided into extrinsic and intrinsic, depending on where they insert and originate from

EXTRINSIC MUSCLES
They insert into the bones of the upper limbs and of the girdle, but they originate inside the trunk. The thoracoappendicular muscles and the spinoappendicular muscles ►62 make up these muscles; particularly the muscles of the shoulder, which all originate inside the thoracic girdle, and insert into the humerus. A set of

different mucous bursae is attached to these muscles. These bursae facilitate the flow of muscular planes and of tendons. This group is made up by:
-A deltoid muscle; this is a flat and triangular muscle, which covers the lateral part of the articulation of the shoulder. Its fascicles converge on the bottom and they insert into the humerus.

-The supraspinatus muscle; it has a prismatic shape and it connects to the fibrous capsule of the joint.
-The subspinatus muscle; it is a flat triangular shaped muscle, and it inserts into the articular capsule.
-The small round muscle; it is an elongated and flat muscle and it runs close to axillary margin;
-The large round muscle is the same as the small round one, but it is deeper.
-The subscapular muscle; it is a flat and triangular shaped muscle.

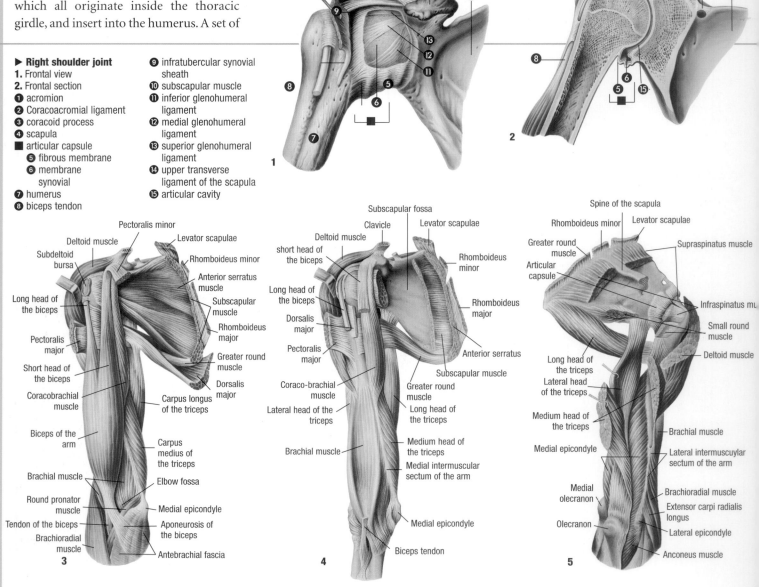

► Right shoulder joint
1. Frontal view
2. Frontal section
❶ acromion
❷ Coracoacromial ligament
❸ coracoid process
❹ scapula
■ articular capsule
  ❺ fibrous membrane
  ❻ membrane synovial
❼ humerus
❽ biceps tendon
❾ infratubercular synovial sheath
❿ subscapular muscle
⓫ inferior glenohumeral ligament
⓬ medial glenohumeral ligament
⓭ superior glenohumeral ligament
⓮ upper transverse ligament of the scapula
⓯ articular cavity

INTRINSIC MUSCLES

They are divided into:

-Anterior muscles of the arm they include the brachial bicep muscles, the coracobrachial and the anterior brachial;

-Posterior muscles of the forearm: there are 8 and are arranged in 4 consecutive layers. The round pronator makes up the superficial layer with the flexor carpi radialis, the long palmar and with the ulnar flexor of the carpus. The second layer is made up of the superficial flexor of the fingers (flexor digitorum superficialis) and the flexor longus pollicis

(thumb). The quadrate pronator muscle is found in the deep layer.

-Lateral muscles of the forearm: they are the brachiocardial, the extensor carpi radialis longus, and the extensor carpi radialis brevis.

-Posterior muscles of the forearm: there are 9 and they are arranged on two layers. On the superficial layer we find the extensor digitorum (fingers), the

extensor digiti minimi, the extensor carpi ulnaris and the anconeus. In the deeper layer, we have the supinator muscles, the longus pollici adductor, the extensor pollicis brevis and the extensor indicis.

-Muscles of the hand: they are all found on the palmar side of the hand and are divided into three groups: lateral, medial, intermediate ▶ ϵ8-69.

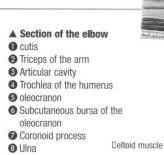

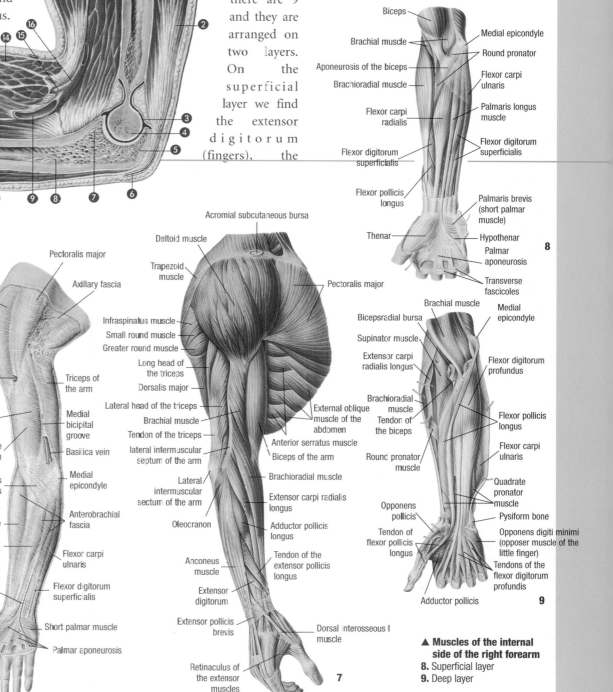

▲ Section of the elbow
1. cutis
2. Triceps of the arm
3. Articular cavity
4. Trochlea of the humerus
5. oleocranon
6. Subcutaneous bursa of the oleocranon
7. Coronoid process
8. Ulna
9. Ulnar artery
10. Extensor carpi ulnaris
11. Flexor digitorum profundus
12. Flexor digitorum superficialis
13. Flexor carpi radialis
14. Round pronator muscle
15. Radial artery
16. Brachial artery
17. Biceps of the arm
18. Brachial muscle
19. Humeri

◄ Muscles of the right shoulder
3. Frontal view, second layer
4. Frontal view, third layer
5. Dorsal view, deep muscles

▶ Muscles of the upper right arm
6. Internal frontal view of superficial muscles
7. Frontal view, first layer

▲ Muscles of the internal side of the right forearm
8. Superficial layer
9. Deep layer

Labels (Section of the elbow)
- cutis
- Triceps of the arm
- Articular cavity
- Trochlea of the humerus
- oleocranon
- Subcutaneous bursa of the oleocranon
- Coronoid process
- Ulna
- Ulnar artery
- Extensor carpi ulnaris
- Flexor digitorum profundus
- Flexor digitorum superficialis
- Flexor carpi radialis
- Round pronator muscle
- Radial artery
- Brachial artery
- Biceps of the arm
- Brachial muscle
- Humeri

Labels (figure 6)
- Pectoralis major
- Axillary fascia
- Deltoid muscle
- Triceps of the arm
- Cephalic vein
- Brachial fascia
- Medial bicipital groove
- Basilica vein
- Biceps of the arm
- Medial epicondyle
- Biceps aponeurosis
- Brachioradial muscle
- Anterobrachial fascia
- Flexor carpi radialis
- Flexor carpi ulnaris
- Tendon of the flexor digitorum superficialis
- Thenar
- Short palmar muscle
- Palmar aponeurosis

Labels (figure 7)
- Acromial subcutaneous bursa
- Deltoid muscle
- Trapezoid muscle
- Infraspinatus muscle
- Small round muscle
- Greater round muscle
- Long head of the triceps
- Dorsalis major
- Lateral head of the triceps
- Brachial muscle
- Tendon of the triceps
- lateral intermuscular septum of the arm
- Lateral intermuscular sectum of the arm
- Oleocranon
- Anconeus muscle
- Extensor digitorum
- Extensor pollicis brevis
- Retinaculus of the extensor muscles
- Pectoralis major
- External oblique muscle of the abdomen
- Anterior serratus muscle
- Biceps of the arm
- Brachioradial muscle
- Extensor carpi radialis longus
- Adductor pollicis longus
- Tendon of the extensor pollicis longus
- Dorsal interosseous I muscle

Labels (figure 8)
- Biceps
- Brachial muscle
- Aponeurosis of the biceps
- Brachioradial muscle
- Flexor carpi radialis
- Flexor digitorum superficialis
- Flexor pollicis longus
- Thenar
- Medial epicondyle
- Round pronator
- Flexor carpi ulnaris
- Palmaris longus muscle
- Flexor digitorum superficialis
- Palmaris brevis (short palmar muscle)
- Hypothenar
- Palmar aponeurosis
- Transverse fascicoles

Labels (figure 9)
- Brachial muscle
- Bicepsradial bursa
- Supinator muscle
- Extensor carpi radialis longus
- Brachioradial muscle
- Tendon of the biceps
- Round pronator muscle
- Opponens pollicis
- Tendon of flexor pollicis longus
- Adductor pollicis
- Medial epicondyle
- Flexor digitorum profundus
- Flexor pollicis longus
- Flexor carpi ulnaris
- Quadrate pronator muscle
- Pysiform bone
- Opponens digiti minimi (opposer muscle of the little finger)
- Tendons of the flexor digitorum profundis

6 7 8 9

PELVIS AND LOWER LIMBS

They are divided into 4 groups, depending on part the limb from which they insert.

HIP MUSCLES

They are divided into:
-*Internal:* the iliopsoas muscle and the small iliopsoas muscle.
-*External:* the gluteus maximus (the most superficial and the vastest, the largest muscle in the entire body), the gluteus medius and minor, the pyriform muscle, the inferior gemellus muscle, the external and the internal obturator muscle and the quadratus femuris

THIGH MUSCLES

They are divided into:
-*Anterolateral:* the tensor fasciae latae, the sartorius, and the quadriceps femuralis.
-*Posteromedial:* the gracilis muscle, the pectineous, the adductor longus, brevis and minimus; the biceps femuralis, the semitendineous and semi membranous muscle.

LEG MUSCLES

They are divided into 3 groups:
-*Anterior:* the anterior tibialis, the extensor digitorum longus, extensor hallucis longus and the anterior fibular;
-*Lateral:* the fibular longus and brevis.
-*Posterior:* these muscles are arranged on two layers: in the superficial layer we can find the triceps of the suture and the plantar muscle; deeper, there is the poplyteous muscle, the posterior tibialis, the flexor digitorum longus and the flexor hallucis longus.

▶ **Superficial muscles of the right thigh**
External lateral view:
❶ External oblique muscle of the abdomen;
❷ Gluteus medium;
❸ Tensor of the fasciae latae;
❹ Startorius;
❺ Iliotibial tract;
❻ Rectus femoris;
❼ Vastus lateralis;
❽ Ligament of the patella;
❾ Head of the tibia;
❿ Gastrocnemius

muscle;
⓫ Semimembranous muscle;
⓬ Biceps of the femus;
⓭ Gluteus maximus;
⓮ Iliac crest
⓯ Popliteal crest
Internal lateral view:
① Pyriform muscle;
② Internal obturator muscle.;
③ Sacrospinal ligament;
④ Gluteus maximus;
⑤ Sacrotuberal ligament;
⑥ Adductor magnus;

⑦ Gracilis muscle;
⑧ Semitendineous muscle;
⑨ Semimembranous muscle;
⑩ Gastrocnemius muscle;
⑪ Medial major muscle;
⑫ Sartorius muscle;
⑬ Rectus femoris;
⑭ Adductor longus;
⑮ Pectineal muscle;
⑯ Iliac muscle;
⑰ Psoas major

▼ **Muscles of the right thigh**
Frontal view:
1. First layer
2. Deep muscles
Dorsal view:
3. Superficial muscles
4. Deep muscles

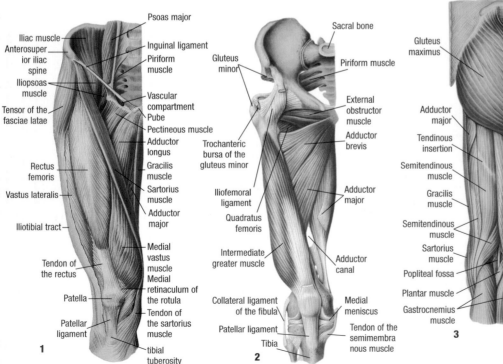

1
Iliac muscle
Anterosuperior iliac spine
Iliopsoas muscle
Tensor of the fasciae latae
Rectus femoris
Vastus lateralis
Iliotibial tract
Tendon of the rectus
Patella
Patellar ligament

Psoas major
Inguinal ligament
Piriform muscle
Vascular compartment
Pube
Pectineous muscle
Adductor longus
Gracilis muscle
Sartorius muscle
Adductor major
Medial vastus muscle
Medial retinaculum of the rotula
Tendon of the sartorius muscle
tibial tuberosity

2
Gluteus minor
Sacral bone
Piriform muscle
External obstructor muscle
Adductor brevis
Trochanteric bursa of the gluteus minor
Iliofemoral ligament
Quadratus femoris
Adductor major
Intermediate greater muscle
Adductor canal
Collateral ligament of the fibula
Patellar ligament
Tibia
Medial meniscus
Tendon of the semimembranous muscle

3
Gluteus maximus
Adductor major
Tendinous insertion
Semitendinous muscle
Gracilis muscle
Semitendinous muscle
Sartorius muscle
Popliteal fossa
Plantar muscle
Gastrocnemius muscle
Ischial foramen major
Sacrospinal ligament
Internal obturator muscle
Ischial bursa of the internal obturator muscle
Ischial tuber
Iliotibial tract
Biceps of the femur
Collateral ligament of the tibia
Posterior cruciate ligament

4
Gluteus medius
Gluteus maximus
Gluteus minor
Bursa of the piriform muscle
Trochanteric bursa of the gluteus medius
Twin (gemellus) muscles
External obturator muscle
Gluteus maximus
Adductor major
Lateral intermuscular sectum of the femur
Vastus lateralis
Vastus medial
Ligament of the meniscus lateralis
Popliteal muscle
Collateral ligament of the fibula

FOOT MUSCLES

The plantar region is covered by the plantar and subcutaneous aponeurosis. This is a fibrous formation similar to that found on the top of the hand, but a much thicker one. It separates all the muscles of the foot and of the skin (cutis), and it is divided into areas named after the underlying muscles. Such areas are the
-*Dorsal:* extensor digitorum or pedidius brevis.
-*Plantar:* divided into medial (they move the hallux: abductor, flexor brevis and the adductor), intermediate (flexor digitorum brevis, quadratus of the foot, 4 lumbrical, and 7 interosseous), and lateral (they move the 5th toe: abductor, flexor brevis and opposer of the 5th toe).

▶ **Right foot**
Plantar view.

▶ **Muscles of the right foot**
Right lateral view.

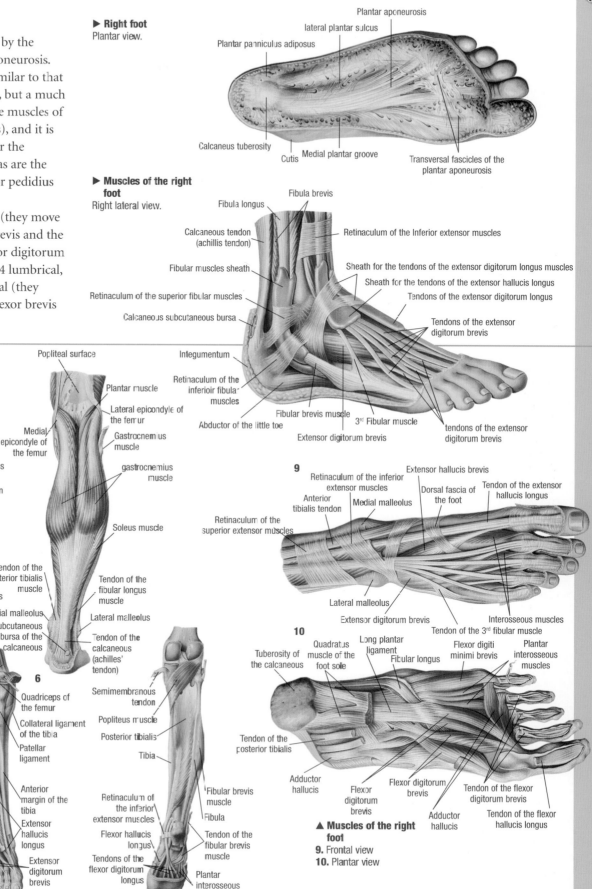

Plantar aponeurosis
lateral plantar sulcus
Plantar panniculus adiposus
Calcaneus tuberosity
Cutis   Medial plantar groove
Transversal fascicles of the plantar aponeurosis

Fibula brevis
Fibula longus
Calcaneous tendon (achillis tendon)
Fibular muscles sheath
Retinaculum of the superior fibular muscles
Calcaneous subcutaneous bursa
Integumentum
Retinaculum of the inferioir fibular muscles
Abductor of the little toe
Fibular brevis muscle
3ʳᵈ Fibular muscle
Extensor digitorum brevis
Retinaculum of the Inferior extensor muscles
Sheath for the tendons of the extensor digitorum longus muscles
Sheath for the tendons of the extensor hallucis longus
Tendons of the extensor digitorum longus
Tendons of the extensor digitorum brevis
tendons of the extensor digitorum brevis

Quadriceps of the femur
Patellar ligament
Iliotibial tract
Articular capsule of the knee
Fibular longus muscle
Fibular brevis muscle
Gastrocnemius muscle
Anterior margin of the tibia
Extensor digitorum longus
Soleus muscle
Retinaculum of the superior extensor muscles
Anterior tibialis muscle
Retinaculum of the inferior extensor muscles
Tendon of the extensor hallucis longus

Popliteal surface
Plantar muscle
Lateral epicondyle of the femur
Medial epicondyle of the femur
Gastrocnemius muscle
gastrocnemius muscle
Soleus muscle
Tendon of the posterior tibialis muscle
Tendon of the fibular longus muscle
Medial malleolus
Lateral malleolus
Subcutaneous bursa of the calcaneous
Tendon of the calcaneous (achilles' tendon)

9
Retinaculum of the inferior extensor muscles
Anterior tibialis tendon
Retinaculum of the superior extensor muscles
Extensor hallucis brevis
Dorsal fascia of the foot
Medial malleolus
Tendon of the extensor hallucis longus
Lateral malleolus
Extensor digitorum brevis
Interosseous muscles
Tendon of the 3ʳᵈ fibular muscle

10
Tuberosity of the calcaneus
Quadratus muscle of the foot sole
Long plantar ligament
Fibular longus
Flexor digiti minimi brevis
Plantar interosseous muscles
Tendon of the posterior tibialis
Adductor hallucis
Flexor digitorum brevis
Flexor digitorum brevis
Tendon of the flexor digitorum brevis
Adductor hallucis
Tendon of the flexor hallucis longus

▲ **Muscles of the right foot**
9. Frontal view
10. Plantar view

5
Articular muscles of the knee
Collateral ligament of the fibula
Interosseous membrane
Quadriceps of the femur
Collateral ligament of the tibia
Patellar ligament
Anterior margin of the tibia
Extensor hallucis longus
Extensor digitorum brevis
Fibular brevis muscle
Fibula
Sheath of the extensor digitorum longus

6
Semimembranous tendon
Popliteus muscle
Posterior tibialis
Tibia
Fibular brevis muscle
Fibula
Tendon of the fibular brevis muscle
Plantar interosseous muscles
Retinaculum of the inferior extensor muscles
Flexor hallucis longus
Tendons of the flexor digitorum longus

▲▶ **Muscles of the right leg**
Superficial:
5. Frontal view
6. Dorsal view
Deep:
7. Frontal view
8. Dorsal view

7

8

AN ORGAN TIPICALLY HUMAN: THE HAND

Even though the characteristics of our hand are similar to those of primates (flat nails, an opposable thumb capable of touching the fingers of the same hand), we are the only species able to use it, thanks to a brain (82), which regulates its finer and more complex movements. The thumb, with its saddle articulation (trapezius) that allows it to flatten on the palm of the hand and to meet the base of all the other fingers, is the most essential element of the prehensile ability, and consequently of our cultural development. Even our

palm is extremely mobile: it is made up by 13 bones (8 of the carpus, 5 of the metacarpus), that articulate with each other and with the forearm, in a very complex way. The bones of the hand are moved by muscles, which originate in the forearm, and by 3 groups of muscles of the hand:

-*Lateral, or of the thenar eminence*: it is made up by 4 muscles, which move the thumb (adductor brevis, flexor brevis, opponent and adductor pollicis);

-*Medial or of hypothenar eminence*: it is made up by 4 muscles, which contribute to the movement of the little finger (palmar brevis, adductor, flexor brevis, opponent of the little finger).

-*Intermediate or palmar muscles*: this group is made up of a large number of small bones located in the central region of the hand. They are divided into the lumbricoid muscles (4 muscles, placed between tendons and flexor digitorum muscles) and the interosseous palmar and dorsal muscles, located in the spaces between the bones of the metacarpus.

The palmar aponeurosis is a very strong subcutaneous fibrous membrane, and it covers the palm of the hand

1 **2** **3** **4**

▲ Comparison
Bones of the index finger and of the thumb of a gorilla's hand (1), compared to the corresponding bones of a human hand (2). Structure of a human hand (3) and of a gorilla's (4).

Extensor carpi radialis brevis

Extensor pollicis longus

Extensor indicis muscle

Extensor carpi radialis longus

Extensor digitorum muscle

Intertendinous connection

Extensor pollicis brevis

Dorsal interosseous muscle of the hand

Fibrous sheath of the fingers

Interphalanx articular capsule

Lumbricoid I muscle

Crossing of the tendons

Interosseous I muscle

Adductor pollicis longus

Retinacula of the tendons

Adductor pollicis brevis

Adductor pollicis

▲ Bones, tendons and muscles
The anatomical elements involved in the precise movements of the opposer muscle of the thumb.

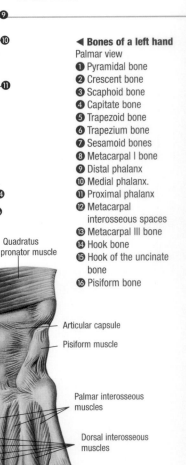

Quadratus pronator muscle

Tendon of the Brachioradial muscle

Tendon of the adductor pollicis longus

Tendon of the flexor carpi radialis

Pollicis Opponent muscle

flexor pollicis brevis

Adductor pollicis brevis

Adductor pollicis

Tendon of the flexor pollicis long

Fibrous sheaths of the flexor digitorum superficialis and profundis

Articular capsule

Pisiform muscle

Palmar interosseous muscles

Dorsal interosseous muscles

◄ Deep muscles of the palm of a hand

◄ Bones of a left hand
Palmar view
❶ Pyramidal bone
❷ Crescent bone
❸ Scaphoid bone
❹ Capitate bone
❺ Trapezoid bone
❻ Trapezium bone
❼ Sesamoid bones
❽ Metacarpal I bone
❾ Distal phalanx
❿ Medial phalanx.
⓫ Proximal phalanx
⓬ Metacarpal interosseous spaces
⓭ Metacarpal III bone
⓮ Hook bone
⓯ Hook of the uncinate bone
⓰ Pisiform bone

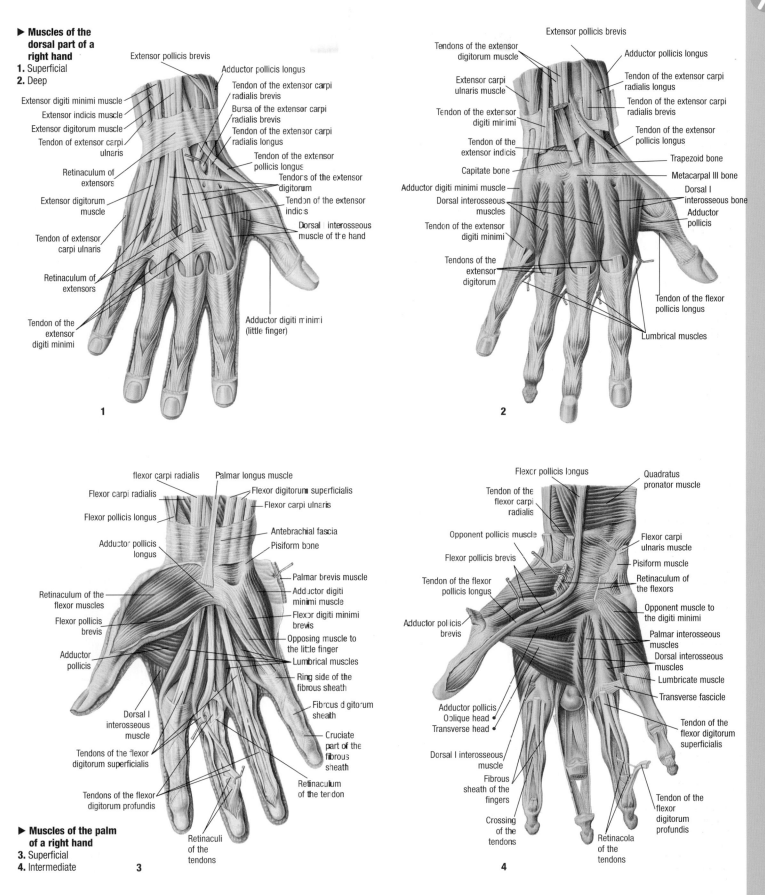

▶ Muscles of the dorsal part of a right hand
1. Superficial
2. Deep

Extensor pollicis brevis
Adductor pollicis longus
Tendon of the extensor carpi radialis brevis
Bursa of the extensor carpi radialis brevis
Tendon of the extensor carpi radialis longus
Tendon of the extensor pollicis longus
Tendons of the extensor digitorum
Tendon of the extensor indicis
Dorsal interosseous muscle of the hand
Adductor digiti minimi (little finger)

Extensor digiti minimi muscle
Extensor indicis muscle
Extensor digitorum muscle
Tendon of extensor carpi ulnaris
Retinaculum of extensors
Extensor digitorum muscle
Tendon of extensor carpi ulnaris
Retinaculum of extensors
Tendon of the extensor digiti minimi

1

Extensor pollicis brevis
Tendons of the extensor digitorum muscle
Adductor pollicis longus
Tendon of the extensor carpi radialis longus
Extensor carpi ulnaris muscle
Tendon of the extensor carpi radialis brevis
Tendon of the extensor digiti minimi
Tendon of the extensor pollicis longus
Tendon of the extensor indicis
Trapezoid bone
Capitate bone
Metacarpal III bone
Adductor digiti minimi muscle
Dorsal interosseous muscles
Dorsal I interosseous bone
Tendon of the extensor digiti minimi
Adductor pollicis
Tendons of the extensor digitorum
Tendon of the flexor pollicis longus
Lumbrical muscles

2

▶ Muscles of the palm of a right hand
3. Superficial
4. Intermediate

flexor carpi radialis
Palmar longus muscle
Flexor carpi radialis
Flexor digitorum superficialis
Flexor carpi ulnaris
Flexor pollicis longus
Antebrachial fascia
Adductor pollicis longus
Pisiform bone
Palmar brevis muscle
Adductor digiti minimi muscle
Retinaculum of the flexor muscles
Flexor digiti minimi brevis
Flexor pollicis brevis
Opposing muscle to the little finger
Adductor pollicis
Lumbrical muscles
Ring side of the fibrous sheath
Dorsal I interosseous muscle
Fibrous digitorum sheath
Tendons of the flexor digitorum superficialis
Cruciate part of the fibrous sheath
Tendons of the flexor digitorum profundis
Retinaculum of the tendon
Retinaculi of the tendons

3

Flexor pollicis longus
Quadratus pronator muscle
Tendon of the flexor carpi radialis
Opponent pollicis muscle
Flexor carpi ulnaris muscle
Flexor pollicis brevis
Pisiform muscle
Tendon of the flexor pollicis longus
Retinaculum of the flexors
Adductor pollicis brevis
Opponent muscle to the digiti minimi
Palmar interosseous muscles
Dorsal interosseous muscles
Lumbricate muscle
Transverse fascicle
Adductor pollicis Oblique head
Transverse head
Dorsal I interosseous muscle
Tendon of the flexor digitorum superficialis
Fibrous sheath of the fingers
Crossing of the tendons
Retinacola of the tendons
Tendon of the flexor digitorum profundis

4

69

THE KEY POINTS OF AN ERECT POSTURE: HIP, KNEE, FOOT

Compared to any other mammal, the way humans move is unique. However an erect posture is not the most efficient way to move around: a slight push is sufficient to make us loose our equilibrium. In fact, the barycenter is a point in the body, which offers us a particularly unstable equilibrium: walking is nothing but a "constant recovery from a fall". However, metabolic tests show that we burn more energy while we sleep than when we go for a stroll. This is possible because the forces are perfectly balanced. This balancing is produced by the anatomical structures involved in maintaining an erect posture. The spinal column plays an essential and very important role, but the hip, the knee and the foot are the body parts that have to support most of the mechanical stressors.

These three anatomical structures have undergone the most marked evolutionary changes of the entire body.

THE SPINAL COLUMN

The three physiological curvatures (cervical, thoracic and lumbar) balance each other, by supporting the body. In fact, because of them, the spinal column's

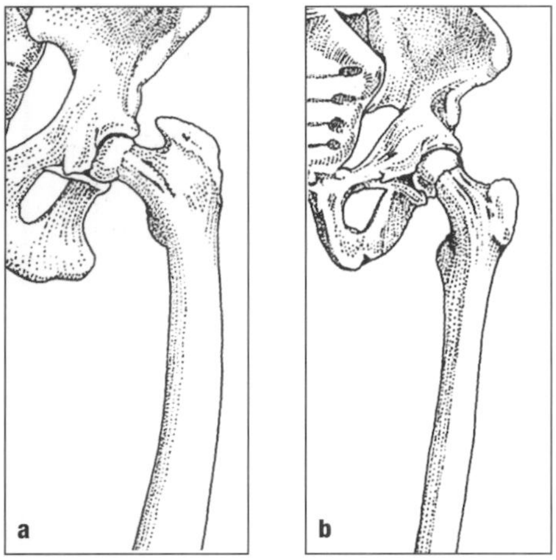

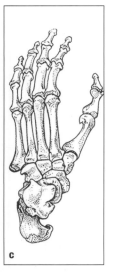

◀ **Femurs: a comparison**
a. Gorilla
b. Human being

▼◀ **Bones of the foot: a comparison**
c. Gorilla
d. Human being

▲ **Comparison between the stride of a gorilla and that of a human being**
Leg and hip bone disposition. Muscles involved in locomotion. The erect posture has

evolved because it offers a series of advantages, one of which is the opportunity for an exceptional development of the cranial volume and of a progressively more advanced brain.

Furthermore, an erect posture allows for the hands to remain free. This is an essential feature of our species l

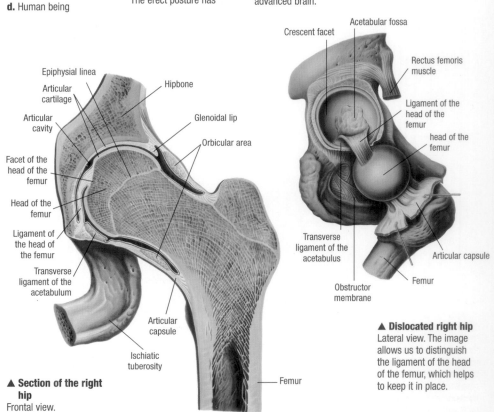

Epiphysial linea
Articular cartilage
Articular cavity
Facet of the head of the femur
Head of the femur
Ligament of the head of the femur
Transverse ligament of the acetabulum
Hipbone
Glenoidal lip
Orbicular area
Articular capsule
Ischiatic tuberosity

▲ **Section of the right hip**
Frontal view.

Crescent facet
Acetabular fossa
Rectus femoris muscle
Ligament of the head of the femur
head of the femur
Transverse ligament of the acetabulus
Obstructor membrane
Articular capsule
Femur

Femur

▲ **Dislocated right hip**
Lateral view. The image allows us to distinguish the ligament of the head of the femur, which helps to keep it in place.

resistance against longitudinal pressures, such as those caused by an overload, is ten times stronger than a hypothetical straight column. The stability of the physiological curvatures is guaranteed by the vertebral ligaments and by the different muscle tones of intrinsic muscles, especially the deep (profundus) spinodorsal muscles ▶ 60: any mechanical movement relaxes them, triggering the adjusting reflex, which re-balances the spinal column.

THE PELVIS

Having acquired a supporting role, it has become a very strong structure: ileum, ischium and pubis fuse with each other, and firmly articulate to the spinal column. The pelvis has a very important role to maintain the static equilibrium of the spinal column, changing the angle of the sacral plane, in relation to the horizontal one. It is stabilized by the ligaments of the hip and by some muscles.

HIP JOINT

Also known as coxofemoral, this joint resolves static and dynamic problems. The head of the femur rests on a deep groove (acetabulus), which is made up of the ischium, the ileum and the pubis. A cartilaginous junction keeps it in place, and strong ligaments and muscles solidify it, by supporting it from the outside. Furthermore, a special internal ligament (teres) connects the head of the femur to the pelvis. The perfect contraction of the joint and a precise architecture of the articular heads, makes this "resting position" very functional: the acetabulum must be at 41° degrees inclination compared to the horizontal plane, the

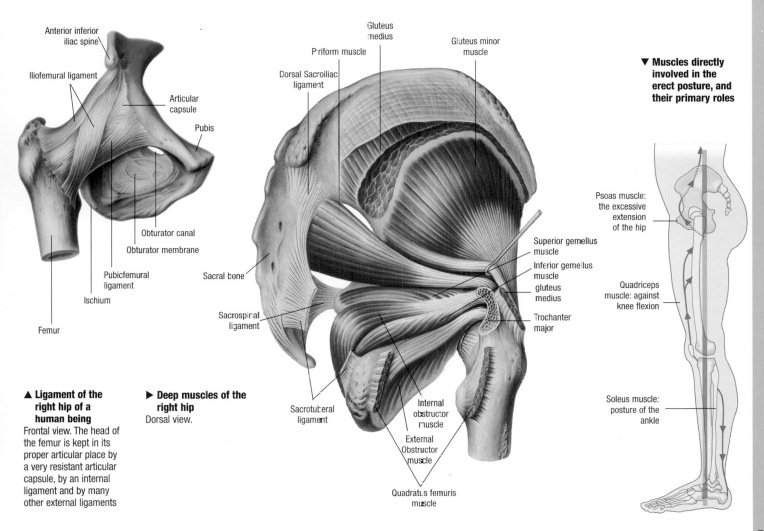

▲ Ligament of the right hip of a human being
Frontal view. The head of the femur is kept in its proper articular place by a very resistant articular capsule, by an internal ligament and by many other external ligaments

▶ Deep muscles of the right hip
Dorsal view.

▼ Muscles directly involved in the erect posture, and their primary roles

71

inclination angle of the neck of the femur must be at 125°, while the inclination of the head of the femur on the neck must be between at 12° and 30°.

Knee joint

Even this joint plays an essential role in static and dynamic functions, by supporting and moving the body. Stability is maintained as follows:
-Transversally, by the medial collateral ligaments. These offer resistance to the internal sliding of the condyles. Stability is also maintained thanks to the capsular fasciae and some muscles such as the tensor fasciae latae.
- Anteroposteriorly, by the posterior fasciae of the capsule, the posterior ligamentous complex and by the quadriceps. In a rotating direction, by the cruciate and the collateral ligaments. The meniscus is a cartilaginous body and it increases the unloading surface of the femur on the tibia, and it creates stability. If such stability is missing, the weight is released on a limited area of the articular plate of the tibia, causing severe damage to the cartilage. The femur and the tibia make up an external angle of 174° (more accentuated in the female, due to a wider pelvis ▶ 66). Because the supporting line of the limb is straight (from the center of the coxofemoral joint at the center of the intermalleolar of the foot), the wear and tear of the lateral condyles and of their ligaments increases. Finally, to absorb shocks, bursae ▶ 57 are located inside the knee, the malleolus, the heel and inside other major joints.

The talocrural joint

The ankle releases the weight of the body on the plantar arches of the foot: here, the stability is mainly kept through a constant involvement of the gastrocnemius and the soleus muscles.

The foot

Having lost its prehensile function, the foot has become both a 3 point support mechanism, and a device for movement that works like a lever, increasing the propulsive strength of the leg: when moving, the vault of the plantar arches flattens and lifts up again, distributing the weight on the entire external arch of the foot. Despite all these skeletal-muscular mechanisms, there is still a problem with equilibrium. In order to avoid falling, there are some sophisticated sensory systems ▶ 82,88, which constantly control the posture and the relationship between the body and the three-dimensional environment. They elaborate stimuli and transform them into a set of coordinated signals, which continually "adjust" the muscles that control for posture.

◀ **Knee high, longitudinal section of the inferior limb**

Tendon of the quadriceps femuris
Articular muscle of the knee
Triceps femuris
Cutis
Overpatellar bursa
Articular face of the patella
Patella
Popliteal vein
Popliteal artery
Prepatellar subcutaneous bursa
Medial Condyle
Articular capsule
Anterior cruciate ligament of the knee
Posterior cruciate ligament of the knee
Popliteus
Infrapatellar synovial fold
Infrapatellar adipose body
Deep infrapatellar bursa
Gastrocnemious muscle
Patellar ligament
Tibia

▼ **Knee joint**
Frontal view. The patella was removed, and is hanging from its ligament.

Patellar surface
Collateral ligament of the patella
Lateral meniscus
Medial condyle
Posterior cruciate ligament
Anterior cruciate ligament
Anterior meniscofemoral ligament
Transverse ligament of the knee
Medial meniscus
Tendon of the biceps femuris
Tibiofibular joint
Patellar head
Collateral ligament of the tibia
Deep Infrapatellar bursa
Patellar ligament
Articular surface of the patella
Interosseous membrane

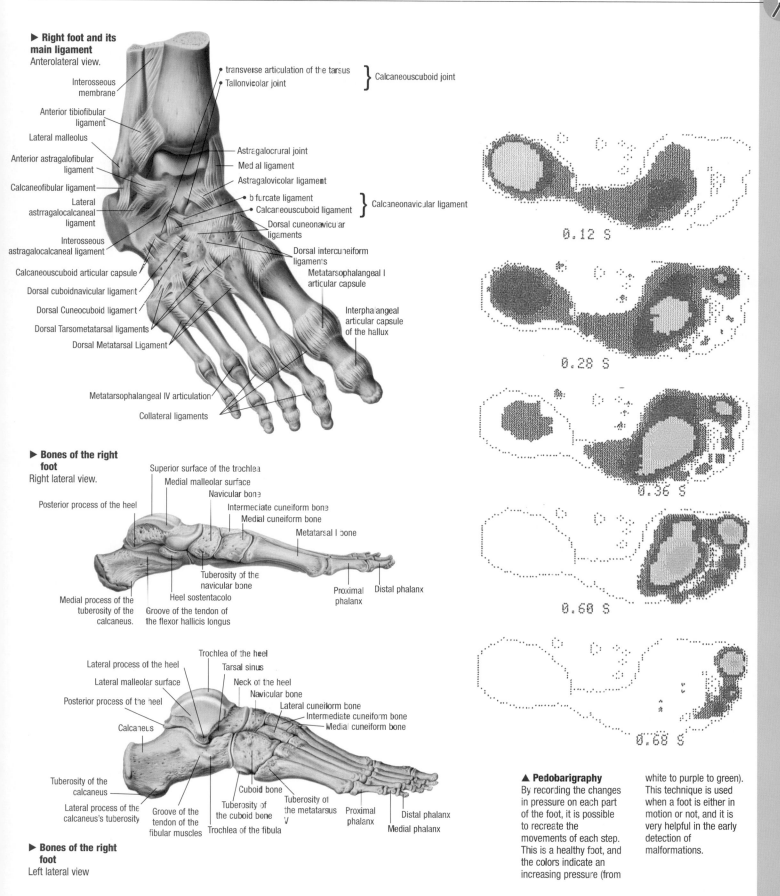

▶ Right foot and its main ligament
Anterolateral view.

Interosseous membrane

Anterior tibiofibular ligament

Lateral malleolus

Anterior astragalofibular ligament

Calcaneofibular ligament

Lateral astragalocalcaneal ligament

Interosseous astragalocalcaneal ligament

Calcaneouscuboid articular capsule

Dorsal cuboidnavicular ligament

Dorsal Cuneocuboid ligament

Dorsal Tarsometatarsal ligaments

Dorsal Metatarsal Ligament

Metatarsophalangeal IV articulation

Collateral ligaments

transverse articulation of the tarsus
Tallonvicolar joint
} Calcaneouscuboid joint

Astragalocrural joint
Medial ligament
Astragalovicolar ligament

bifurcate ligament
Calcaneouscuboid ligament
} Calcaneonavicular ligament

Dorsal cuneonavicular ligaments

Dorsal intercuneiform ligaments

Metatarsophalangeal I articular capsule

Interphalangeal articular capsule of the hallux

▶ Bones of the right foot
Right lateral view.

Superior surface of the trochlea
Medial malleolar surface
Navicular bone
Intermediate cuneiform bone
Medial cuneiform bone
Metatarsal I bone

Posterior process of the heel

Medial process of the tuberosity of the calcaneus.

Groove of the tendon of the flexor hallicis longus

Heel sostentacolo

Tuberosity of the navicular bone

Proximal phalanx

Distal phalanx

Lateral process of the heel
Trochlea of the heel
Tarsal sinus
Neck of the heel
Navicular bone
Lateral malleolar surface
Lateral cuneiform bone
Intermediate cuneiform bone
Medial cuneiform bone
Posterior process of the heel

Calcaneus

Tuberosity of the calcaneus

Lateral process of the calcaneus's tuberosity

Groove of the tendon of the fibular muscles

Tuberosity of the cuboid bone

Trochlea of the fibula

Cuboid bone

Tuberosity of the metatarsus V

Proximal phalanx

Medial phalanx

Distal phalanx

▶ Bones of the right foot
Left lateral view

0.12 S

0.28 S

0.36 S

0.60 S

0.68 S

▲ Pedobarigraphy
By recording the changes in pressure on each part of the foot, it is possible to recreate the movements of each step. This is a healthy foot, and the colors indicate an increasing pressure (from white to purple to green). This technique is used when a foot is either in motion or not, and it is very helpful in the early detection of malformations.

✚ FRACTURES

①

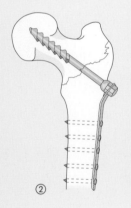

②

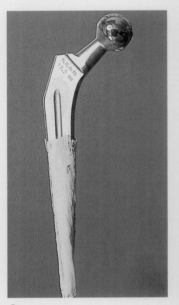

③

The most common problem in the field of orthopedics is a full or partial fracture of a bone. Fractures can be of different types:

-**Spontaneous**: if a physiological weakening of the osseous framework causes them.

-**Traumatic**: if an external agent, which acts on the organism in a violent and rapid way, causes them. Traumatic fractures are further divided into direct fractures: they occur on the area where the trauma has occurred

-**Indirect fractures**: they occur away from where the trauma has occurred, or they are a consequence of violent movements or of sudden tractions of ligaments. Depending on the type of bone fracture, they are divided into:

-**Green stick fracture**, if the bone is not completely broken.

-**Transverse fractures**, if the fracture line is perpendicular to the longitudinal axis of the bone.

-**Spiral fracture**, if the fracture line has a spiral or helicoid course around the bone.

-**Oblique or diagonal fracture**, if the fracture line is on an incline compared to the axis of the bone.

-**Longitudinal fracture**, if the fracture line is parallel to the longitudinal axis of the bone.

-**Splinter fractures**, if the bone is fragmented.

Fractures can also be

Simple or closed- it occurs when the bone has fractured, without creating a lesion in the soft tissue that surrounds it (it hasn't perforated the skin).

-**Complex or exposed**, it occurs when the bone has lacerated the soft tissue, and has created a wound that exposes it. This type of fracture carries a higher risk of infection and a more difficult recovery.

The fracture is diagnosed quite easily, even if a radiological investigation is necessary to guide the precise intervention of the orthopedic surgeon. Generally speaking this occurs in two different stages:

1.**REDUCTION OF THE FRACTURE**

The two surfaces of the broken bone are re-united and their original shape is recreated. This happens with a close reduction, meaning that the bone undergoes a manual or an instrumental traction; or with a bloody reduction, which involves a surgical intervention, in which the fractured elements are reconnected using screws, threads, plaques or different types of nail.

2.**RETENTION OF THE FRACTURE**

The part of the fractured bone is immobilized so that the separated parts of the bone remain in the correct position until proper knitting has been obtained. To guarantee the immobility of the area, a plaster cast or a rigid splint, placed on the joints above and below the fracture, is utilized: for example, if the ulna fractures, the wrist and the elbow are placed in a plaster cast as well (48).

The broken bone regenerates new bone tissue (the so called "callus") along the fracture line. In order to facilitate the regeneration of the bone tissue, it is necessary to integrate a normal diet with vitamins and minerals, more specifically with Vitamin D, calcium, phosphorus and magnesium.

If the fractured bone re-seals well, the bone re-gains its full strength.

◀ **Three types of surgical fissions**
In case of a surgical fracture of the proximal epiphysis femuris, very common with osteoporosis:
① The fracture is very close to the head of the femur. In this case a metallic nail is used;
② The fracture is further away. Besides a screw, extra support is necessary;
③ If the head of the femur is very damaged, it is necessary to substitute it with a metal prosthesis ④ covered with inert plastic materials. The "anima" which supports the leg, is inserted inside the marrow cavity.

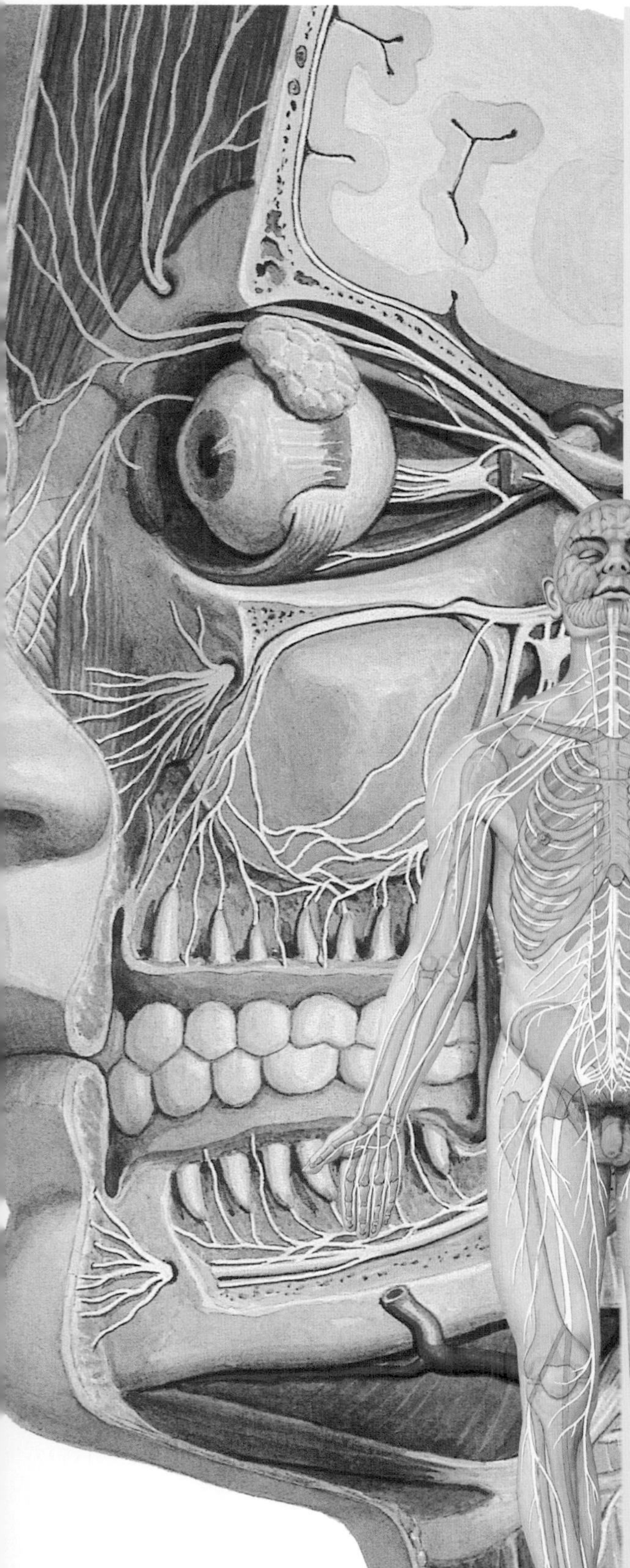

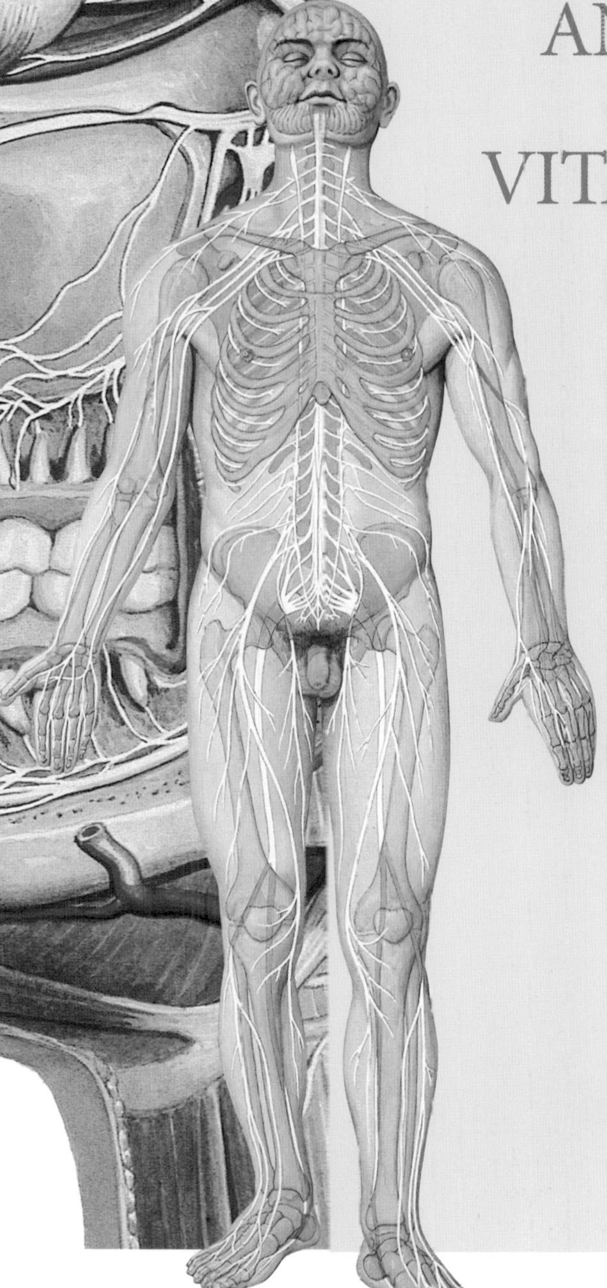

NERVES AND ENDOCRINE GLANDS

THE THINKING PROCESS AND CONTROL OF THE VITAL FUNCTIONS

*Thanks to a complex
evolution over
thousands of years,
man's brain has reached
abilities without equal,
and he has learned
to interact with
the environment
allowing him to speak.*

WETHER WE ARE AWAKE OR ASLEEP, THE COMPLEX AND ENORMOUS CELLULAR NETWORK OF THE NERVOUS SYSTEM CONTROL EACH ACTIVITY OF OUR BODY: THE NERVOUS SYSTEM IS COMPRISED OF THOUSANDS OF NERVOUS FIBERS AND SENSORY ENDINGS TIGHTLY CONNECTED TO EACH OTHER.

THE NERVOUS SYSTEM

Every activity of our body is kept under control by an enormous and complex cellular network: thousands of nervous fibers and sensorial endings collect data allowing them to be recognized, collected and elaborated. Other thousands of nervous fibers send the body, all the commands elaborated by the central nervous units, which are organized inside the complex encephalic structures.

Even though the human brain is very similar to the brain of anthropoid apes, it has developed much further, from a dimension (a little over 400 cm to a little less than 1300 cm) and an operational standpoint. It is unquestionable that a human being is the only technological animal, and that our evolution, from the first stone tool to the discovery of fire, has been deeply influenced by psychism.

Let's take a look at the general characteristics of the biological system, which organizes and coordinates the vital functions of our organism. Let's look at those characteristics, which, at the same time, allow it to interact with the world around it, and also to elaborate each sensation into memories, emotions, self-knowledge, dreams, feelings, logic, creativity and intelligence.

THE STRUCTURE OF THE NERVOUS SYSTEM

The term "nervous system" refers to all the organs in charge of this complex group of coordinated activities. As in all the other mammals, it is divided into parts that work closely with each other:

-A **system of sensors**: this registers environmental changes (inside and outside the body) and transforms them into nervous stimuli. It allows the organism to interact with the outside environment, and allows the body to have a sense of self-perception.

-A **network of nerves**: They reach every part of the body. It connects the sensors and the other elements of the body with the organs that elaborate the stimuli and that give nervous "orders".

-A **centralized neuronal system**: depending on the need, this system includes organs, which elaborate, memorize and produce a response to the stimuli through the network. The central system includes the encephalon, about 1300 g of compact neuronal mass located inside the skull cap (this is exactly the part of the nervous system that, in human beings, has undergone a fast and complex evolution) and the spinal marrow ▶ 106. An extension of the encephalon, the canal made up of the vertebrae of the spinal column.

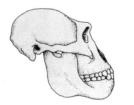 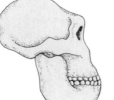 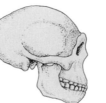 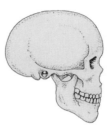

◀ **Volumetric increment**
In human evolution, the increasing volume occupied by the encephalon, corresponded to the development of psychic and technological capabilities.
From left to right, the fossilized skulls of an Astralopythecus afarensis (about 3 million years ago: 400 cm_), a Homo erectus (about 1 million of years ago: 1250 cm_), and a homo sapiens (about 100.000 years ago: 1300 cm_).

▼ The nervous system

This complex network of cells, which branches out into the entire body, is distinguished into two main parts, with very different roles:
- The *central nervous system*, made up of the encephalon, the spinal marrow, a network of 12 pairs of cranial nerves and 32 pairs of spinal nerves.
- The *autonomous nervous system*, made up of nervous fibers of the sympathetic and of the parasympathetic.

-A *peripheral nervous system* or *autonomous nervous system* or *vegetative system*: includes all the nerves centers called ganglia, connected to the central system, which elaborate the involuntary nervous stimuli. The nerves which transmit all those impulses that regulate the physiological activities, independently from our will, are distinguished by two subgroups, which are often act in an antagonistic fashion:

- The *orthosympathetic nervous system*, or simply *sympathetic*: includes two long chains of ganglia, even and symmetrical to the spinal column, made up of fibers that are organized into plexuses. They are distributed to all the organs by following the arterious pathway. This system plays a coordination and control role similar and often opposite to that of the parasympathetic system.

- The *parasympathetic nervous system*: it is essentially made up of the vagus nerve, which not only controls homeostasis ▶ 205, but also all of the functions of the internal organs (for example, it reduces heart and respiratory rate, increases the acidic secretion inside the stomach, and the intestinal peristaltic movements ▶ 156).

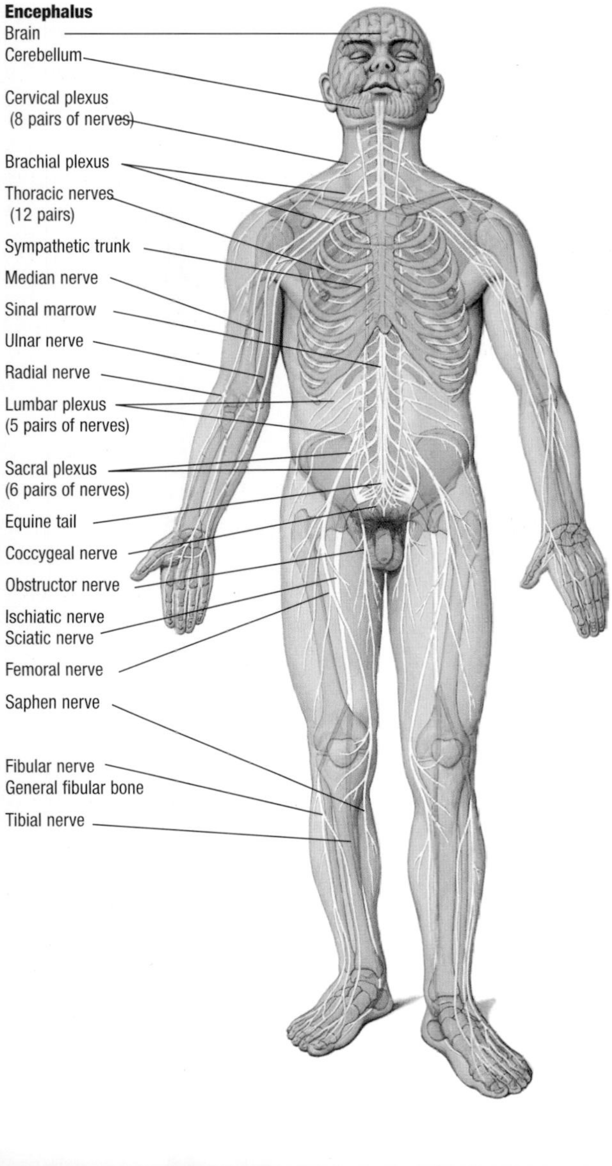

Encephalus
Brain
Cerebellum
Cervical plexus (8 pairs of nerves)
Brachial plexus
Thoracic nerves (12 pairs)
Sympathetic trunk
Median nerve
Sinal marrow
Ulnar nerve
Radial nerve
Lumbar plexus (5 pairs of nerves)
Sacral plexus (6 pairs of nerves)
Equine tail
Coccygeal nerve
Obstructor nerve
Ischiatic nerve
Sciatic nerve
Femoral nerve
Saphen nerve
Fibular nerve
General fibular bone
Tibial nerve

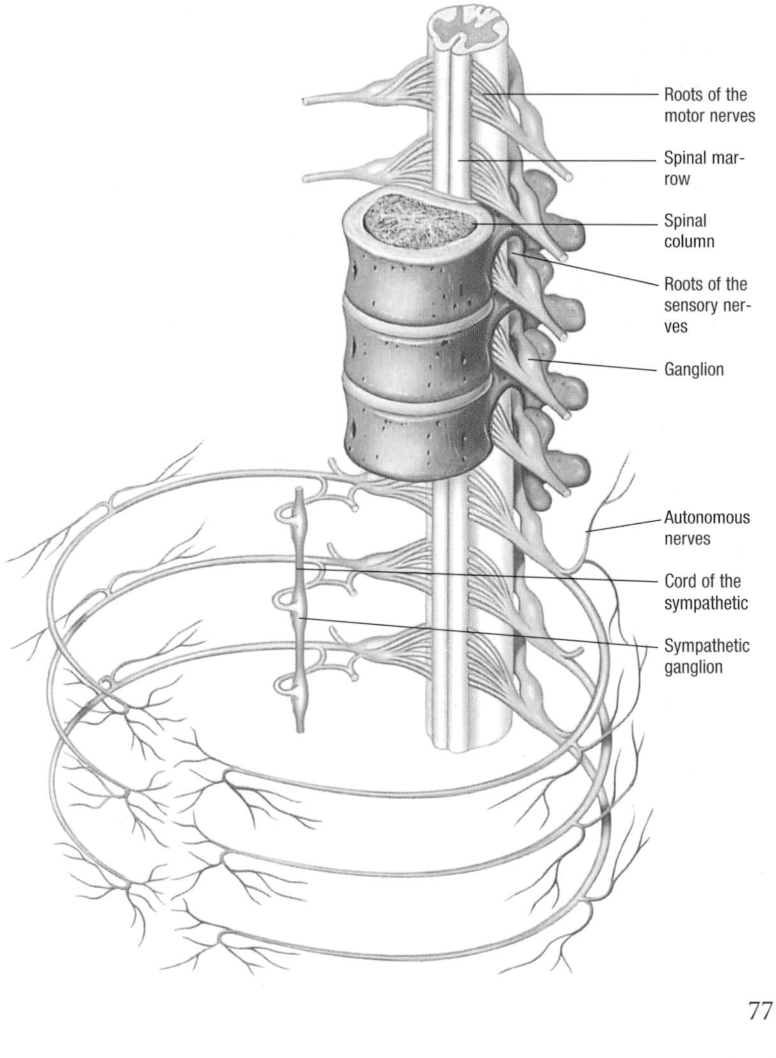

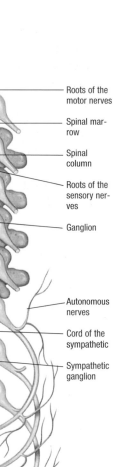

Roots of the motor nerves
Spinal marrow
Spinal column
Roots of the sensory nerves
Ganglion
Autonomous nerves
Cord of the sympathetic
Sympathetic ganglion

▲ Purkinje neurons
They are characteristic of the cerebellum

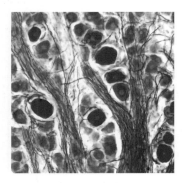

▲ Spinal ganglion neurons

NERVES

They are fasciae of cells, which transmit nervous impulses ▶ 30, guaranteeing the communication between different parts of the body. They are distinguished into:
- *Afferent nerves* or *sensory*, which transfer messages that have been collected from the sensory receptors, to the central system;
- *Efferent nerves* or *motor*, which transfer nervous impulses from the central nervous system to the organs of the body (for example a muscle, or an endocrine gland).

NERVOUS CIRCUITS

All of the nervous fibers (or neurons) are directly or indirectly interconnected. The fibers of the central nervous system, for example, tend to gather into specific parts of the body, and to organize in a way that they are able to strengthen the wave of stimulation, which generally tends to fade by passing from one fiber to the next. Therefore, inside the cerebral cortex, different cells (pyramidal, stellate) layered and connected to each other through various dendrites, create neuronic circuits of various complexity. On the other hand, the fibers of the spinal marrow ▶ 106 come in contact with the somatic muscles, whose activity is also controlled by other neurons connected to different areas of the nervous system (spinal or dorsal ganglia, cerebral cortex). Even the fibers of the peripheral nervous system are tightly connected to each other: those of the sympathetic one often make up such complex networks (plexuses), that they make it very hard to be able to follow precisely the course of a single nerve. The nerve connections can have different effects on the transmitted nervous impulse:

1. When a nervous fiber comes in contact with many other fibers, and sends its nervous impulse in a "cascading" fashion, a divergent nervous circuit or amplifier is created. This impulse allows the organism to respond to even the slightest or circumscribed impulse.
2. When a nervous signal is collected by numerous neurons that are connected to each other in a "pyramid" order, a convergent circuit is formed: the response is guaranteed, rapid and uniform, even if the signals are produced by different stimuli.
3. When a nervous signal, which runs through a chain of neurons, encounters a branch that causes it to "go back", a recurring circuit or reverberating is created. From a simple impulse, the nervous message transforms itself into a continuous discharge; this is necessary, for example, in the case of some smooth muscles.
4. When a neuron contemporarily stimulates other neurons (through different endings), a parallel circuit is created: in response to a sole stimulus, the terminal neuron receives a close discharge of impulses, which stimulates it for a long period of time. These are the operating schemes of the most important neuronal circuits: they intertwine inside our body in very different ways, adding to each other, inhibiting, and influencing one another. Their activity

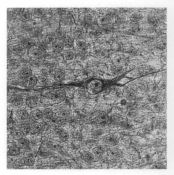

▲ Bipolar neuron
It's characteristic of the retina

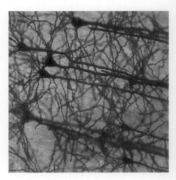

▲ Pyramidal neurons
They are characteristic of the cerebral cortex

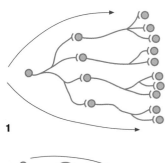

1

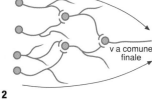

v a comune
finale

2

is regulated not only by the disposition of the neurons and the type of connection that they have, but also by different excitability thresholds ▶ ³⁰, by their possible spontaneous activity, and by the diffuse and continuous presence of thousands of sensory stimuli. It is practically impossible to follow the nervous activity of the body, still less foresee it, other than at a macroscopic level: if we want to pick up an apple, we know which impulses go to the encephalon, where are they elaborated, which areas of the encephalon the signals for the right movements come from, and in which order they occur. However, if what we are trying to obtain is a complete cellular analysis of a nervous transmission, it is impossible to even describe a single blinking of the eye.

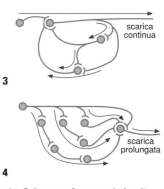

scarica
continua

3

scarica
prolungata

4

◄▲ Schemes of neuronal circuits
1. Divergent circuit (ex. Efferent fibers)
2. Convergent circuit (ex: afferent fibers))
3. Recurring circuit (continuous impulse)
4. Parallel circuit (intense an prolonged discharge)

THE EVOLUTION OF THE HUMAN BRAIN

When we talk about human evolution, and particularly about the evolution of the encephalon, the most complex and mysterious organ in our body, it is difficult to identify a single selective cause, which could justify its rapid development. More than a "cause", it would be better to refer to it as "retroaction rings": a chain of events that by acting positively or negatively on the initial phenomenon can accelerate or inhibit the entire process.

By living in the forest, primates had developed an opposing thumb, which allowed them both fine motor manipulation, and a grip. However, as important as it may be, such change was not the only important one in relation to our most primitive "relatives". For

example, in the forest, there were no predators, and it was vitally important to be able to recognize edible fruits and being able to grab them before they would fall on the ground. As a result, primates whose eyes were positioned more on the front were favored in the natural selection. Thanks to the superimposition of the visual field they were able to have a three-dimensional view. Not only this, but while their ancestors were nocturnal, primates became diurnal, sensitive to a wide range of colors. At the same time, their sense of smell became less and less useful: in the forest, there were too many smells already! So the face started to flatten and due to some complicated retroaction systems, primates took on the look that is

familiar to us today. However, not only did the outward appearance change: thanks to constant manipulation of different objects, the amount of tactile stimuli received by the brain, began to increase, compared with the amount produced by the leg.

Furthermore, the inflow of visual and auditory messages became enormous and much more articulated: to elaborate this river of stimuli the brain had to transform and become much faster, and to be able to deal with data rapidly and effectively. However the neuronal system of the brain cannot sustain more than a certain amount of information at one time: if what needs to be handled surpasses a certain threshold, the brain goes into a short circuit. The increasing amount of perceptive stimuli forced the neuronal network to reorganize itself, exercising a selective pressure which brought on the development of a higher quality

brain, bigger, and better organized: differently from the norm, the evolution of the brain is not linked to any event "outside" the organism, but instead, to internal changes produced by the physical changes occurred with primates.

The area of the brain in which the sensory perceptions were elaborated (the neocortex ▶⁸²⁻⁸⁷) began to expand further. In turns, the development of the brain brought about a rapid evolution of behavior. This factor indirectly contributed to the preservation of the morphology of other parts of the body. If behavioral changes are enough to face and overcome survival problems, the body doesn't undergo any more natural selection pressures. In the long run this affected the cultural development.

▶ Comparison of encephala
During its development, the anterior and superficial cerebral structures were the ones to develop the most:
1. Macaque;
2. Chimpanzee;
3. Gorilla;
4. Man.

☐ Frontal lobe
☐ Parietal lobe
☐ Temporal lobe
☐ Occipital lobe
☐ Cerebellum
☐ Brainstem

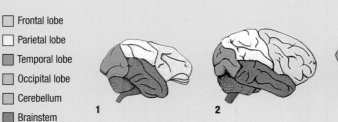

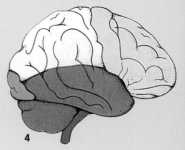

The central nervous system is made up of the encephalon, the spinal marrow, 12 pairs of cranial nerves, and 31 pairs of spinal nerves. It collects, elaborates and memorizes external and internal stimuli, and it reacts to them with nervous impulses.

THE CENTRAL NERVOUS SYSTEM AND THE SENSORY ORGANS

The central nervous system is a very complex anatomical structure, which collects millions of stimuli per second. These stimuli are constantly elaborated and memorized by the central nervous system, which is able to adapt the responses of the body to internal and external conditions.

It is divided into three different parts, depending on their structure and function: each part distinguishes further into elements, which are named after their function. Particularly, the encephalon is further divided into three parts: the Romboencephalon (which includes the medulla oblongata, the cerebellum and the bridge); the mesencephalon (tectum mesencephalic) and the proencephalon (brain, thalamus, hypothalamus).

COMPONENTS OF THE CENTRAL NERVOUS SYSTEM

THE ENCEPHALON
The encephalon is made up on average, of over 10 billion nervous cells, surrounded by the glia cells (from the Greek "glue"), ten times more numerous. The neurons and the glia make up the soft, gel-like cerebral tissue, which maintains its shape by being contained inside the skullcap. The encephalon is wrapped into three membranes, the meninx (protective and nutritive function), the dura mater (more external and solid); the arachnoid (reticulated and run through by numerous canals with cephalorachidian liquid); and the pia mater, which is the thinnest. They coat the entire spinal marrow ▶106. Blood vessels and nervous fibers are located in between. They are immersed in cephalorachidian liquid, which also circulates inside the ventricles (cavities) of the encephalon and inside the thin canal located in the middle of the spinal marrow. The encephalon is divided into different parts, which correspond to anatomically and functionally different compartments:

THE BRAIN: it includes most of the encephalon (it weighs about 1200 g); besides receiving and elaborating the stimuli, it hosts the most advanced human psychic functions.

THE CEREBELLUM: it directly controls precise muscular movements.

THE LIMBIC SYSTEM: it elaborates and memorizes emotions.

THE HYPOTHALAMUS: it controls the pituitary gland (the biggest endocrine gland) and many other vital functions of the body.

THE BRAINSTEM: it also includes the thalamus, which "sorts" the incoming and outgoing messages from the other encephalic areas. It extends into the spinal marrow ▶106 and controls for some internal conditions such as blood pressure and breathing rate, adapting them to the different physiological needs of the body

THE SPINAL MARROW
It includes various types of neurons, which derive from or end in the encephalon. Such neurons reside inside the spinal canal, the cavity that longitudinally crosses the spinal column. About 45cm long, it has a cylinder-like shape and a stratified

THE SENSORY ORGANS

Even if they are not strictly considered as a part of the central nervous system, they are the primary centers in which stimuli are collected: the brain receives essential data from the nerve endings of sight, hearing, smell, taste, and of the proprioceptives of the equilibrium. These data are essential for the development of the brain itself and also for the survival of the entire organism. The sensory organs collect different types of stimuli, and transmit them to the brain through specific nerves. The brain then integrates the signals, memorizes them, recognizes and codes them, before elaborating the most appropriate response. The cerebral activity is constantly stimulated by the inflow of data: it has been demonstrated that the higher the inflow of data to be elaborated, the higher the number of interconnections between cerebral cells, the higher the development of intellectual capabilities will be. On this basis, some anthropologists believe that the intelligence of our species is the direct consequence of a higher inflow of stimuli produced by both the use of hands, and the improvement of language abilities. Besides sensitivity to touch, heat, pain and proprioceptive sensitivity, typical of the skin and found on the entire body, we can say that the sensory organs, which are essential for the interaction with the world, the self preservation of the body and the cerebral activity, are located all inside the head, and are protected by the bones of the skull.

◄ Sectional view of the meninges
❶ Cutis-hair, ❷ Aponeurosis, ❸ External lamina, ❹ Arachnoidal granules , ❺ Internal lamina, ❻ Lateral space (lacuna), ❼ Subdural space, ❽ Dura mater, ❾ Arachnoidal meninges, ❿ Subarachnoidal cavity, ⓫ Encephalon, ⓬ Cerebral falx, ⓭ Pia mater, ⓮ Superior sagittal sinus., ⓯ Diploe, ⓰ Emissary vein.

▼ The brain and the cranial nerves

structure similar to the cerebral one. Its role is to transmit data from and to the encephalon; furthermore, it contains important centers for the regulation of the autonomous nervous system. Finally, the "reflected" nervous responses to the stimuli are elaborated at the level of spinal marrow ►106 without the intervention of the brain.

THE CRANIAL NERVES

12 cranial nerves depart form the inferior surface of the encephalon. These nerves send and receive information to the head, the neck and to the majority of the internal organs. Usually they are indicated with a roman number.
Out of all these nerves, three are exclusively afferent, which means that they carry information that comes from sensory organs, to the encephalon:
-Olphactory nerve (I)
-Optical nerve (II)

-Acoustic nerve (VIII)
Two of them are exclusively motor nerves:
-The accessory nerve (XI), which sends information to the muscles of the neck.
-The hypoglossal nerve (XII), which moves the tongue and other small neck muscles involved in talking. The other seven pairs of cranial nerves are made up by motor and sensory fibers:
- The trigeminal nerve (V), which innervates the muscles of mastication (chewing), and transmits sensations coming from the face.
- The facial nerve (VII), which moves the mimic muscles and transmits the sensations coming from 2/3 of the taste buds.
- The glossopharingeal (IX), which transmits tactile and taste information collected by the posterior part of the tongue and by the pharynx, and regulates swallowing.
- The vagus (X), which is connected

to the muscles of the thorax and of the abdomen.
- The trochlear (IV), abducent (VI) and ocular motor (III) nerves, which innervate the external muscles of the ocular globi: thanks to their motor and sensory nature they can constantly adjust the position of the eye.

THE SPINAL NERVES

These are bundles of nerve fibers, which originate from the spinal marrow and exit the spinal column in pairs – one on each side- at regular intervals. Before exiting through the vertebral openings, each nerve is made up of two bundles of fibers called roots of the nerve: a posterior afferent, and an anterior efferent. Distally speaking, the peripheral

nerves originate from the spinal nerves. Peripheral nerves branch out into various parts of the body.

THE BRAIN

It is the main mass of the encephalon: messages arrive here from sensory organs, from proprioceptive and pain nerve endings. The brain elaborates, analyzes and compares information internal and external to the body; it transforms them into sensations, and stores them as memories. All processes that affect the elaboration of thoughts, decision making, motor or endocrine reaction of the body, take place here. However, most of them are still unknown. The fact that even though it weighs only 2% of the total body weight, the brain consumes about 20% of the total cir-

culating oxygen, gives us a clear indication of its intense and constant metabolic activity. It is divided into two cerebral hemispheres separated on three sides by a deep fissure, but united at the basis of the corpus callosum (commissure of the cerebral hemisphere) by a bundle of nervous fibers about 10 cm long, which guarantee the communication between hemispheres. The following substances are found in each hemisphere:

- The *cerebral cortex* or *grey matter*: it is made up of a high percentage of cellular bodies: it contains about 60% of encepha-

lic neurons. It is deeply folded, so that its surface is about 30 times larger than the surface available inside the cranial space. The traces of the folds visible on the outside of the cortex are called convolutions of the cerebral cortex (gyri cerebri), cerebral sulci or cerebral fissures, and often they outline areas that have specific functions.

- The *white matter*: it is more internally located. It's mostly made of myelitic nervous fibers ▶ [30] which arrive to the cortex. Thousands of nervous fibers branch out from the corpus callosum into the white matter. If the corpus callosum is interrupted, the hemispheres are able to function independently.

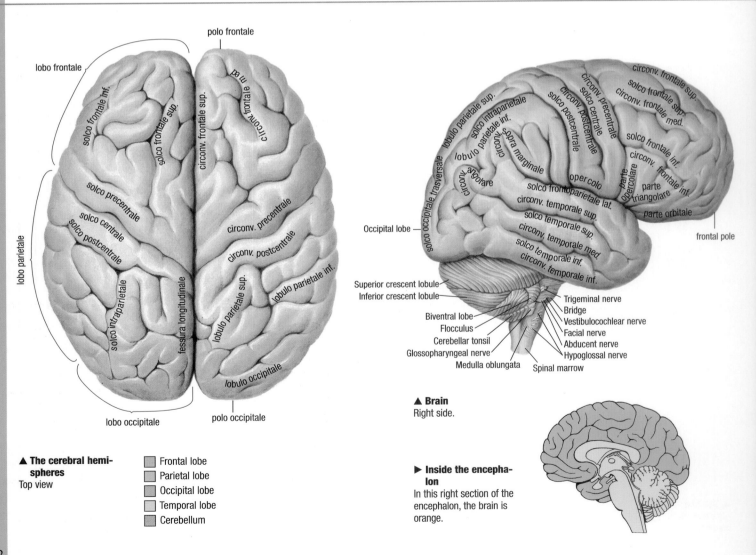

▲ The cerebral hemispheres
Top view

■ Frontal lobe
■ Parietal lobe
■ Occipital lobe
■ Temporal lobe
■ Cerebellum

▲ Brain
Right side.

▶ Inside the encephalon
In this right section of the encephalon, the brain is orange.

CORTEX AND ELABORATION OF SENSORIAL DATA
The term "cortex" must not deceive us in thinking that this is a covering of the brain. In fact the cortex is a live and extremely active tissue, specialized in selecting, comparing, organizing and elaborating the incoming information. It categorizes them as images, thoughts, and emotions and stores them as memories. This part of the brain is made up of about 8 million neurons, which are all located inside a little layer, only few cm thick, and immersed inside the glia, a gelatinous substance made of a number of cells 8 times higher than the number of the cortical neurons. The longitudinal Rolando's fissure, and the transversal Silvio's fissure, outlines the four cerebral lobes, which make up the cerebral cortex. These lobes are separated and symmetrical in both hemispheres. Each lobe is made up of neurons which are both ready to receive and to transmit information. The lobes take on the name of the corresponding cranial bones. Each lobe has a different function: the role of the parietal lobes is to receive and elaborate tactile stimuli; the occipital lobes are the centers of vision; the temporal lobes are the centers for hearing, and so on. It is very difficult for the data coming from only one type of sensory nerves to give us a clear view of the internal and external situation of the body, since all of the impulses that reach the cortex are integrated, modified and elaborated at the same time as all the other incoming information. The areas, in which this rapid modification and integration process of sensory messages take place, are called association areas, and they are distributed along the entire cortex. Also, before being sent to the cortex, the hypothalamus ►94 collects the information arriving from the sensory organs. The grey nuclei of the base, which are clusters of grey matter enclosed between the two hemispheres, are also involved in the process of sorting out information. The sensory cortex is the part of the cortex in which sensations are elaborated. Except for smell, all of the other sensations reach the cerebral cortex through the thalamus, a small structure located in the middle of the encephalon. The thalamus controls, integrates and coordinates all of the impulses, which are directed into those nervous fibers, which connect the hemispheres, collecting sensations from the cerebrospinal axis.

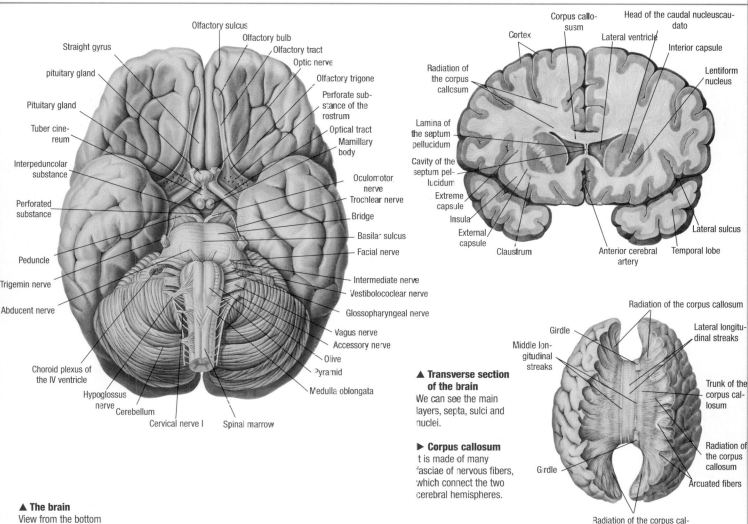

▲ The brain
View from the bottom (ventral)

▲ Transverse section of the brain
We can see the main layers, septa, sulci and nuclei.

► Corpus callosum
It is made of many fasciae of nervous fibers, which connect the two cerebral hemispheres.

THE VISUAL CORTEX

Sight has had a very important role in our evolution: as a result the number of sensory cells involved in it, is the highest of any organ. The images received by the eye ▶96 are transmitted by each optic nerve to the brain. Here they are transformed into multicolor moving images, which can be recognized and recalled by memory. Colors are developed inside the visual cortex: they are not physical properties of the objects themselves, but a sensation produced inside the brain by different combinations of impulses which are generated by different wavelengths of light. Here, another extraordinary process occurs: the transformation of thousands of nervous impulses produced by bi-dimensional and flipped over images, direct images, three-dimensional, and "integrated" images of reality. The brain allows us to see objects as they truly are, despite their possible distortions due to different perspective, distance or other factors. Our mind integrates information with our memory, with the correct images that we have already encountered throughout our life. The constant movements of the eyes are necessary to obtain an adaptable perception of depth, making sure that the image lasts for a long time. If one of the eyes gets "stuck" for a long time, soon the image disappears. Therefore sight is much more than just a simple summation of information collected by the eyes: it requires a wealth of information acquired also through other senses. It is perhaps these complex interactions that create what we call "optical illusions". If the perceptual signs of the image are ambiguous, the brain integrates them according to its experience. The objects are distorted, split, believed to be three-dimensional when they are not, and real when they are actually unrealistic. The mechanisms behind these visual creations of the mind are still unknown, and scientists are still trying to understand how much of this process is innate or learned. The mechanisms of the general visual perception are also a mystery. However, we know that by stimulating the cerebral cortex somewhere in the middle of the retina ▶96-97, we obtain luminous flashes, and by progressively stimulating the more peripheral areas, precise images of objects

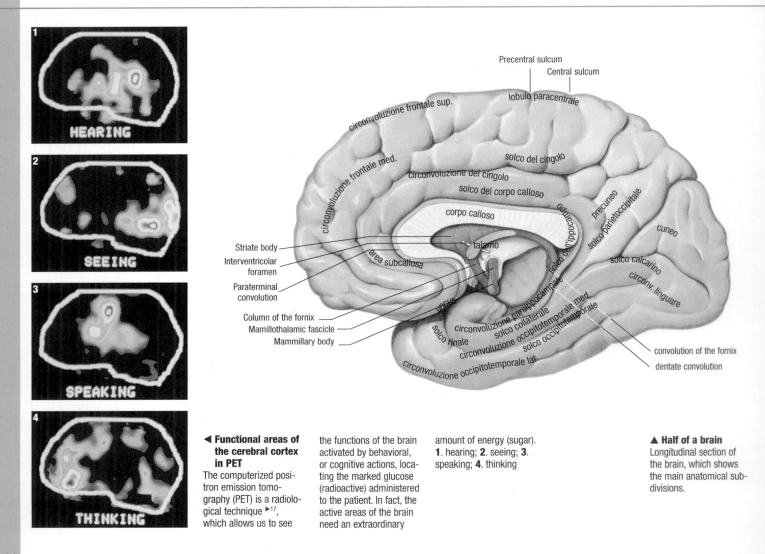

◀ **Functional areas of the cerebral cortex in PET**
The computerized positron emission tomography (PET) is a radiological technique ▶17, which allows us to see the functions of the brain activated by behavioral, or cognitive actions, locating the marked glucose (radioactive) administered to the patient. In fact, the active areas of the brain need an extraordinary amount of energy (sugar). **1**. hearing; **2**. seeing; **3**. speaking; **4**. thinking

▲ **Half of a brain**
Longitudinal section of the brain, which shows the main anatomical subdivisions.

HEARING

SEEING

SPEAKING

THINKING

Precentral sulcus
Central sulcus
lobulo paracentrale
circonvoluzione frontale sup.
solco del cingolo
circonvoluzione frontale med.
Circonvoluzione del cingolo
solco del corpo calloso
precuneo
solco parietoccipitale
corpo calloso
Striate body
Interventricolar foramen
talamo
cuneo
area subcallosa
solco calcarino
Paraterminal convolution
circonv. linguare
Column of the fornix
uncus
Mamillothalamic fascicle
circonvoluzione paraippocampale
solco collaterale
Mammillary body
solco occipitotemporale med.
solco rinale
circonvoluzione occipitotemporale med.
solco occipitotemporale
convolution of the fornix
circonvoluzione occipitotemporale lat.
dentate convolution

and shapes are created, and ultimately even images of scenes from past experiences. It is possible that, even inside the visual cortex, there is a "subdivision" of the cerebral functions, and that some cells recognize only precise objects or movements directed toward preferential areas of the visual field.

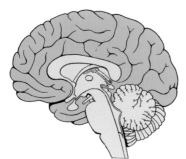

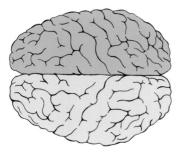

▲ Right cerebral hemisphere
Location of the encephalon, and areas with specific activities. We know a lot less about the right hemisphere, compared to the left.

✚ PROBLEMS ASSOCIATED WITH: BRAIN, MEMORY AND DEGENERATIVE DISEASES

Like any other organ, even the brain ages: on average, after reaching 40, its weight can decrease of 9g each year. This total loss of active cells corresponds to a progressive cerebral degeneration, which can cause a more or less consistent loss of intellective capacity. Different forms of Senile dementia affect about 4% of the population over sixty-five. This percentage increases up to 20%, once the person reaches eighty. This aging-related problem depends also on the overall physical condition of a person: arteriosclerosis (the thickening of the walls of the blood vessels, which decreases the blood inflow and increases blood pressure) is one of the factors, which mostly affect the onset of this disease. On the other hand, exercising the mind, following a good diet and a non-smoking lifestyle, are all preventive factors.

A much more devastating disease is Alzheimer's disease. In people over 50 years old, about 50% of all dementia cases suffers from it. This disease has a clear genetic component, and it not only causes problems with memory and thought formation, but also with personality, language, and in the worst cases, even with posture. These are the consequences of plaque and fibrils, built from the activity of some enzymes, starting from protein elements produced by healthy cerebral cells.

The elaboration of a vaccine and of new drugs, should allow us to resolve or to con-
tain the problem within few years. In fact, today's pharmacology, has not yet found a solution to the progressive cerebral degeneration caused by this disease.

Parkinson's disease can develop at a much earlier time, and it doesn't have a genetic component, even though its onset, just like Alzheimer, is quite slow. Differently from Alzheimer, Parkinson's disease doesn't jeopardize mental faculties until much later. In fact, this degenerative disease is caused by the death of specific neurons located inside the substantia nigra of the mesencephalon. These neurons produce dopamine, a neurotransmitter (30) very important in the modulation of movement. People affected by Parkinson's are easily recognized for their muscular rigidity, a persistent muscular tremor that also occurs while they are asleep, and for a lack of facial expression.

Many drugs are already available on the market; however, drug therapy must be integrated with a regular and specific physical activity.

① Positron emission tomography of a healthy brain and ②) of a brain affected by Alzheimer: the plaques of dead brain tissue are visible (false lighter colors).

THE AUDITORY CORTEX

Even the auditory impulses are integrated with other memorized information by the cortex: visual, olphactory, tactile memories, along with memories of other sounds contribute to the formation of a complete "image" of the registered sound. Memory is essential in recognizing sounds: when the center of memory is damaged, we cannot decode sounds anymore, even though we are still able to perceive them. Since birth, the brain stores memories of sounds: it is believed that it is able to recognize up to half a million of sound signals. However, these are not the only ones: thanks to electrodes, electrical impulses that can be transformed into acoustic signals, recognizable as those sent to the ear ▶98 (functioning as a microphone), "travel" inside an auditory nerve. Furthermore, nervous messages, which carry integrative information (amplitude, pitch, frequency), also travel inside the auditory nerve. It is still a mystery how the auditory cortex decodes such messages.

THE TACTILE CORTEX

The tactile signals ▶104, which come from every part of the body, distribute themselves in different areas of the sensory cortex. Experimental studies have allowed us to trace a "map" of the tactile areas of the cortex, and, as expected, the most sensitive areas of the body are afferent to a higher number of cortical neurons (which means a larger surface of the sensory cortex). For example, the fingers of the hand send stimuli to an area of the cortex equal to that which elaborates the stimuli of all the rest of the body. Furthermore, since the nervous fibers cross each other at the level of the brainstem ▶90, the right sensory cortex is innervated by the fibers of the left half of the body, and vice versa.

COORDINATING MOVEMENTS

The motor cortex is the part of the brain, which organizes and determines the voluntary movements of our body. It is very similar to the sensory cortex by way of its anatomical positioning, and an internal organization. In fact, both the sensory and motor cortex is located in the upper part of the brain hemisphere. Also, both of them are divided into areas, which correspond to a specific part of the body: even in this case, it was possible to draw a "map of the motor

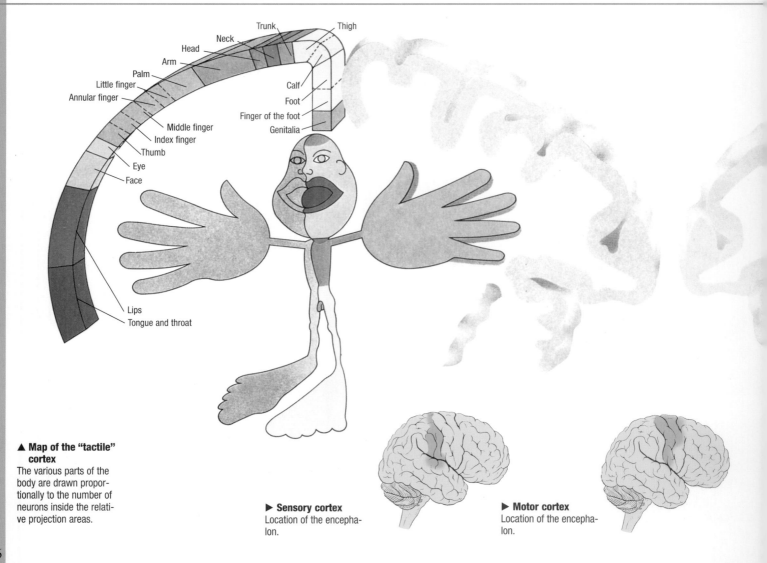

▲ **Map of the "tactile" cortex**
The various parts of the body are drawn proportionally to the number of neurons inside the relative projection areas.

▶ **Sensory cortex**
Location of the encephalon.

▶ **Motor cortex**
Location of the encephalon.

areas". The more extended ones, and those richer with neurons, correspond to more active areas of the body. However, there is a profound functional difference between these two types of cortex. While the stimuli inside the sensory cortex are incoming, in the motor cortex, they depart from the cortical neurons, and they arrive at the muscles, through the efferent fibers. In order to function, the motor cortex needs the body to be "ready to move". Besides the cerebellum ▶88, many structures collaborate in creating and maintaining conditions that favor movement:

-The *spinal marrow*: through reflex movements, it contributes to the relaxation of every antagonist muscle, whenever a muscle contracts.

-The *brainstem*: by maintaining the muscle tone, it allows muscles to react fast.

-The *thalamus*: it guarantees that once movement starts, it occurs smoothly, and in a gradual and progressive way.

In order for a movement to occur, it first needs to be "thought", and then carried out. These two distinct moments occur in two different parts of the motor area: the premotor and the primary area. The elaboration of data and the coordination of motor impulses take place in the premotor area. This is why this area is located behind the frontal lobe; here, the main cerebral processes associated with ideation, elaboration of complex schemes, planning, reason and personality control, take place.

Furthermore, the premotor areas of both hemispheres are connected to each other in order to elaborate a "global motor pro-gram". However, in the primary area, the commands are combined into a "general and coordinated instruction" to the muscle; the two primary areas are almost all disconnected. This allows movements to be independent in both halves of the body. The stimuli reach the spinal marrow, and then the voluntary muscles. The fibers of the autonomous nervous system depart also from the motor roots of the spinal marrow. These fibers reach a chain of ganglia, adjacent to the spinal column, and carry those impulses, which activate the involuntary movements: they are not produced by the motor cortex, but, in some cases, they can be influenced by them (for example respiration ▶162-163).

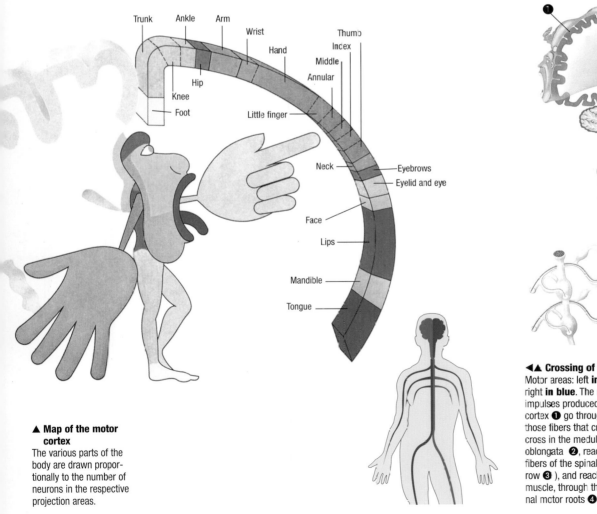

▲ Map of the motor cortex
The various parts of the body are drawn proportionally to the number of neurons in the respective projection areas.

◀▲ Crossing of fibers
Motor areas: left **in red**, right **in blue**. The nerve impulses produced by the cortex ❶ go through those fibers that crisscross in the medulla oblongata ❷, reach the fibers of the spinal marrow ❸), and reach the muscle, through the spinal motor roots ❹ and the spinal nerves ❺ The movements of the involuntary muscles are controlled by the fibers of the autonomous nervous system (**in green**). From there, they reach the chain of nervous ganglia ❻ located adjacent to the spinal column.

THE CEREBELLUM

Located at the basis of the encephalon, the cerebellum, from birth until the age of 2, grows at a faster rate than the brain, rapidly reaching its final volume, which is 11% of the total weight of the encephalon. The way in which it is able to memorize motor schemes is still unknown to us. Such motor schemes are gradually learned as "memories of work", and the cerebellum can access them very quickly: such memories are like a "data bank" of optimal mechanical schemes. The cerebellum refers to them and checks for the exact procedure of each movement. The cerebellum is like a miniature brain. It is divided into 2 lobes (cerebellar hemispheres), and has a surface, which folds into lamellae. Each section of a lamella shows a superficial part of grey matter (cortex), below which, we can find the white matter, made of afferent and efferent nerve fibers. This, however, is the only similarity between the two. Unlike what happens in the brain, the neurons of the cerebellar cortex are orderly distributed on 3 layers. All three of them have a different structure and function:

-The **molecular layer**: it is the most external one, its stellate and crest-shaped cells elaborate information.

- The **intermediate layer**: it is made of Purkinje cells, which carry all the information relative to the movement of the body outside the cerebellum.

-The granular layer: it is the most inner layer, and it is made of granular and Golgi cells, which filter incoming information. Furthermore, the cerebellum has an inhibitory function: thanks to its activity, the rapid impulses produced by the cerebral cortex are arranged and coordinated in such a way that they obtain the correct development of a movement.

This doesn't mean that the cerebellum is slow: on the contrary, it can elaborate data in less than 1/10 of a second! Its cells "do not waste time" in exchanging information: the incoming impulse can be "right"

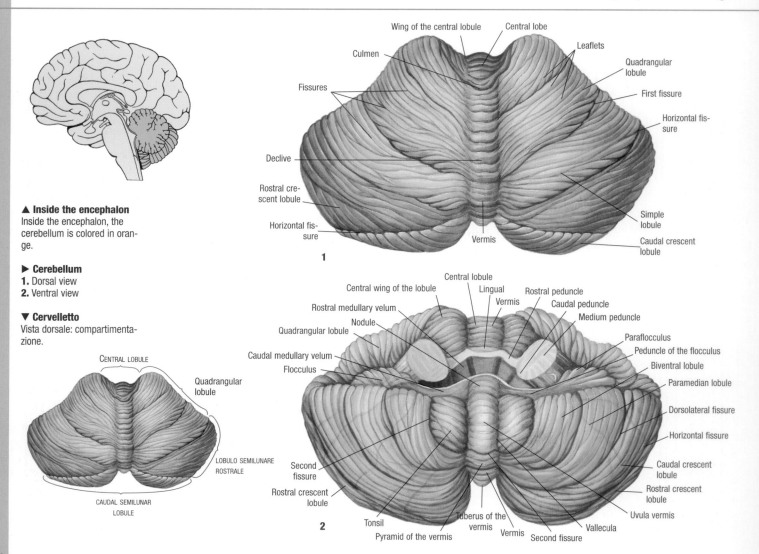

▲ Inside the encephalon
Inside the encephalon, the cerebellum is colored in orange.

▶ Cerebellum
1. Dorsal view
2. Ventral view

▼ Cervelletto
Vista dorsale: compartimentazione.

CENTRAL LOBULE

Quadrangular lobule

LOBULO SEMILUNARE ROSTRALE

CAUDAL SEMILUNAR LOBULE

1

Wing of the central lobule

Central lobe

Culmen

Leaflets

Quadrangular lobule

Fissures

First fissure

Horizontal fissure

Declive

Rostral crescent lobule

Horizontal fissure

Simple lobule

Vermis

Caudal crescent lobule

2

Central lobule

Central wing of the lobule

Lingual

Rostral peduncle

Rostral medullary velum

Vermis

Caudal peduncle

Nodule

Medium peduncle

Quadrangular lobule

Paraflocculus

Caudal medullary velum

Peduncle of the flocculus

Flocculus

Biventral lobule

Paramedian lobule

Dorsolateral fissure

Horizontal fissure

Second fissure

Caudal crescent lobule

Rostral crescent lobule

Rostral crescent lobule

Uvula vermis

Tonsil

Tuberus of the vermis

Vallecula

Pyramid of the vermis

Vermis

Second fissure

or "wrong", is like an electronic calculator. The cerebellum modulates the movements of the body with a "yes" (the impulse is allowed to pass through) or with a "No" (the impulse is stopped). Let's follow a motor sequence:

1. The premotor cortex of the left cerebral hemisphere elaborates the idea of a movement: "lift the right hand and grab the apple".

2. The primary motor cortex of the cerebral hemisphere receives this impulse and transforms it into a complex signal destined to stimulate the muscles of the arm, the forearm, the wrist and hand.

3. Leaving the primary cortex, the signal reaches the cerebral trunk: while some signals move toward the arm, others are sent to the right cerebellar hemisphere,

and within 1/15 of a second from leaving the cortex, it receives complete motor information.

4. As the arm begins to move, the cerebellum compares the information received about the movement, with the information stored in its "data bank". Based on what was memorized, the cerebellum changes the message, allowing it to move on.

As the arm continues moving, the cerebellum receives all the information from the cerebral trunk arriving from the balance receptors, from the receptors of the spatial position of the body, and from the arm. This information describes the amplitude and the speed of the movement. The brain keeps comparing the information received with those memorized, and it modifies the

signals for the upcoming movement.

So, the arm is lifted without any sudden jumps, the forearm progressively moves up, the hand extends and the fingers close around the apple.

The cerebellum is constantly working: it controls every movement and it guarantees that the body is in a perfect position from a gravitational standpoint, and in relation to the space around it. It also controls balance while the body is in a standing or sitting position. In order to be able to fulfill this role, the cerebellum receives continuous information from the spinal cord (106) through voluntary and involuntary muscles, through the organs of balance and all the proprioceptors scattered on the skin.

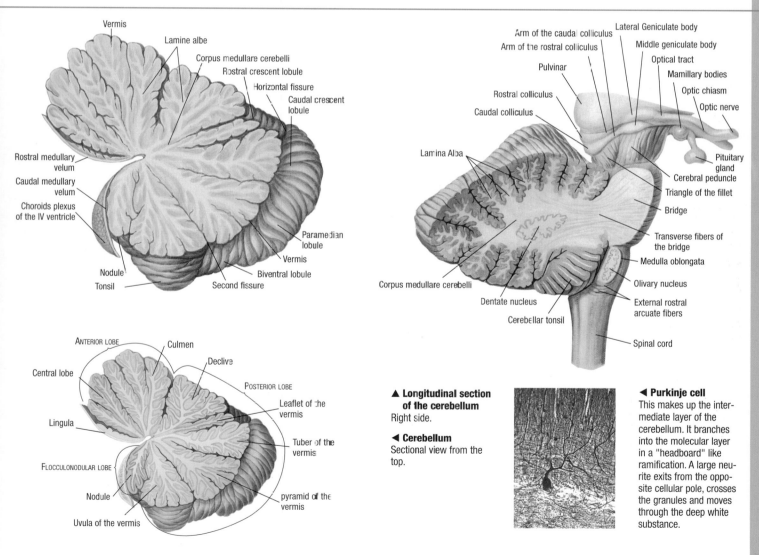

▲ Longitudinal section of the cerebellum
Right side.

◄ Cerebellum
Sectional view from the top.

◄ Purkinje cell
This makes up the intermediate layer of the cerebellum. It branches into the molecular layer in a "headboard" like ramification. A large neurite exits from the opposite cellular pole, crosses the granules and moves through the deep white substance.

THE CEREBRAL TRUNK

Not all the cerebral functions have the same importance: vital functions, such as respiration, cardiac rhythm and blood pressure, are much more important. During the evolution of animals, these abilities were the first to develop, and even in men; they are located inside the most "archaic" portion of the encephalon: the cerebral trunk.

This elongated structure is located under the cerebral hemispheres, in front of the cerebellum, and it is connected to all of the parts of the encephalon. The main motor and sensory pathways are found inside the cerebral trunk, and pass from and to the cerebral centers: it is here, that many of them criss-cross in such a way that each hemisphere controls the opposite side of the body. In the anterior part of the cerebral trunk, the choroid plexus secretes the cerebro-spinal fluid, which is collected inside the cerebral ventricles and inside the central cavity of the encephalon.

From here, the choriod plexus flows into the ventricle above the cerebral trunk, through the Silvio's aqueduct, and it wets the external surface of the encephalon and of the spinal cord ▶106.

The reticular formation is located inside the cerebral trunk. This is an interlacement of thousands of fibers without a predefined nerve pathway. Its role is to maintain the vital functions and to regulate various levels of conscience: any variation in conscience and in the vigilance

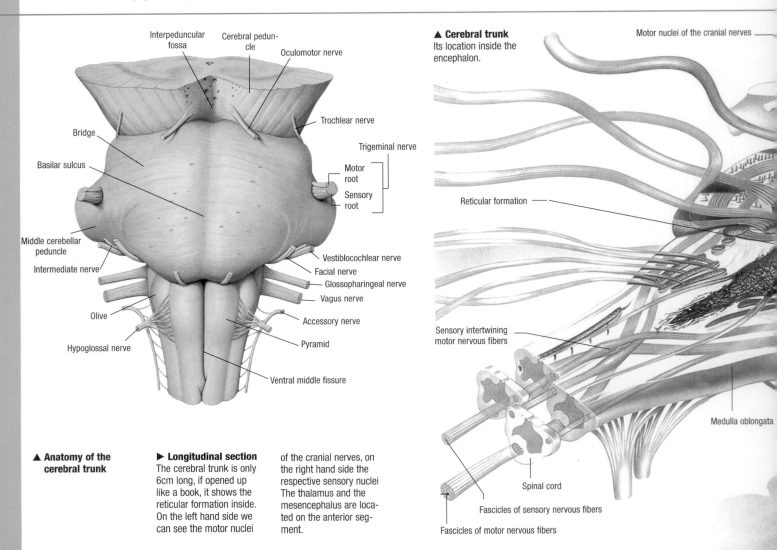

▲ **Cerebral trunk**
Its location inside the encephalon.

Interpeduncular fossa
Cerebral peduncle
Oculomotor nerve
Trochlear nerve
Trigeminal nerve
Motor root
Sensory root
Bridge
Basilar sulcus
Middle cerebellar peduncle
Intermediate nerve
Olive
Hypoglossal nerve
Vestiblocochlear nerve
Facial nerve
Glossopharingeal nerve
Vagus nerve
Accessory nerve
Pyramid
Ventral middle fissure

Motor nuclei of the cranial nerves
Reticular formation
Sensory intertwining motor nervous fibers
Spinal cord
Medulla oblongata
Fascicles of sensory nervous fibers
Fascicles of motor nervous fibers

▲ **Anatomy of the cerebral trunk**

▶ **Longitudinal section**
The cerebral trunk is only 6cm long, if opened up like a book, it shows the reticular formation inside. On the left hand side we can see the motor nuclei of the cranial nerves, on the right hand side the respective sensory nuclei The thalamus and the mesencephalus are located on the anterior segment.

level, are caused by different excitations of the reticular formation, which keeps functioning even when the body is asleep or is unconscious. Such control is possible thanks to the strategically central location of the reticular formation. Information from the entire body arrives here, and from here all this information branches out to other organs.

In fact, in order to produce even a minimal physical change, it is necessary to constantly adjust heart rate, blood pressure, respiration and digestion.

The reticular formation has that specific role. Furthermore, specific fibers emanate from here and regulate finer movements: this role is extremely important because it allows us to perform those infinitesimal functions that influence regular and coordinated movements.

The thalamus is anterior to the cerebral trunk: when the fibers of this area are stimulated, they excite larger areas of the cerebral cortex, activating their processing function.

Inside the thalamus we can find nuclei, thalamic bodies, and compact clusters of neurons, all of which have specific functions and correspond to well-defined areas of the cerebral cortex.

The mesencephalus is found under the thalamus, and it controls the ocular movements and the dilation of the pupil.

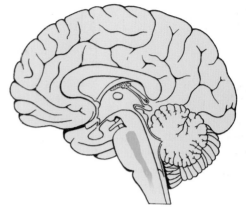

▲ **Activating reticular formation**
Location inside the encephalon.

▲ **Informative routes**
The sensory signals ❶ belonging to the sensory cortex ❷ stimulate the reticular formation (in **yellow**) of the cerebral trunk, before reaching the brain. So the reticular formation stimulates the activity and the vigilance of the entire cortex. In turn, the motor messages leaving the motor cortex ❸, pass through the efferent nervous fibers ❹ and stimulates the reticular formation.

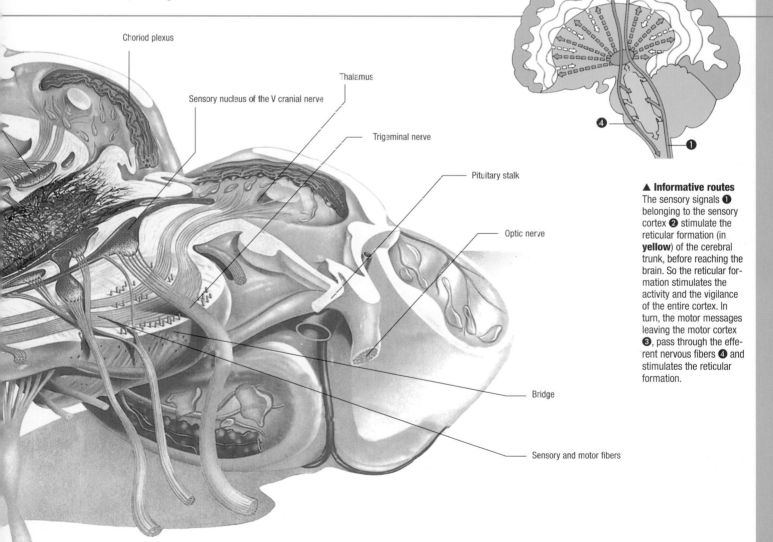

Choriod plexus

Sensory nucleus of the V cranial nerve

Thalamus

Trigeminal nerve

Pituitary stalk

Optic nerve

Bridge

Sensory and motor fibers

THE LIMBIC SYSTEM

This structure influences both hemi-spheres and it is located right over the cerebral trunk, to which it is tightly connected. Its role is to regulate both the stereotipate (or instinctive) behavior, and the vital biological functions and rhythms. Once believed to be tightly connected to olfactory perception (and for this reason called rhinencephalon), the limbic system has complex nervous and biochemical interactions with the cerebral cortex, and today it is considered to be the encephalic element in charge of memory, emotions, attention and lear-

ning. Patients who have been hit in the hippocampus, the closest part of the cerebral trunk, have displayed emotional problems, problems with concentration, with focusing, thinking and perceiving. Also, by electrically stimulating some areas of the limbic system (amygdala, transparent septum, hippocampus), reactions of anger, anxiety, excitement, sexual interest,

colorful visions, deep thoughts and relaxation have been observed. Since the limbic system functions interdependently from the cerebral cortex, it seems possible that a disorder at this level could unleash some mental diseases: the sensory information which usually goes through this structure can get distorted up to the point of completely loosing a sense of reality.

The hippocampus also plays a very important role in relation to emotional responses. It continuously compares sensory data

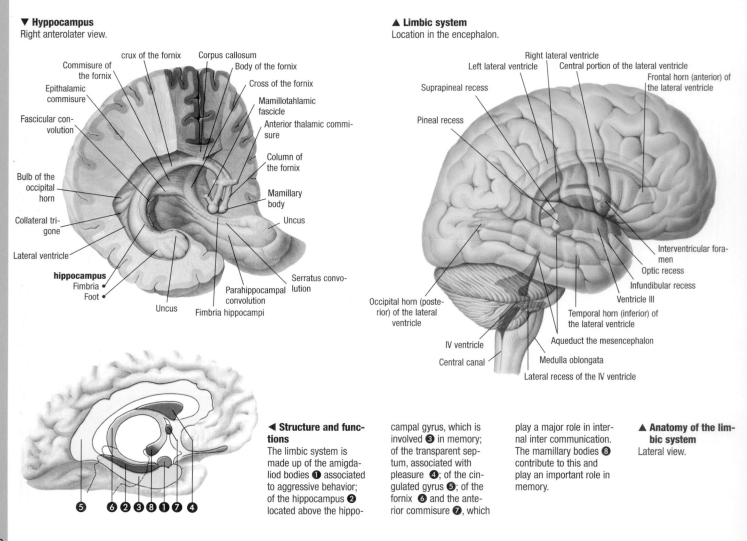

▼ Hyppocampus
Right anterolater view.

crux of the fornix
Commisure of the fornix
Corpus callosum
Body of the fornix
Epithalamic commisure
Cross of the fornix
Fascicular convolution
Mamillotahlamic fascicle
Anterior thalamic commisure
Bulb of the occipital horn
Column of the fornix
Mamillary body
Collateral trigone
Uncus
Lateral ventricle
Serratus convolution
hippocampus
Fimbria
Foot
Uncus
Parahippocampal convolution
Fimbria hippocampi

▲ Limbic system
Location in the encephalon.

Right lateral ventricle
Left lateral ventricle
Central portion of the lateral ventricle
Suprapineal recess
Frontal horn (anterior) of the lateral ventricle
Pineal recess
Interventricular foramen
Optic recess
Infundibular recess
Ventricle III
Occipital horn (posterior) of the lateral ventricle
Temporal horn (inferior) of the lateral ventricle
Aqueduct the mesencephalon
IV ventricle
Medulla oblongata
Central canal
Lateral recess of the IV ventricle

◄ Structure and functions
The limbic system is made up of the amigdaliod bodies ❶ associated to aggressive behavior; of the hippocampus ❷ located above the hippo-campal gyrus, which is involved ❸ in memory; of the transparent septum, associated with pleasure ❹; of the cingulated gyrus ❺; of the fornix ❻ and the anterior commisure ❼, which play a major role in internal inter communication. The mamillary bodies ❽ contribute to this and play an important role in memory.

▲ Anatomy of the limbic system
Lateral view.

to a recalled model, allowing it to detect any environmental change. As environmental conditions change, it interrupts its inhibiting action on the reticular formation, stimulating the vigilance of the organism, allowing us to filter the important elements, among the mass of stimuli or memories that constantly bombard us.

The limbic system is also the place where the "reward" and the "punishment" centers are located. Here our actions are evaluated. Among the most important memories selected in the hippocampal memory, the amigdala allows us to distinguish positive from the negative memories.

THE EFFECT OF DRUGS

Consciousness can be altered by using drugs and pharmaceuticals: substances that can directly or indirectly interfere with nervous transmission and that can cause the development of a habit or a drug addiction. Some of these substances can also induce immediate mental alterations and hallucinations (LSD, cocaine, amphetamines).

Others can cause permanent physiological damages (empathic cirrhosis, cardiovascular diseases, and progressive destruction of the cerebral cells…) if taken for a long period of time. Also, if a habit develops, the dose must increase in order to reach the same effect: this can lead to an acute intoxication due to "overdose". Drugs are divided into three groups:

A. Popular drugs (such as alcohol, tobacco, caffeine…), sold without any limitation, regardless of the fact that their abuse can cause fatal problems.

B. Pharmaceutics (such as stimulants, sedatives, sleeping pills, pain killers…..), sold in pharmacies with a medical prescription. They act directly on the central nervous system.

C. Prohibited substances such as heroin, cocaine, amphetamines, cannabis (ashis and marijuana), hallucinogens (such as LSD, mescal, ecstasy…) act on the central nervous system and are only sold illegally.

The reason for the success of drugs, originally developed for medical purposes, is the immediate feeling of well being they afford. However, all of them can cause tolerance or addiction. In the case of tolerance, higher doses of the drug are necessary to obtain the same result. As for addiction, if the dose in not taken regularly, it can cause severe physiological and psychological problems.

▼ Anatomy of the limbic system
Dorsal view.

Ventricle III

Occipital horn (posterior) of the left lateral ventricle

Frontal horn (anterior) ventricle IV

Temporal horn (inferior) of the left lateral ventricle

Precentral convolution

Central sulcus

Occipital horn (posterior) of the left lateral ventricle

Lateral recess of the ventricle IV

Ventricle IV

Superior parietal lobule

Postcentral convolution

Aqueduct mesencephalon

▲ Action areas of various drugs
☐ antidepressant: mesencephalon
☐ tranquilizers: limbic system and reticular formation
☐ stimulants: reticular formation and hypothalamus
☐ sedatives: reticular formation and cortex

▶ Action of amphetamines
The PET ►30, 84 of a normal brain (1) and of a brain after having taken increasingly higher doses of Methamphetamines (2, 3). We can see an increase in brain activity (yellow areas).

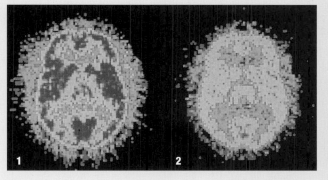

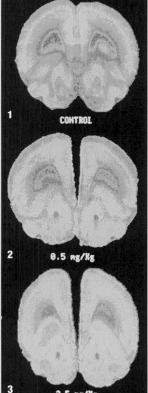

▲ Action of cocaine
The PET ►30, 84 of a normal brain (1) and of a brain after having taken cocaine (2) shows a decrease in cerebral activity (red areas).

HYPOTHALAMUS

The hypothalamus is located under the thalamus and the pituitary gland [128], in the middle of the inferior cerebral surface. The hypothalamus is an uneven and median nervous formation. It is connected to the brain, the cerebral trunk, the limbic system and the spinal cord [106] though many nervous tracts. The optic chiasm, made up of a criss-cross of optic nerves [81], is part of the hypothalamus. The optic tracts are two white cords that embrace the tuber cinereum (this makes up the floor of the cerebral ventricle), and the pituitary stalk and the pituitary gland.

The mamillary bodies are also part of the hypothalamus. They are two round prominences, with a diameter of 5mm; also the median part of the superior surface of the hypothalamus is the floor of the cerebral ventricle III. Inside the hypothalamus there are different functional areas called hypothalamic nuclei, which constantly regulate the basic pulsions and the internal conditions of the organism (homeostasis, nutrients' level, temperature). Furthermore, the hypothalamus is involved in the elaboration of emotions,

and the sensations of pleasure and pain (even sadness); it is also in charge of the female menstrual cycle [130-134].

In addition, this part of the encephalon responds "automatically" to the chemical stimuli of the body: So, for example, the osmoreceptors, which are located in the supraoptical region of the hypothala-

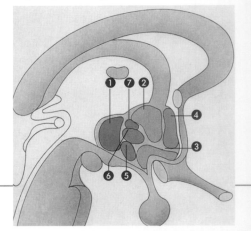

▲ Structure and functions

The hypothalamus is divided into different areas, which regulate the body's different activities; the posterior area ❶ controls the sexual urges; the anterior area ❷ is in charge of thirst sensations, and, along with the supra-optical nuclei ❸, it regulates the use of water; the preoptical nucleus ❹ regulates temperature inside the body. The ventromedial nucleus (also known as "appetistato") ❺ controls the hunger stimulus; the dorsomedial nucleus ❻ controls aggressive behavior, and the dorsal area ❼ is probably the center for "pleasure".

▶ Cerebral centre and endocrine center

The hypothalamus is connected to the brain and the spinal cord. It is the most important meeting point between the central nervous system and the endocrine system, thanks to the direct control it has on the hormonal production of the anterior pituitary. It functions automatically, and it presides over the autonomic nervous system, the metabolism of the organism, the defensive reactions in case of an emergency, and the menstrual cycle.

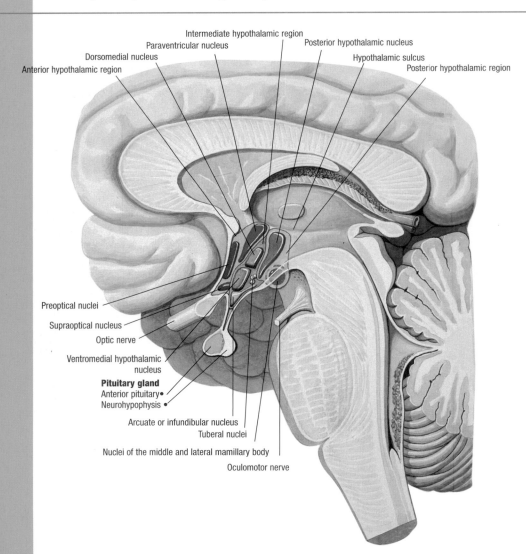

Anterior hypothalamic region
Dorsomedial nucleus
Paraventricular nucleus
Intermediate hypothalamic region
Posterior hypothalamic nucleus
Hypothalamic sulcus
Posterior hypothalamic region

Preoptical nuclei
Supraoptical nucleus
Optic nerve
Ventromedial hypothalamic nucleus
Pituitary gland
Anterior pituitary•
Neurohypophysis •
Arcuate or infundibular nucleus
Tuberal nuclei
Nuclei of the middle and lateral mamillary body
Oculomotor nerve

◀ Hypothalamus
Lateral section: anatomical elements.

mus, register a shortage of water and consequently trigger the sensation of thirst; likewise, as a result of a close relationship between the ventromedial nucleus and the lateral hypothalamus, the encephalon is able to signal a shortage of nutrients (low blood glucose) triggering the sensation of hunger. Also, the hypothalamus connects the central nervous system to the endocrine system [126]: in fact, the supraoptical nucleus, the paraventricular nucleus and the so called medial eminence, are all made up of neurosecreting cells, which produce hormones [127-129].

It is not a novelty that neurons are able to produce chemically active substances: the neurotransmitters, necessary for chemical synapses [30], are produced by almost all nervous cells. On the other hand, the hormones that make up the hypothalamus, unlike the neurotransmitters, are not released near the interneuronal spaces, but they are transported inside the cell along the axons of the hypothalamic-pituitary tract toward the neurohypophysis [130-133]. They first accumulate here, and then they are either released into the blood flow, or they stimulate specific pituitary endocrine cells. This way, thanks to this unique neuronal activity of the hypothalamus, the encephalon also directly controls the pituitary gland, one of the "key" glands of the endocrine system, and the chemical control system of the body.

In turns, the activity of the pituitary gland, and more generally, the basal endocrine situation of the body, acts as a modulator of the neurosecreting activity of the hypothalamus [94].

▲ **Hypothalamus**
Location inside the encephalon.

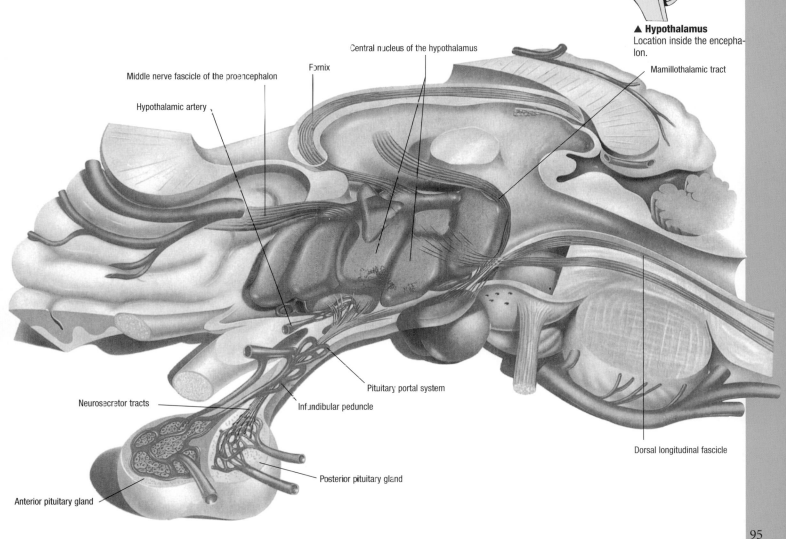

Middle nerve fascicle of the proencephalon

Hypothalamic artery

Fornix

Central nucleus of the hypothalamus

Mamillothalamic tract

Neurosecretor tracts

Infundibular peduncle

Pituitary portal system

Posterior pituitary gland

Anterior pituitary gland

Dorsal longitudinal fascicle

THE EYE AND VISION

Ones visual apparatus is made up of two equal and symmetrical organs, anterior to the skull: the eyes, or the ocular bulbs, are directly connected to the encephalon through the optic nerves. Each ocular bulb is internally divided into 3 fluid filled spaces, and some parts have different tissues, structure and functions:

-The **Outer fibrous tunic**: it is the external membrane, divided into a perfectly transparent (cornea) anterior part, which lacks lymphatic and blood vessels, and a posterior white one (sclera), poorly vascularized, but very robust. Its role is to support and protect. Tendons of the extrinsic muscles of the eye join here, and the most anterior part is coated with a thin transparent membrane (conjunctiva).

-The **vascular tunic** (uvea): it is the intermediate membrane, divided into a posterior part (choriodeal), rich with blood vessels, an intermediate one (ciliary body) where the ciliary muscle, which indirectly moves the chrystalline, attaches; and an anterior (iris) one, perforated by the papillary foramen or pupil. On the front part, its color varies depending on the degree of pigmentation, while on the back it has a black and velvety color. This part is richly vascularized and innervated (parasympathetic fibers of the sphincter muscle of the pupil and of the dilator muscle of the pupil).

- The **nervous tunic or retina**: it is the most internal membrane and it is made up of two things (external or pigmented epithelium, and internal) and is divided into a posterior part (optic), which is the center for photoreceptors, and an anterior part (pars ceca retinae);

-The **crystalline**: it is an element that acts as a lens, connected to the ciliary body inside the iris;

-the **anterior chamber**: it is located

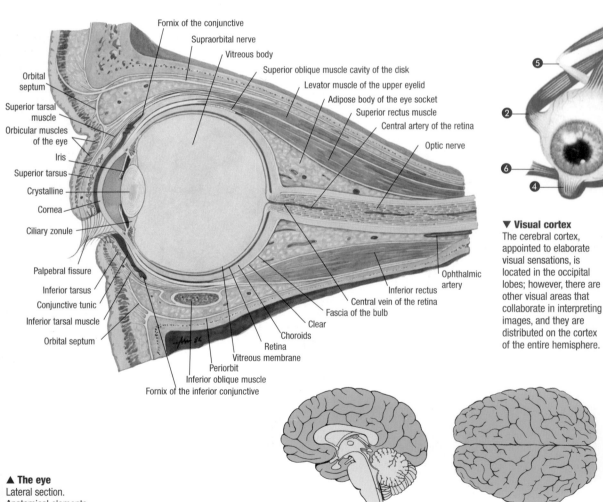

Fornix of the conjunctive
Supraorbital nerve
Vitreous body
Superior oblique muscle cavity of the disk
Levator muscle of the upper eyelid
Adipose body of the eye socket
Superior rectus muscle
Central artery of the retina
Optic nerve

Orbital septum
Superior tarsal muscle
Orbicular muscles of the eye
Iris
Superior tarsus
Crystalline
Cornea
Ciliary zonule

Palpebral fissure
Inferior tarsus
Conjunctive tunic
Inferior tarsal muscle
Orbital septum

Ophthalmic artery
Inferior rectus
Central vein of the retina
Fascia of the bulb
Clear
Choroids
Retina
Vitreous membrane
Periorbit
Inferior oblique muscle
Fornix of the inferior conjunctive

▲ **The eye**
Lateral section.
Anatomical elements

▼ **Visual cortex**
The cerebral cortex, appointed to elaborate visual sensations, is located in the occipital lobes; however, there are other visual areas that collaborate in interpreting images, and they are distributed on the cortex of the entire hemisphere.

▲ **Oculomotor muscles**
There are 6, and they allow the oculary bulb to rotate.
❶ lateral rectus muscle: outward horizontal movement.
❷ median rectus muscle: horizontal movement toward the medial line of the body.
❸ superior rectus muscle: upward movement.
❹ inferior rectus muscle: down ward movement.
❺ inferior oblique muscle: downward outer rotation.
❻ Superior oblique muscle: upward external rotation.

between the cornea and the iris.

-The **posterior chamber**: it is located between the iris and the crystalline; The aqueous humor is found inside the anterior and posterior chambers, while the vitreous body is found inside the vitreous chamber.

Eyes are moved by their own motor apparatus, made up of a rich mixture of muscles, managed by specific encephalic areas. Each eye "works" a little like a camera: the crystalline and the cornea act like a lens, which projects the images that pass though the iris, onto the photosensitive surface of the retina.

The iris, just like a photographic diaphragm, regulates the incoming amount of light that hits the retina, and it contri-

butes, along with the crystalline, to the clarity of the image focused on the retina. Inside the retina, light hits special receptors, which transform the luminous images into nerve stimuli. These stimuli reach the cerebral cortex ▶82-87, crossing the fibers of the optic nerves. Thanks to the optic chiasm, part of the signals collected by the right eye, reach the left occipital lobe, and vice versa. Optic nerves are 50 mm long, and are divided into portions: intratubular, orbital, canalicular and intracranial. The orbital, canalicular and intracranial portions are covered by the meninges; particularly, the intracranial portion is coated by the arachnoid membrane of the pia mater.

✚ VISION PROBLEMS

The most common vision problems are so-called "refraction" difficulties, which can be corrected with glasses, contact lenses of specific surgical operations. In such cases, a defect of the ocular globe or of the crystalline causes the projected image to be out of focus.

① Myopia is generally caused by an excessively long ocular bulb or by the opacification of the crystalline. The image in front of the retina is in focus

② Hypermetropia is caused by an excessively short ocular bulb: the image is in focus behind the retina.

③ Astigmatism is caused by a defected cornea and/or crystalline: light is refracted differently depending on the wavelength, and images double.

④ Presbyopia or "senile vision" is caused by a reduced elasticity of the crystalline: the ability to focus decreases: the image is in focus behind the retina.

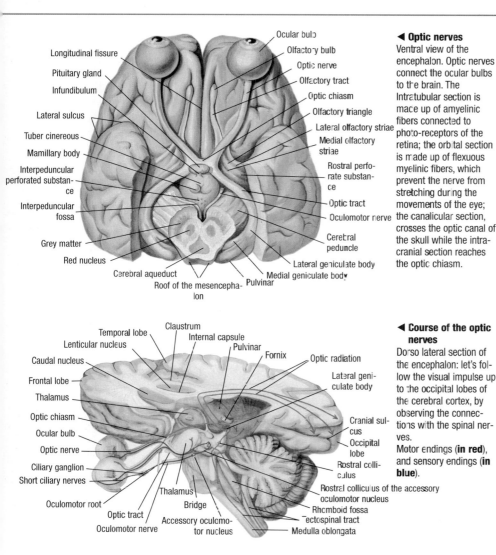

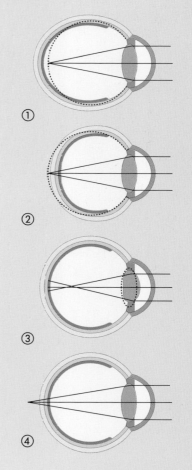

◀ Optic nerves

Ventral view of the encephalon. Optic nerves connect the ocular bulbs to the brain. The Intratubular section is made up of amyelinic fibers connected to photo-receptors of the retina; the orbital section is made up of flexuous myelinic fibers, which prevent the nerve from stretching during the movements of the eye; the canalicular section, crosses the optic canal of the skull while the intracranial section reaches the optic chiasm.

◀ Course of the optic nerves

Dorso lateral section of the encephalon: let's follow the visual impulse up to the occipital lobes of the cerebral cortex, by observing the connections with the spinal nerves.

Motor endings (**in red**), and sensory endings (**in blue**).

THE EAR: HEARING AND EQUILIBRIUM

The ear plays three different perceptive roles: besides transforming sound waves (change of pressure) into nervous stimuli, it informs the brain of the position of the body compared to its vertical position, and of the three-dimensional space that surrounds it. Almost all of the structures that make up the ear are contained inside the thickness of the temporal bone of the skull; they are grouped in the three parts that divide the ear into the following:

- **external ear**: includes the auricle, and the external acoustic canal; its job is to collect sound waves and to transport them to the tympanum;
- **middle ear**: includes the tympanic cavity, a bony cavity which houses the chain of the little bones of the ear (malleus, incus, and stirrup) and communicates, via the auditory tube (Eustachian tube, 35-45mm long), along with the pharynx, the tympanic membrane (made up of three layers, and external one of the tympanic mucouse, with a fibrous middle

one, and an internal one, which extends into the coating of the internal cavity of the ear); the mastoid apparatus made up of a special cavities which communicate with the tympanic cavity and are filled with air;

- **internal ear**: it includes the osseous labyrinth (a complex system of cavities of the temporal bone) and the membranous labyrinth , which fills the spaces, separated from the osseous labyrinth by a perilymphatic space, filled with a fluid called perilymph. The osseous labyrinth is divided into an anterior or acoustic part,

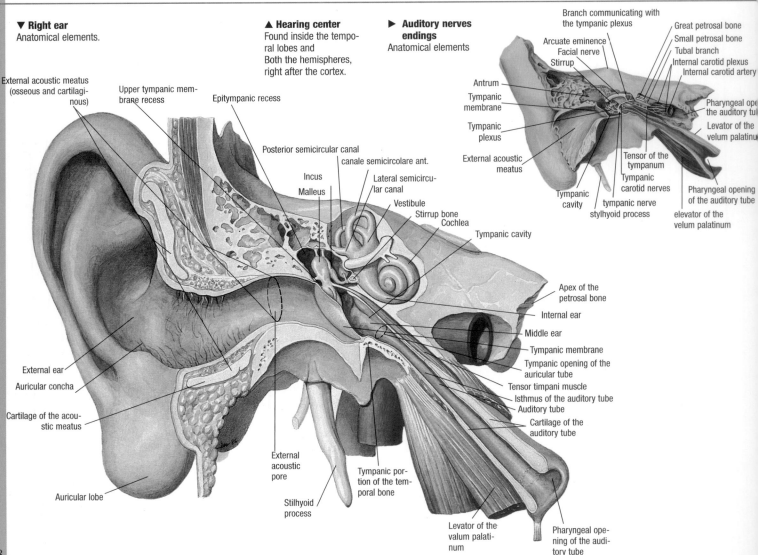

▼ **Right ear**
Anatomical elements.

▲ **Hearing center**
Found inside the temporal lobes and Both the hemispheres, right after the cortex.

► **Auditory nerves endings**
Anatomical elements

Branch communicating with the tympanic plexus
Arcuate eminence
Facial nerve
Stirrup
Great petrosal bone
Small petrosal bone
Tubal branch
Internal carotid plexus
Internal carotid artery
Antrum
Tympanic membrane
Tympanic plexus
External acoustic meatus
Pharyngeal ope the auditory tu
Levator of the velum palatinu
Tensor of the tympanum
Tympanic carotid nerves
Tympanic cavity
tympanic nerve
stylhyoid process
Pharyngeal opening of the auditory tube
elevator of the velum palatinum

External acoustic meatus (osseous and cartilaginous)
Upper tympanic membrane recess
Epitympanic recess
Posterior semicircular canal
canale semicircolare ant.
Incus
Malleus
Lateral semicircular canal
Vestibule
Stirrup bone
Cochlea
Tympanic cavity
Apex of the petrosal bone
Internal ear
Middle ear
Tympanic membrane
Tympanic opening of the auricular tube
Tensor timpani muscle
Isthmus of the auditory tube
Auditory tube
Cartilage of the auditory tube
External ear
Auricular concha
Cartilage of the acoustic meatus
Auricular lobe
External acoustic pore
Stilhyoid process
Tympanic portion of the temporal bone
Levator of the valum palatinum
Pharyngeal opening of the auditory tube

made up of the cochlear aqueduct and the cochlea, which is the site for the acoustic receptors (Corti's organ) and is filled with fluid (endolymph); There is also a posterior or vestibular part, which includes the vestibulum, the semicircular canals (which are the state-kinetic receptors), and the vestibular aqueduct.

THE EQUILIBRIUM AND THE PERCEPTIONS OF GRAVITY

The equilibrium signals are produced by the movement of the endolymph, which circulates inside the 3 semicircular canals: it moves the tufts of the receptor cilia located inside the ampullae, at the base of the canals, in preferential directions. These canals, once stimulated, send signals to the cerebellum, which translates them into a three dimensional sensation. While the movements of the head along the horizontal axis produce signals that come from the horizontal canal, the movements on the vertical and diagonal plane cause different and specific movements, which come from the other two canals, according to the specific direction of the movement. When we make a sudden move with the head and then stop, there is a temporary feeling of disorientation and a loss of equilibrium; in fact, the visual perception (and all the proprioceptive stimuli which derive from it) sends a still image to the brain. Such image clashes with the messages arriving from the ear, in which the endolymph takes more time to return to it original resting state. In such conditions, the brain cannot register a coherent motor response. Gravity precision is produced by the movement of the otoliths, which are located inside the internal pockets of the ear (utricle and saccule); in fact, under the action of gravity, minuscule calcium carbonate crystals press on to the ciliated receptors of the internal surface of the pockets.

When the head moves, the corpuscles change position based on gravitational needs, and press against a different area of the pockets: the signals sent to the cerebellum allow us to distinguish the "top" from the "bottom", even with our eyes closed or in an upside down position

Unlike the auditory nerves, the receptors' fibers of equilibrium and of spatial position, do not reach the brain, even though they come from the internal ear. In fact,

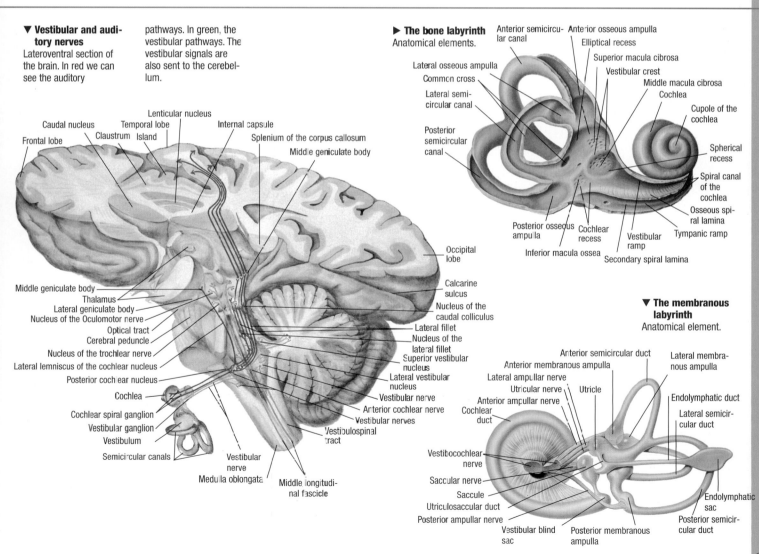

▼ **Vestibular and auditory nerves**
Lateroventral section of the brain. In red we can see the auditory pathways. In green, the vestibular pathways. The vestibular signals are also sent to the cerebellum.

▶ **The bone labyrinth**
Anatomical elements.

▼ **The membranous labyrinth**
Anatomical element.

the nerves that originate from the vestibular semicircular canals (equilibrium), the utricle and the vestibular saccule (position), end in the vestibular nuclei of the cerebellum ▶88, and not into the cerebral cortex. Here they are integrated and developed in such a way that they can create motor coordination, by producing stimuli that activate the muscles that control posture.

HEARING

The tympanic membrane is moved by air vibrations, which are transported by the auricule of the external auditory meatus. The tympanic vibrations have different frequencies, depending on different sounds. They are transmitted to the chain of ossicles located inside the tympanic membrane, and suspended on the upper wall of the middle ear. They amplify the tympanic vibrations transforming them into shorter and more powerful movements: the stirrup bone, located at the end of the chain, vibrates at the same frequency as the tympanum, but 20 times stronger. As a result of this, it is able to transmit the vibrations to the fluid that fills the cochlea, on which it distally rests. The cochlea is occupied in its entire length by the Corti's organ: no more than 2,5 cm long, it contains over 2500 ciliated cells, arranged in a parallel fashion, and sensitive to the vibration of the basal membrane. In fact, the basal membrane moves thanks to the vibrations transmitted by the endolymph, and it presses on them unlike the tectory membrane, stimulating the cellular cilia.

The nervous fibers depart from the Corti's cells and transmit the impulse to the auricular nervous fiber. The pressure changes of the endolymph found inside the cochlea, translate into nervous impulses that cross the nerve, and reach the sensory cortex, where they are then decoded ▶84. The auditory receptors of an adult react to sounds with frequencies ranging between 16000 and 20000 cycles per second, even though the optimal capacities occur at frequencies ranging between 1000 and 2000 cycles per second. Children can hear higher pitch sounds than adults, but by the time they reach sexual maturity this ability has decreased, while sensitivity to lower sounds remains constant during ones entire life.

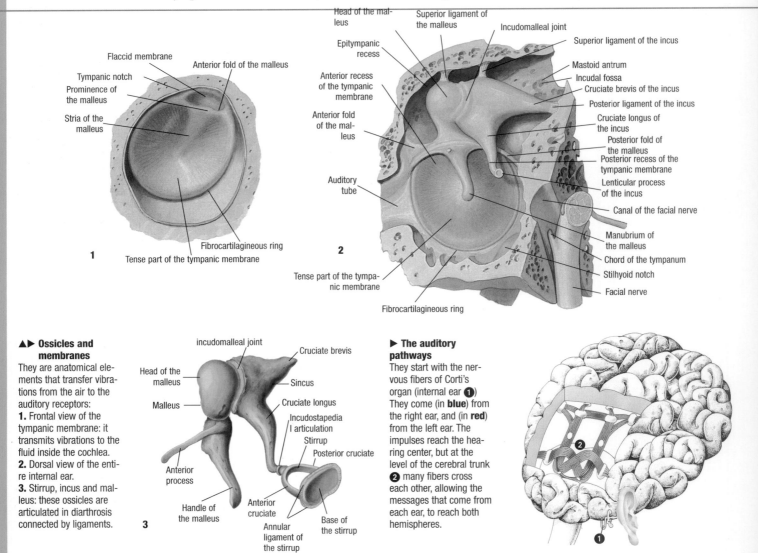

1
Flaccid membrane
Anterior fold of the malleus
Tympanic notch
Prominence of the malleus
Stria of the malleus
Fibrocartilagineous ring
Tense part of the tympanic membrane

2
Head of the malleus
Superior ligament of the malleus
Incudomalleal joint
Superior ligament of the incus
Epitympanic recess
Mastoid antrum
Anterior recess of the tympanic membrane
Incudal fossa
Cruciate brevis of the incus
Posterior ligament of the incus
Anterior fold of the malleus
Cruciate longus of the incus
Posterior fold of the malleus
Posterior recess of the tympanic membrane
Lenticular process of the incus
Auditory tube
Canal of the facial nerve
Manubrium of the malleus
Chord of the tympanum
Stilhyoid notch
Facial nerve
Tense part of the tympanic membrane
Fibrocartilagineous ring

▲▶ Ossicles and membranes
They are anatomical elements that transfer vibrations from the air to the auditory receptors:
1. Frontal view of the tympanic membrane: it transmits vibrations to the fluid inside the cochlea.
2. Dorsal view of the entire internal ear.
3. Stirrup, incus and malleus: these ossicles are articulated in diarthrosis connected by ligaments.

3
incudomalleal joint
Cruciate brevis
Head of the malleus
Sincus
Malleus
Cruciate longus
Incudostapedial articulation
Stirrup
Posterior cruciate
Anterior process
Handle of the malleus
Anterior cruciate
Annular ligament of the stirrup
Base of the stirrup

▶ The auditory pathways
They start with the nervous fibers of Corti's organ (internal ear ❶) They come (in **blue**) from the right ear, and (in **red**) from the left ear. The impulses reach the hearing center, but at the level of the cerebral trunk ❷ many fibers cross each other, allowing the messages that come from each ear, to reach both hemispheres.

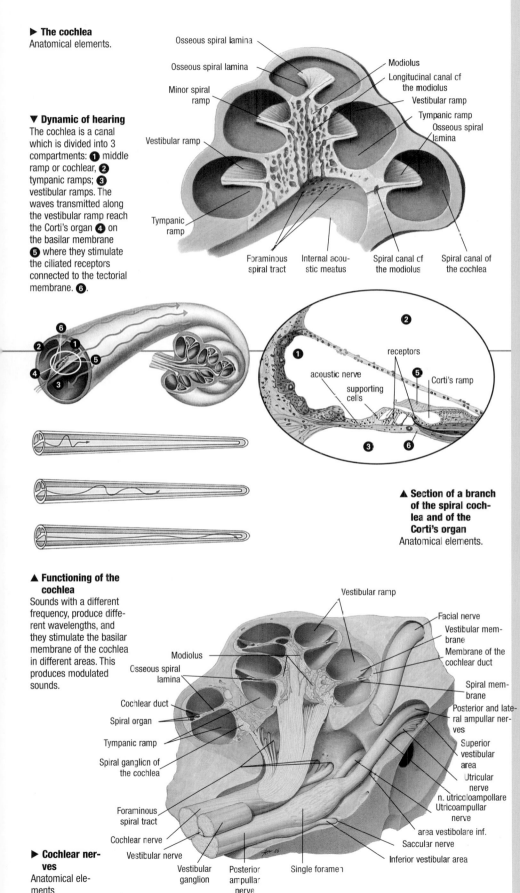

► **The cochlea**
Anatomical elements.

Osseous spiral lamina
Osseous spiral lamina
Minor spiral ramp
Vestibular ramp
Tympanic ramp
Modiolus
Longitucinal canal cf the modiolus
Vestibular ramp
Tympanic ramp
Osseous spiral lamina
Foraminous spiral tract
Internal acoustic meatus
Spiral canal cf the modiolus
Spiral canal of the cochlea

▼ **Dynamic of hearing**
The cochlea is a canal which is divided into 3 compartments: ❶ middle ramp or cochlear, ❷ tympanic ramps; ❸ vestibular ramps. The waves transmitted along the vestibular ramp reach the Corti's organ ❹ on the basilar membrane ❺ where they stimulate the ciliated receptors connected to the tectorial membrane. ❻.

receptors
acoustic nerve
supporting cells
Corti's ramp

▲ **Section of a branch of the spiral cochlea and of the Corti's organ**
Anatomical elements.

▲ **Functioning of the cochlea**
Sounds with a different frequency, produce different wavelengths, and they stimulate the basilar membrane of the cochlea in different areas. This produces modulated sounds.

Vestibular ramp
Facial nerve
Vestibular membrane
Membrane of the cochlear duct
Spiral membrane
Posterior and lateral ampullar nerves
Superior vestibular area
Utricular nerve
n. utricoloampullare
Utricoampullar nerve
area vestibolare inf.
Saccular nerve
Inferior vestibular area
Single forame7
Posterior ampullar nerve
Vestibular ganglion
Vestibular nerve
Cochlear nerve
Foraminous spiral tract
Spiral ganglicn of the cochlea
Tympanic ramp
Spiral organ
Cochlear duct
Osseous spiral lamina
Modiolus

► **Cochlear nerves**
Anatomical elements

✚ HEARING PROBLEMS

Audiometric tests are very precise in showing the sensitivity of each ear to different sound waves and different frequencies, and they allow us to determine the possible degree of deafness or hypoacusys (reduced acoustic perception). In 50% of the cases deafness is genetically based. In the remaining 50% of the cases, it can depend on different causes, which can all alter the production, the transmission and the elaboration of the auditory nervous message. As a result there are many different types of deafness: congenital deafness (genetically determined), central deafness (caused by cerebral problems), perceptual deafness (caused by an alteration of the internal ear and of the acoustic nerve), and transmission deafness, caused by a disease of the middle ear and of the auricular duct. The presbyacusus and socioacusis are two terms that indicate a hypoacusys caused, respectively, by the aging of the acoustic structures and by an acoustic trauma. Sudden and prolonged auditory stimuli over 120 or 100 decibels (dB), and for over 8 hours, can reduce hearing. If the exposure is not prolonged, the damage can be temporary, but it can become permanent if the exposure is prolonged.

dB	SUONI	OSSERVAZIONI
10-20	rustling	
30-40	quiet road theater play	
50-60	louder voice telephone high volume radio and TV	
70-80	alarm clock bus sewing machine traffic	growing discomfort
90-100	high traffic heavy engine train foundry	
110-120	circular saw planer motorcycle horn race car bell	pain threshold: protective earphones necessary
130-140	cannon airplane	
150-170	jet machine gun	
180	missile foundry	

NOSE AND MOUTH: TASTING AND SMELLING

The nose and the mouth have receptors that are able to perceive different chemical stimuli: the nose receptors produce the olfactory sensations; those of the mouth produce tasting sensations. The taste buds of an adult, grouped in sets of 100-200, make up about 9000 papillae, mostly spread on the top surface of the tongue, but also on the palate, the pharynx and the tonsils. During childhood, their number is very high, but then it decreases with age. Each taste bud is made up of cells that are connected to a number of nervous fibers. In turn, each fiber can be connected to more than one taste bud: this makes it very hard to understand the mechanism which creates a specific taste. Traditionally, we distinguish between 4 basic flavors: salty, sweet, sour, and bitter. By examining the reactions of the tongue to various chemical solutions, we were able to establish that it is sensitive to different flavors (gustatory areas); for example, the sweet and salty sensations are detected mainly on the tip of the tongue, while the acid ones on the middle part. Nerves transmit thermal, tactile, pain sensations: they reach the brain separately, but are elaborated together. So, hot milk tastes different from cold milk, and dry bread tastes different from wet bread. It is still not quite clear why substances have different tastes: we suspect that different

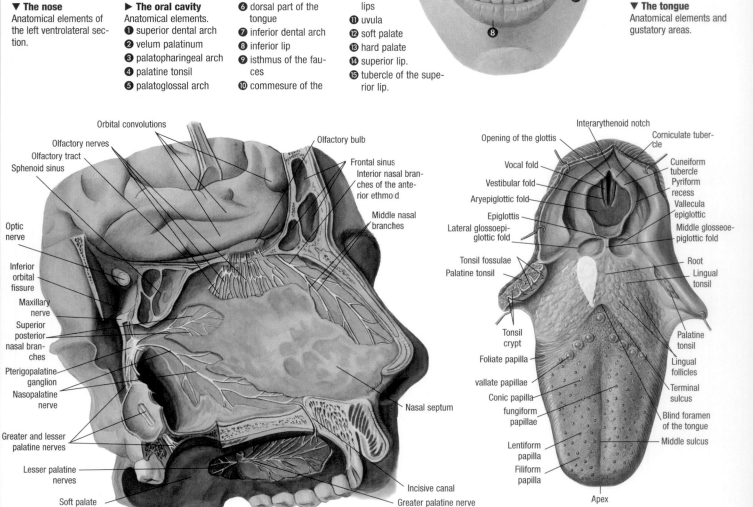

▼ **The nose**
Anatomical elements of the left ventrolateral section.

▶ **The oral cavity**
Anatomical elements.
❶ superior dental arch
❷ velum palatinum
❸ palatopharingeal arch
❹ palatine tonsil
❺ palatoglossal arch
❻ dorsal part of the tongue
❼ inferior dental arch
❽ inferior lip
❾ isthmus of the fauces
❿ commesure of the lips
⓫ uvula
⓬ soft palate
⓭ hard palate
⓮ superior lip.
⓯ tubercle of the superior lip.

▼ **The tongue**
Anatomical elements and gustatory areas.

Orbital convolutions
Olfactory nerves
Olfactory tract
Sphenoid sinus
Olfactory bulb
Frontal sinus
Interior nasal branches of the anterior ethmoid
Middle nasal branches
Optic nerve
Inferior orbital fissure
Maxillary nerve
Superior posterior nasal branches
Pterigopalatine ganglion
Nasopalatine nerve
Greater and lesser palatine nerves
Lesser palatine nerves
Soft palate
Nasal septum
Incisive canal
Greater palatine nerve

Interarythenoid notch
Opening of the glottis
Corniculate tubercle
Vocal fold
Cuneiform tubercle
Pyriform recess
Vestibular fold
Aryepiglottic fold
Epiglottis
Vallecula epiglottic
Lateral glossoepiglottic fold
Middle glosseoepiglottic fold
Tonsil fossulae
Palatine tonsil
Root
Lingual tonsil
Tonsil crypt
Foliate papilla
vallate papillae
Conic papilla
fungiform papillae
Lentiform papilla
Filiform papilla
Palatine tonsil
Lingual follicles
Terminal sulcus
Blind foramen of the tongue
Middle sulcus
Apex

chemical components act on taste buds, altering their metabolism and triggering the nervous stimulus. In any case, flavors can be detected only when the substance has been dissolved in water and that often in order to taste something, it is necessary to smell it also. What we define as a "flavor" maybe is nothing else than the brain's elaboration of an olfactory stimulus. In fact, compared to the gustatory receptors, the olfactory receptors can detect substances up to 25000 times smaller. Also, an adult with an exceptional smell is able to detect up to 1000 different smells, and at times he can recognize a single molecule per square foot, and be able to smell it. After all, a clear example of how important smell is in the process of tasting is when we get a cold or when our nose is stuffed up, and we cannot taste anything! The olfactory receptors are also able to only detect substances in a solution: In fact, they must dissolve inside the humid film that coats the nasal fossae. It is still unknown to us how these minuscule particles can stimulate smell. They are all gathered in a 5 cm_ area of the upper part of the nasal cavity, and they produce nervous signals that reach the root of the nose where the nervous fibers gather in the olfactory bulbs. The olfactory pathways continue through the limbic system (92) up to the cortex of the frontal lobes of the brain, where the impulse is elaborated and the smell is recognized. Smell is probably one of the most primitive senses, and it is also the one most directly connected to the subconscious layers of the psyche and of memory, and for many it is the one that triggers memories the most: sometimes a simple smell is enough to relive a scene from the past and to feel all the emotions associated with it. Most probably the limbic system, influenced by the olfactory stimuli that travel the cortex, contributes to this. On the other hand, the gustatory nerves of the tongue, cross each other inside the marrow, and reach the gustatory cortex, through the thalamus. Therefore, those flavors which come from the right side of the tongue are elaborated by the cerebral cortex of the left hemisphere, and vice versa.

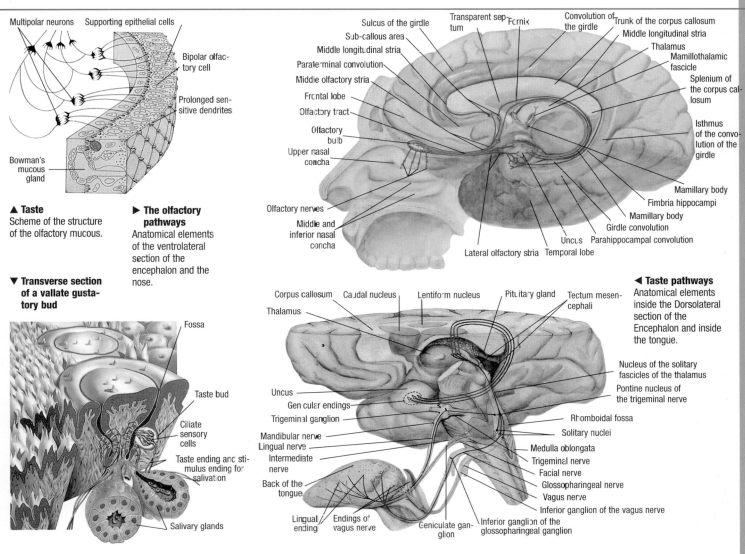

▲ Taste
Scheme of the structure of the olfactory mucous.

► The olfactory pathways
Anatomical elements of the ventrolateral section of the encephalon and the nose.

▼ Transverse section of a vallate gustatory bud

◄ Taste pathways
Anatomical elements inside the Dorsolateral section of the Encephalon and inside the tongue.

THE SENSATIONS OF THE SKIN

On the surface of the body, tactile senses collect a wide range of information about the environment and the world outside our body. However, besides the tactile ones, the skin hosts other receptors as well. These play a complementary role, by sending information about the pressure and temperature of the skin to the sensory cortex. When necessary, they also send pain information, which inform us of an imminent danger. These receptors are distributed on the entire surface of the body, and collaborate with those found inside the tendons, muscles and joints, in making sure

that the brain is able, at all times, to keep the entire body "under control". The receptors are found in different layers of the skin (epidermis, dermis), are different in look and function, and often they take on the name of illustrious anatomists who have studied them:

-**Krause's terminal bulbs**: they are located in the dermis, and are made up of a thin capsule which contains a nerve ending. They are sensitive to cold.

-**Pacini's corpuscles**:

they are located in the dermis; they are made up of concentric rings of capsular cells, which enclose nervous endings. They are pressure sensitive.

-**Ruffini's corpuscles**: are located in the

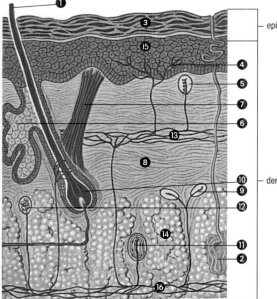

▲ The skin
Structure and receptors
❶ hair
❷ sudoriparous gland

Epidermis:
❸ corneal layer

Dermis:
❹ free nerve endings (pain)
❺ Meissner's corpuscle
❻ Sebaceous gland
❼ Erector muscle of the hair
❽ Connective tissue

❾ Ruffini's corpuscle
❿ hair bulb
⓫ Pacini's corpuscle
⓬ Krause's corpuscle
⓭ Sub papillary nervous plexus
⓮ Subcutaneous cellular tissue
⓯ Malpighi's layer
⓰ Hypodermic nervous plexus

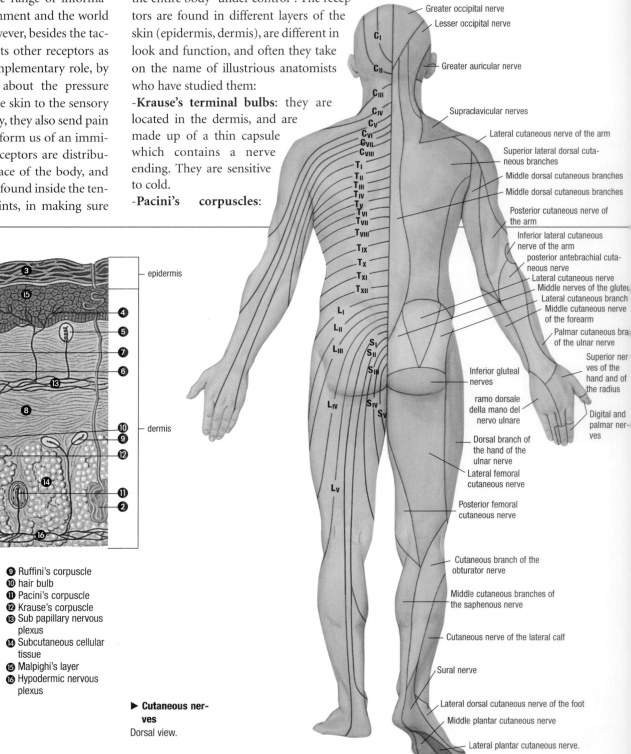

▶ Cutaneous nerves
Dorsal view.

Greater occipital nerve
Lesser occipital nerve
Greater auricular nerve
Supraclavicular nerves
Lateral cutaneous nerve of the arm
Superior lateral dorsal cutaneous branches
Middle dorsal cutaneous branches
Middle dorsal cutaneous branches
Posterior cutaneous nerve of the arm
Inferior lateral cutaneous nerve of the arm
posterior antebrachial cutaneous nerve
Lateral cutaneous nerve
Middle nerves of the glutei
Lateral cutaneous branch
Middle cutaneous nerve of the forearm
Palmar cutaneous bra of the ulnar nerve
Superior nerves of the hand and of the radius
Digital and palmar nerves
Inferior gluteal nerves
ramo dorsale della mano del nervo ulnare
Dorsal branch of the hand of the ulnar nerve
Lateral femoral cutaneous nerve
Posterior femoral cutaneous nerve
Cutaneous branch of the obturator nerve
Middle cutaneous branches of the saphenous nerve
Cutaneous nerve of the lateral calf
Sural nerve
Lateral dorsal cutaneous nerve of the foot
Middle plantar cutaneous nerve
Lateral plantar cutaneous nerve.

deep layer of the dermis. These are ramifications of flattened nervous fibers that are enclosed inside layers of capsular cells; they are heat sensitive.

-**Merkle disks**: are located inside the dermis; they are made up of a sheath that encloses a biconvex disk connected to a nerve ending. They are sensitive to continuous tactile stimulation.

Meissner's corpuscles: they are located on the outer surface of the dermis, and are made up of a cluster of nerve endings enclosed by a sheath. They are touch sensitive.

The following also have receptive qualities:
-**Free nerve endings**: are generally located in the epidermis and also in the dermis (lesser amount). They are pain and touch sensitive.

-**Nerve endings of the hair follicles**: they are located inside the deep dermis and they cover the hair follicle; they are stimulated by each movement of the hair, even with minimal contact.

The sensory nerve fibers emanating from the receptors have different lengths: from the few cm long cranial ones, to the 2 m long fibers that connect the tip of the toe to the sensory cortex.

Starting from the surface of the skin, they pass through the spinal cord (106) and arrive at the cerebral trunk (90) in bundles. Here they cross, pass through the thalamus and then end in the cortex. The response to pain is often reflected (108-109): because they have to produce a movement that quickly neutralizes the pain, they emanate directly from the spinal cord, often before the brain can perceive the sensation of pain.

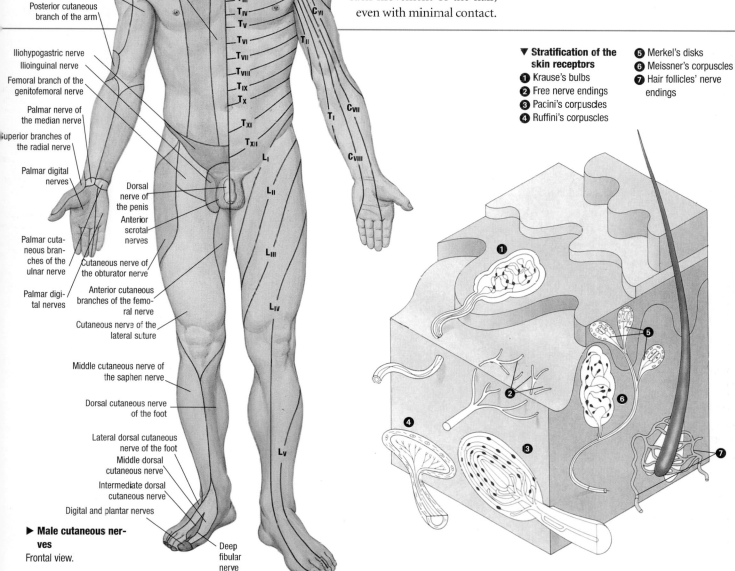

▼ **Stratification of the skin receptors**
1 Krause's bulbs
2 Free nerve endings
3 Pacini's corpuscles
4 Ruffini's corpuscles
5 Merkel's disks
6 Meissner's corpuscles
7 Hair follicles' nerve endings

▶ **Male cutaneous nerves**
Frontal view.

105

SPINAL CORD AND NERVES

The spinal cord is also part of the central nervous system. Different neurons are subdivided in different types (sensory, motor, and associative). This subdivision is similar to that of the brain: the grey matter is found here, and the white matter (made up of longitudinal bundles of myelitic fibers) makes up the surrounding layer. Because of this, the spinal cord is more flexible, elastic, and consequently it has a higher consistency than the brain. Protected by the meninges and by the cephalo-spinal fluid, the spinal cord is found inside the spinal canal and it fol-

lows all its curvatures without ever touching the bones. It is a white cylindrical substance, 45 cm long, able to undergo slight stretching, when the trunk is forcefully extended. It is divided into a bulb (as part of the encephalic trunk) and into neuromers, which correspond to the different vertebral sections: cervical, thoracic, lumbar, sacral, and coccygeal. Its primary

function is to collect stimuli from the environment and transmit them to the cerebral cortex, and to re-transmit the centrally elaborated answers all the way to the periphery. Some important elements of the autonomous nervous system ▶110 are also found in the spinal cord. Inside the grey matter, neurons are organized into clusters where all the elements have the same nervous connections. They are called nuclei or columns, depending on how their orientation is studied (transverse or longitudinal to the marrow). They are organized into lamellae. Many of the 9 laminae found inside the grey matter of the spinal cord, correspond to nuclei and columns- each lamina is made up of special neurons that

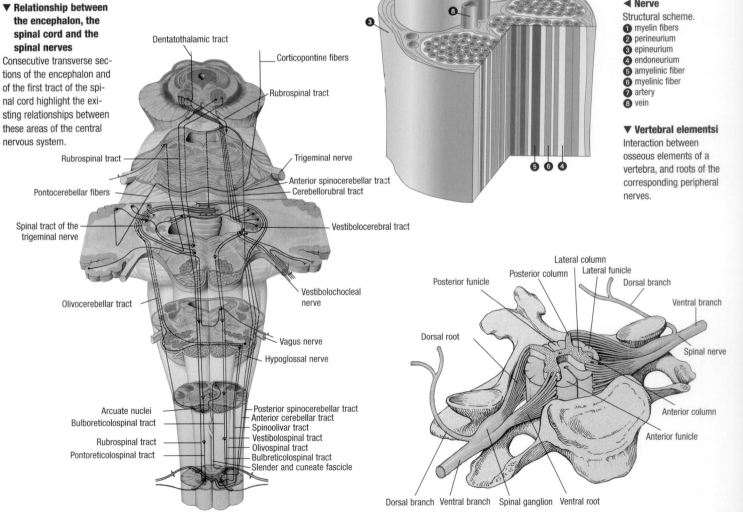

▼ Relationship between the encephalon, the spinal cord and the spinal nerves

Consecutive transverse sections of the encephalon and of the first tract of the spinal cord highlight the existing relationships between these areas of the central nervous system.

Dentatothalamic tract

Corticopontine fibers

Rubrospinal tract

Rubrospinal tract

Trigeminal nerve

Anterior spinocerebellar tract
Cerebellorubral tract

Pontocerebellar fibers

Spinal tract of the trigeminal nerve

Vestibolocerebral tract

Olivocerebellar tract

Vestibolochocleal nerve

Vagus nerve

Hypoglossal nerve

Arcuate nuclei
Bulboreticolospinal tract

Rubrospinal tract

Pontoreticolospinal tract

Posterior spinocerebellar tract
Anterior cerebellar tract
Spinoolivar tract
Vestibolospinal tract
Olivospinal tract
Bulbreticolospinal tract
Slender and cuneate fascicle

◀ Nerve
Structural scheme.
1 myelin fibers
2 perineurium
3 epineurium
4 endoneurium
5 amyelinic fiber
6 myelinic fiber
7 artery
8 vein

▼ Vertebral elementsi
Interaction between osseous elements of a vertebra, and roots of the corresponding peripheral nerves.

Lateral column
Posterior column
Lateral funicle
Posterior funicle
Dorsal branch
Ventral branch
Dorsal root
Spinal nerve
Anterior column
Anterior funicle
Dorsal branch Ventral branch Spinal ganglion Ventral root

have a specific function or that end in a specific part of the body, or by nervous fibers that come from or are directed to the encephalic fasciae. The central canal runs in the middle of the grey matter, which occupies a section of the marrow, is shaped as an "H", and whose length depends on its distance from the encephalon. The central canal runs in its length and contains very little cephalo-spinal fluid. Inside the white matter, the bundles of nervous fibers can be made up of either the extensions of the ganglion neurons, which are attached to the posterior root of the spinal nerves ▶116, or by the extension

of the cells of the spinal grey matter, or even by the neurons that are located on supramedullary axial centers or supraxial centers. Therefore, the fasciae can be crossed by unidirectional nervous impulses (projection fasciae), or by impulses that travel up and down the spinal cord (association fasciae, made up of a different type of fiber, which send unidirectional impulses). The fasciae guarantee a tight connection between the different spinal sections, and they play an essential role in organizing the spinal reflexes.

The spinal cord is connected on the periphery to 33 pairs of spinal nerves:

their roots (33 pairs per side) are divided into an anterior root (or motor) and a posterior root (or sensitive). The anterior root is responsible for transferring the stimuli coming from the encephalon or the centers of the spinal cord, and directed toward the muscles. The posterior root transfers the stimuli from the periphery of the body to the central nervous system. Along the course of each posterior root, there is a spinal ganglion: this is an enlargement made up of cellular bodies that originate from the root itself. The posterior and anterior roots unite and form the spinal nerve, on the side of the ganglion. A peripheral nerve originates far from each nerve, and its ramifications reach the different structures of the body. The spinal

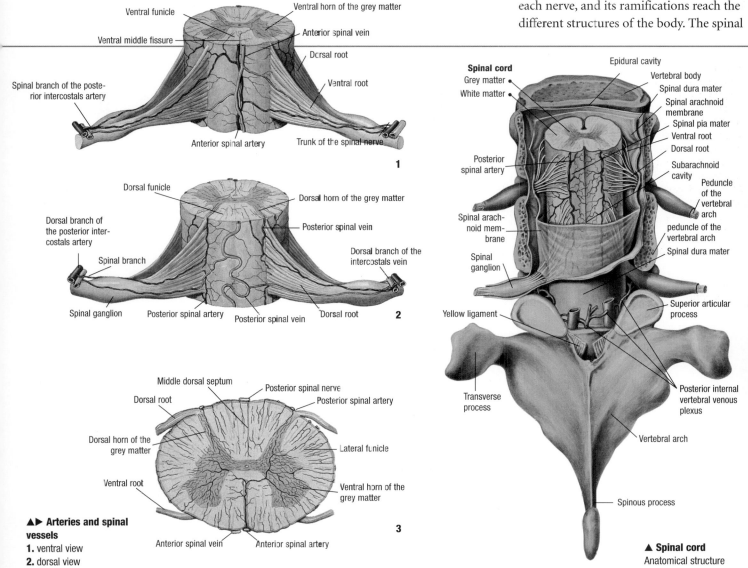

1

Ventral funicle
Ventral middle fissure
Spinal branch of the posterior intercostals artery
Ventral horn of the grey matter
Anterior spinal vein
Dorsal root
Ventral root
Anterior spinal artery
Trunk of the spinal nerve

2

Dorsal funicle
Dorsal branch of the posterior intercostals artery
Spinal branch
Spinal ganglion
Posterior spinal artery
Posterior spinal vein
Dorsal horn of the grey matter
Posterior spinal vein
Dorsal branch of the intercostals vein
Dorsal root

3

Middle dorsal septum
Dorsal root
Dorsal horn of the grey matter
Ventral root
Posterior spinal nerve
Posterior spinal artery
Lateral funicle
Ventral horn of the grey matter
Anterior spinal vein
Anterior spinal artery

▲▶ **Arteries and spinal vessels**
1. ventral view
2. dorsal view
3. transverse section.

Spinal cord
Grey matter
White matter
Posterior spinal artery
Spinal arachnoid membrane
Spinal ganglion
Yellow ligament
Transverse process
Epidural cavity
Vertebral body
Spinal dura mater
Spinal arachnoid membrane
Spinal pia mater
Ventral root
Dorsal root
Subarachnoid cavity
Peduncle of the vertebral arch
peduncle of the vertebral arch
Spinal dura mater
Superior articular process
Posterior internal vertebral venous plexus
Vertebral arch
Spinous process

▲ **Spinal cord**
Anatomical structure

nerves are at times connected into plexuses: there are crossings of fibers connected to each other (for example, the solar plexus or celiac, in which nervous ramifications or efferent nervous branches of the crescent ganglion cross around the celiac trunk and the superior mesenteric artery).

REFLEXED ARCH

This is the name of a complex neuromo-tor structure made of receptors, afferent and efferent peripheral nervous fibers and neurons of the spinal cord. It is able to manage specific muscular reactions, in a reflexive way, in other words without the conscious intervention of the encephalon. The stimulus (for example a painful sensation) reaches the grey matter of the spinal cord, through the receptors and the afferent fibers of the spinal nerves. A "short circuit" directly stimulates the motor neuron responsible for the reflex movement (muscular contraction). This is a typical simple reflex arch, but there are many which are more complex, such as those that control for the secretion of substances, or that involve unconscious levels of memory, such as the conditional reflexes. In many cases, the reflex arch can also evoke automatic responses, through nervous fibers that connect the central system to the sympathetic system.

GAMMA CIRCUIT

This is a complex neuromotor structure able to regulate the muscular tone in a

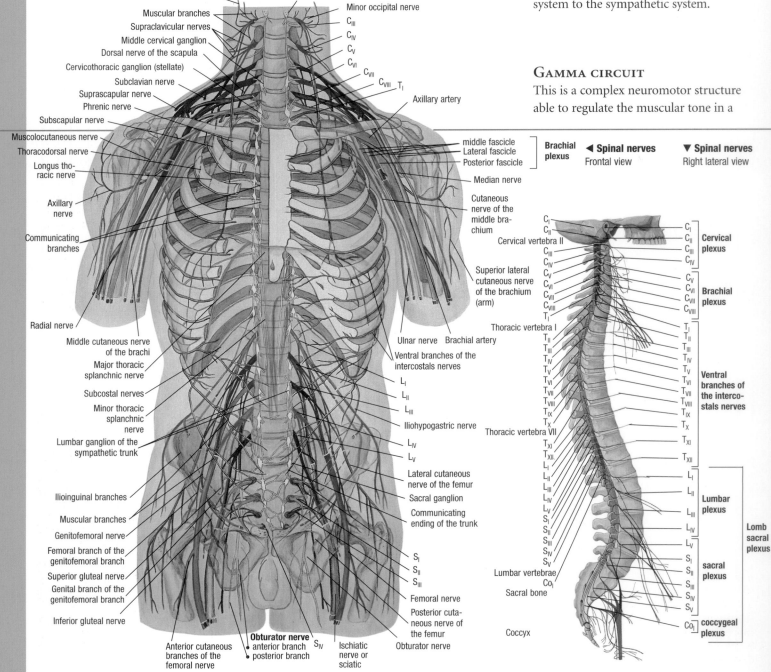

reflex manner. It is made up of neuromuscular spindles, by efferent and afferent peripheral nervous fibers, and by neurons of the spinal cord.

The gamma motor neuron of the lamina 9, and their neuritis, reaches the fibers of the neuromuscular spindles and the motor end plates, through the anterior roots of the spinal nerves, stimulating a muscle contraction. Their activity is induced by peripheral stimuli (articular or cutaneous receptors), but mainly by the cerebellum and by the extra pyramidal system. When the muscle fibers, which have been stimulated by these motor neurons, contract, the surrounding fibers that are not directly stimulated, passively relax. This stimulates the receptors found inside them, which then transfers impulses through the spinal ganglion and the posterior root, all the way up to the alpha motor neurons. These cells of the spinal grey matter directly stimulate all of the fibers of the target muscle. The action of the gamma neurons becomes useless: they go back to a resting state, waiting to respond to a new stimulation.

✚ TRAUMAS OF THE SPINAL COLUMN

By inserting some electrodes inside the motor cortex, we can detect an electrical signal that precedes a movement: this is the signal sent by the cortex to one or more voluntary muscles in order to move them. However, if a spinal cord lesion has occurred, this signal cannot reach the target: the voluntary movement doesn't occur. In order to give back movement to the millions of people who have become paralyzed due to a spinal trauma, medical research is trying to pick up and detect such electrical activity "before" a lesion may occur. The goal is to try to amplify it and to send it to the targeted muscles, by transmitting it to spinal nerves that are still efficient. This type of research has been developing from different perspectives from a cellular to a more technological one. So, while in some laboratories we are trying to stimulate the regeneration of those spinal neurons damaged by the trauma or to grow those in vitro (228), in other laboratories they are trying to connect electrical wires (metal) to muscular and nervous fibers. Both of these routes are very promising, and the second one especially, has already shown some successful results with couple of volunteers, who have been able to walk again after a bad accident.

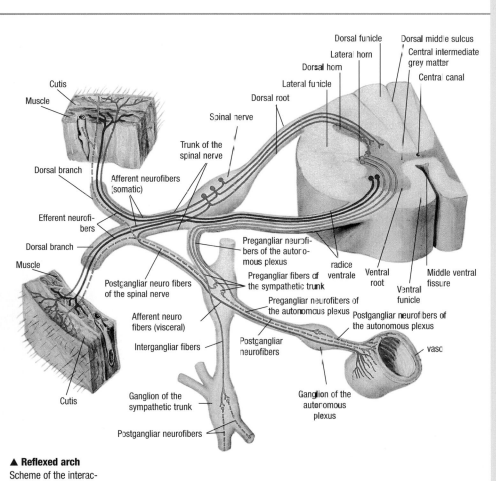

▲ Reflexed arch
Scheme of the interactions between the stimuli, The autonomous nervous signals and the reflexed nervous signals of the spinal cord.

Labels: Cutis, Muscle, Dorsal branch, Efferent neurofibers, Dorsal branch, Muscle, Cutis, Spinal nerve, Trunk of the spinal nerve, Afferent neurofibers (somatic), Postgangliar neuro fibers of the spinal nerve, Afferent neuro fibers (visceral), Interngangliar fibers, Postgangliar neurofibers, Ganglion of the sympathetic trunk, Postgangliar neurofibers, Dorsal root, Dorsal funicle, Lateral horn, Dorsal horn, Lateral funicle, Pregangliar neurofibers of the autoromous plexus, Pregangliar fibers of the sympathetic trunk, Pregangliar neurofibers of the autonomous plexus, Postgangliar neurofibers of the autonomous plexus, Ganglion of the autonomous plexus, radice ventrale, Ventral root, Ventral funicle, Dorsal middle sulcus, Central intermediate grey matter, Central canal, Middle ventral fissure, vaso

The peripheral nervous system is also called autonomous, because it is independent from the central system, and vegetative, since it regulates the "basic" vital functions. It is made up of peripheral nerves and the ganglia, and it is divided into the parasympathetic system and the orthosympathetic system, called the sympathetic

THE PERIPHERAL NERVOUS SYSTEM

The peripheral nervous system guarantees the connection between the various organs of the body and the central nervous system. It is also called the autonomous system because it induces behavior that does not involve either a conscious effort or superior cerebral structures. It is also called the vegetative system, because it automatically regulates the "basic" vital functions of the body. In fact, besides connecting the rest of the body to the central system, the peripheral nervous system keeps some functions of single organs under control. It also regulates the homeostasis of the entire organism, by stimulating or hindering activities such as cardiac and respiratory frequency, acidic secretion of the stomach, intestinal movements, and so on. It is made up of nervous fibers and of ganglia (organs made of neuron aggregates): depending on their characteristics, these nerves and ganglia are divided into two main groups: the parasympathetic, and the orthosympathetic system. Both of them innervate the same organs, and often their action is antagonistic: for example, the vagus nerve (parasympathetic) allows brachial muscles to contract, while the endings of the sympathetic relax them.

THE PARASYMPATHETIC SYSTEM

It is made up of nervous fibers, which originate from the cranial and the bulbar centers. Their course is always mixed in with the fibers of the encephalic or spinal somatic nerves. It is primarily made up of the vagus nerve (10th pair of cranial nerves) and its ramifications, it also includes parts of the 3rd, 7th and 9th pair of nerves, along with some nuclei of the sacral position of the spinal cord. All of the cells are pregangliar neurons: the body is found inside the nucleus of a cranial nerve or inside the spinal cord, and the fiber reaches a ganglion, always located in the thickening of the innervated viscera.

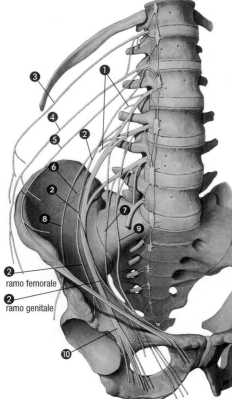

ramo femorale

ramo genitale

◄ **Lumbar plexus**
Vertebral and pelvic lumbar connections between the peripheral nervous system and the sympathetic nervous system.
❶ branches of the great and small psoas muscles; ❷ genitofemoral nerve; ❸ sub costal nerve; ❹ iliohypogastric nerve; ❺ ilioinguinal nerve; ❻ branch of the iliac muscle; ❼ accessory obstructor nerve; ❽ lateral femoral cutaneous nerve; ❾ obstructor nerve; ❿ femoral nerve

▼ **Action of the parasympathetic system**
The main nervous tracts of the parasympathetic system are related to the organs they affect. Often their actions are antagonistic to those of the orthosympathetic system.

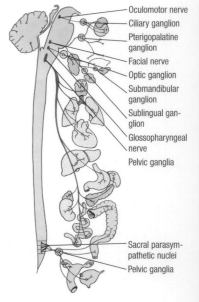

Oculomotor nerve
Ciliary ganglion
Pterigopalatine ganglion
Facial nerve
Optic ganglion
Submandibular ganglion
Sublingual ganglion
Glossopharyngeal nerve
Pelvic ganglia

Sacral parasympathetic nuclei
Pelvic ganglia

▼ Vertebral elements
Components of the central, peripheral and sympathetic nervous system near a vertebra.
Axial session.

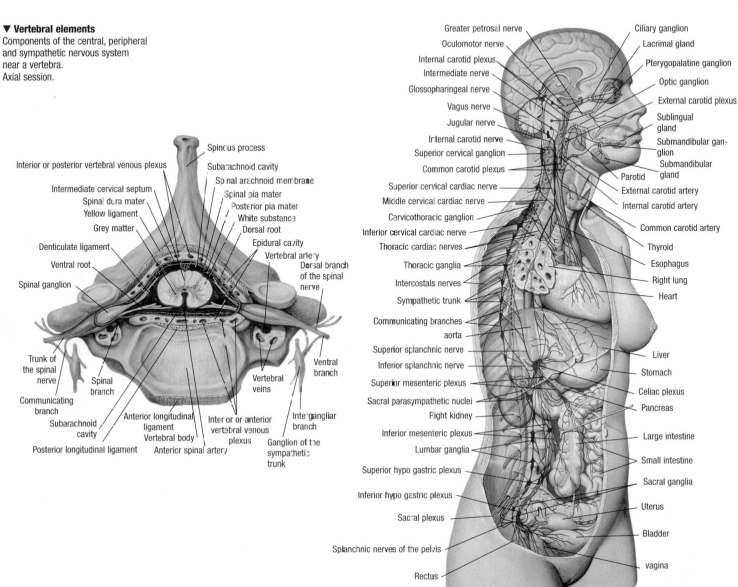

Spincus process
Interior or posterior vertebral venous plexus
Subarachnoid cavity
Spinal arachnoid membrane
Intermediate cervical septum
Spinal dura mater
Yellow ligament
Grey matter
Denticulate ligament
Ventral root
Spinal ganglion
Spinal pia mater
Posterior pia mater
White substance
Dorsal root
Epidural cavity
Vertebral artery
Dorsal branch of the spinal nerve
Trunk of the spinal nerve
Spinal branch
Communicating branch
Subarachnoid cavity
Posterior longitudinal ligament
Anterior longitudinal ligament
Vertebral body
Anterior spinal artery
Interior or anterior vertebral venous plexus
Interganglar branch
Ganglion of the sympathetic trunk
Ventral branch
Vertebral veins

Greater petrosal nerve
Oculomotor nerve
Internal carotid plexus
Intermediate nerve
Glossopharingeal nerve
Vagus nerve
Jugular nerve
Internal carotid nerve
Superior cervical ganglion
Common carotid plexus
Superior cervical cardiac nerve
Middle cervical cardiac nerve
Cervicothoracic ganglion
Inferior cervical cardiac nerve
Thoracic cardiac nerves
Thoracic ganglia
Intercostals nerves
Sympathetic trunk
Communicating branches
aorta
Superior splanchnic nerve
Inferior splanchnic nerve
Superior mesenteric plexus
Sacral parasympathetic nuclei
Fight kidney
Inferior mesenteric plexus
Lumbar ganglia
Superior hypo gastric plexus
Inferior hypo gastric plexus
Sacral plexus
Splanchnic nerves of the pelvis
Rectus

Ciliary ganglion
Lacrimal gland
Pterygopalatine ganglion
Optic ganglion
External carotid plexus
Sublingual gland
Submandibular ganglion
Submandibular gland
Parotid
External carotid artery
Internal carotid artery
Common carotid artery
Thyroid
Esophagus
Right lung
Heart
Liver
Stomach
Celiac plexus
Pancreas
Large intestine
Small intestine
Sacral ganglia
Uterus
Bladder
vagina

THE ORTHOSYMPATHETIC SYSTEM

This system is much more complex than the sympathetic one, and besides being made up of many distinct nerves, it also has many neuronal networks, called plexi.
The fibers of the orthosympatetic, currently called simply "sympathetic", originate from the dorso-lumbar positions of the cord and from ganglia that are made of the cellular bodies of the postgangliar neurons, aligned in front of the vertebral column in two long rows on the sides of the marrow. Nervous fibers called communicating branches (if they reach the spinal nerves) or visceral nerves (if they reach an organ) depart from each ganglion. The visceral nerves cross anastomize and connect to other ganglia, forming the complex reticulate of nervous fibers called plexi.

▲ Involuntary nervous system of a woman
Besides some nerves of the central system, we can also see the main parasympathetic and sympathetic nerves
Right anterolateral section

THE CRANIAL NERVES

They depart form the encephalon or from the spinal cord, on the level of the cervical vertebrae, reaching the head, the neck, part of the trunk and the upper limbs. They are divided into:
-12 pairs of symmetric **encephalic nerves**, which connect the encephalon to numerous peripheral territories (head, neck, thorax and the abdomen). Each pair is indicated by a progressively higher number (in a cranial-caudal direction) or with a name that indicates its function. Each pair is made up of different fibers from the spinal nerves: in fact, the visceroeffectrice fibers of these nerves are only parasympathetic, and

the components of each nerve are quite different. Also, while the somatic and visceral sensitivity of the spinal nerves is transmitted by neurons localized in the same spinal ganglion, such neurons are found in different ganglia, in the encephalic nerves (the geniculate ganglion is an exception). In fact, the encephalic nerves do not only transmit a generalized sensitivity, but also specific sensitive stimuli: gustatory (nerves of the 7th, 9th and 10th pair), vestibular and acoustic (8th pairs), visual (2nd pair) and olfactory (1st pair). Also, in some

cases, some encephalic nerves that transport the same type of impulses are afferent to the same cerebral nucleus: This is the case of those fibers that come from the 5th, 7th, 9th and 10th pair, that transmit the sensitivity of the head and converge in the

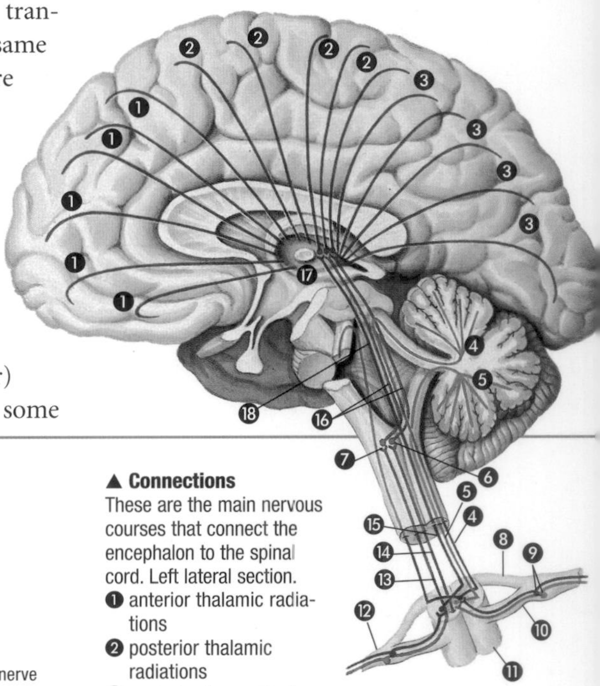

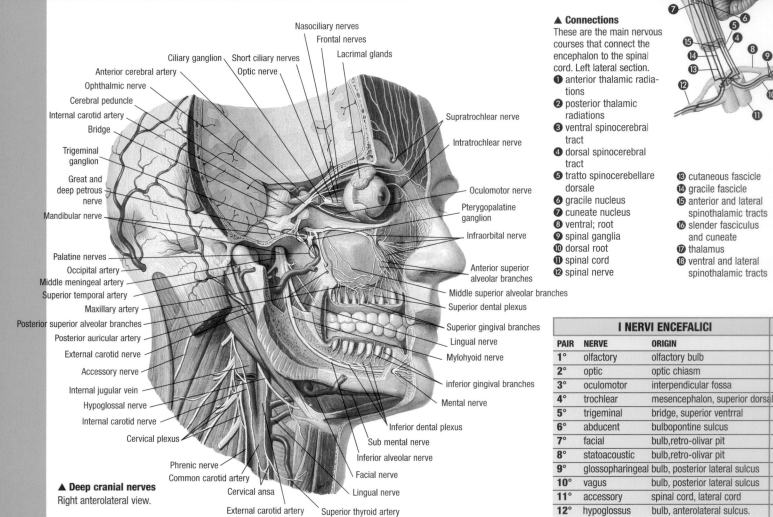

▲ **Connections**
These are the main nervous courses that connect the encephalon to the spinal cord. Left lateral section.
❶ anterior thalamic radiations
❷ posterior thalamic radiations
❸ ventral spinocerebral tract
❹ dorsal spinocerebral tract
❺ tratto spinocerebellare dorsale
❻ gracile nucleus
❼ cuneate nucleus
❽ ventral; root
❾ spinal ganglia
❿ dorsal root
⓫ spinal cord
⓬ spinal nerve

⓭ cutaneous fascicle
⓮ gracile fascicle
⓯ anterior and lateral spinothalamic tracts
⓰ slender fasciculus and cuneate
⓱ thalamus
⓲ ventral and lateral spinothalamic tracts

Deep cranial nerves labels:
Nasociliary nerves
Frontal nerves
Lacrimal glands
Ciliary ganglion
Short ciliary nerves
Optic nerve
Anterior cerebral artery
Ophthalmic nerve
Cerebral peduncle
Internal carotid artery
Bridge
Trigeminal ganglion
Great and deep petrous nerve
Mandibular nerve
Palatine nerves
Occipital artery
Middle meningeal artery
Superior temporal artery
Maxillary artery
Posterior superior alveolar branches
Posterior auricular artery
External carotid nerve
Accessory nerve
Internal jugular vein
Hypoglossal nerve
Internal carotid nerve
Cervical plexus
Supratrochlear nerve
Intratrochlear nerve
Oculomotor nerve
Pterygopalatine ganglion
Infraorbital nerve
Anterior superior alveolar branches
Middle superior alveolar branches
Superior dental plexus
Superior gingival branches
Lingual nerve
Mylohyoid nerve
inferior gingival branches
Mental nerve
Inferior dental plexus
Sub mental nerve
Inferior alveolar nerve
Facial nerve
Lingual nerve
Phrenic nerve
Common carotid artery
Cervical ansa
External carotid artery
Superior thyroid artery

▲ **Deep cranial nerves**
Right anterolateral view.

I NERVI ENCEFALICI		
PAIR	**NERVE**	**ORIGIN**
1°	olfactory	olfactory bulb
2°	optic	optic chiasm
3°	oculomotor	interpendicular fossa
4°	trochlear	mesencephalon, superior dorsal
5°	trigeminal	bridge, superior ventrral
6°	abducent	bulbopontine sulcus
7°	facial	bulb,retro-olivar pit
8°	statoacoustic	bulb,retro-olivar pit
9°	glossopharingeal	bulb, posterior lateral sulcus
10°	vagus	bulb, posterior lateral sulcus
11°	accessory	spinal cord, lateral cord
12°	hypoglossus	bulb, anterolateral sulcus.

sensitive nucleus of the trigeminal; -8 pairs of cervical nerves, all originating from the spinal cord.

Like all the other spinal nerves, the cranial nerves are characterized by 4 different types of fibers: motor somatic visceral effectrice, somatic sensitive and visceral sensitive. Also, in these types of nerves, the somatic and visceral sensitivity is transmitted by neurons which are localized inside the same spinal ganglia. Each nerve originates from the gathering of numerous radicles, united into 2 roots, and near the inter-vertebral foramen from which they depart: an anterior root, made up of somatic motor and visceral effectrice fibers, and a posterior root, made up of sensitive fibers, which form a spinal ganglion. The direction of these roots changes, and their

length increases in a cephalo-caudal direction: the first nerves have an horizontal root, while the direction of the others is progressively more oblique, reaching the typical "horse tail " shape in the sacral and coccygeal roots ▶116-119. Once the spinal nerve has formed, it releases two collateral branches: the meningeal branch (or recurrent), which distributes sensitive fibers to the vertebral structures, and the white communicating branch (lumbar and thoracic nerves) that reaches a ganglion of the sympathetic system with pregangliar medullar fibers. From the sympathetic ganglion, one or more grey communicating branches, made of amyelinic fibers, goes back into the spinal nerve and acts as a distributor in specific territories. At this point the spinal nerve splits into an ante-

rior or ventral branch and a posterior or dorsal branch: usually the posterior branch is the shortest, and it innervates a territory circumscribed by muscles' motor fibers and cutaneous sensitive fibers, conserving its individuality. On the other hand, the anterior branch crosses with anterior branches of other nerves, establishing with them complex nervous relationships and creating complex anatomical structures called plexi. The anterior branches are responsible for the motor and sensitive innervations of the anterolateral region of the neck, the trunk and the upper limbs. Among the spinal nerves, we must remember the particular characteristics of the dorsal branches of the first two cervical nerves (or occipital), the characteristics of the cervical plexus and of the brachial

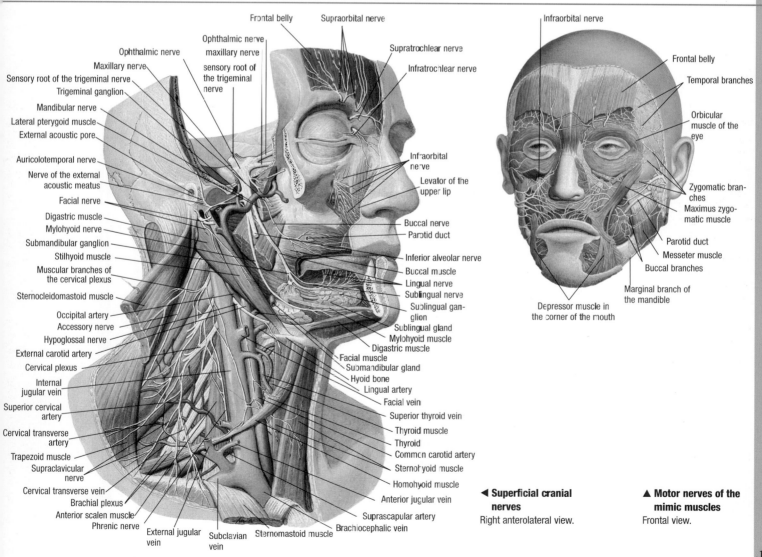

◀ Superficial cranial nerves
Right anterolateral view.

▲ Motor nerves of the mimic muscles
Frontal view.

plexus:

-Dorsal branch of the 1st cervical nerve (suboccipital nerve): larger than the corresponding ventral branch and with an exclusive motor function. It runs across the vertebral canal along with the vertebral artery, and therefore it moves upward toward the head, in which mostly motor fibers arrive (suboccipital muscles). One of the descending branches connects to a corresponding posterior branch of the 2nd occipital nerve;

-Dorsal branch of the 2nd cervical nerve (great occipital nerve): it is the largest among the posterior spinal nerves; it originates from the vertebral canal, between the atlas ▶44-45 and the epistropheus, in an upward direction. On average, it perforates the semispinal muscle and the trapezium,

under the inferior oblique muscle, becoming subcutaneous in the occipital region: it is not by chance that it made up primarily of sensory fibers. In its initial branch, some muscular fibers innervate the inferior oblique muscles, the semispinal muscles, the longissimus capitis musculus, the splenium and the trapezoid muscle, while two ascending and descending fibers attach respectively to the dorsal branches of the 1st and the 3rd cervical nerve.

-Cervical plexus: it is made up of the anterior branches of the 1st, 2nd, 3rd and 4th cervical nerves. In fact, each anterior branch is further divided into two branches (ascending and

descending) which tie to the corresponding branches of the adjacent nerves, forming 3 cervical ansae, layered on top of each other and differentiated into superior, middle and inferior. The plexus is located deep inside the neck, and from it new nervous fibers originate. They are divided into anastomotic branches (Which put other nervous fibers in contact with each other), cutaneous branches (sensory) and muscular branches (motor and sensory). The phrenic nerve is located inside the muscu-

---

▼ Pterygopalatine ganglion

Main nervous courses.
Left lateral section of the skull

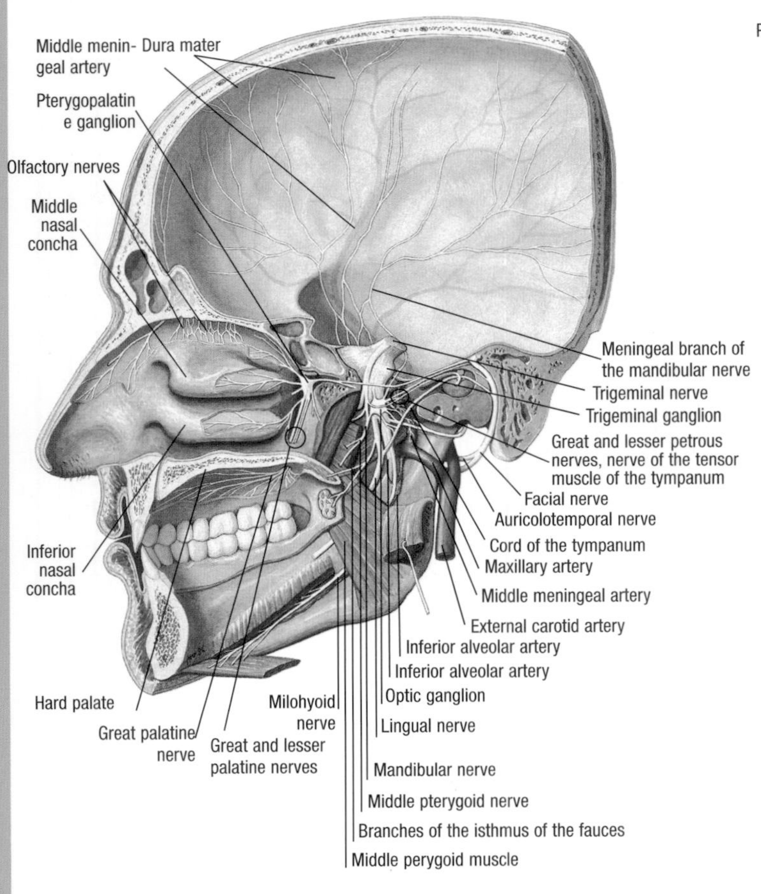

▶ Deep nerves of the neck

Frontal view.

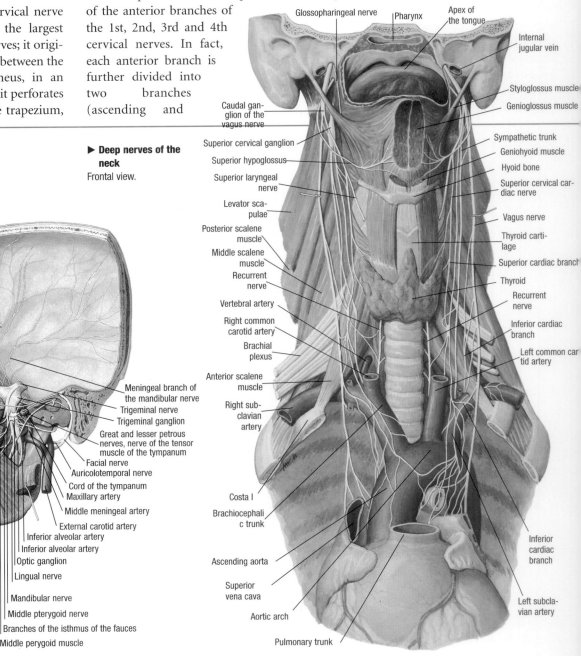

Labels (left illustration):
Middle meningeal artery — Dura mater
Pterygopalatine ganglion
Olfactory nerves
Middle nasal concha
Inferior nasal concha
Hard palate
Great palatine nerve
Great and lesser palatine nerves
Milohyoid nerve
Meningeal branch of the mandibular nerve
Trigeminal nerve
Trigeminal ganglion
Great and lesser petrous nerves, nerve of the tensor muscle of the tympanum
Facial nerve
Auriculotemporal nerve
Cord of the tympanum
Maxillary artery
Middle meningeal artery
External carotid artery
Inferior alveolar artery
Inferior alveolar artery
Optic ganglion
Lingual nerve
Mandibular nerve
Middle pterygoid nerve
Branches of the isthmus of the fauces
Middle perygoid muscle

Labels (right illustration):
Glossopharyngeal nerve
Pharynx
Apex of the tongue
Internal jugular vein
Styloglossus muscle
Genioglossus muscle
Caudal ganglion of the vagus nerve
Sympathetic trunk
Superior cervical ganglion
Geniohyoid muscle
Superior hypoglossus
Hyoid bone
Superior laryngeal nerve
Superior cervical cardiac nerve
Levator scapulae
Vagus nerve
Posterior scalene muscle
Thyroid cartilage
Middle scalene muscle
Superior cardiac branch
Recurrent nerve
Thyroid
Vertebral artery
Recurrent nerve
Right common carotid artery
Inferior cardiac branch
Brachial plexus
Left common carotid artery
Anterior scalene muscle
Right subclavian artery
Costa I
Brachiocephalic trunk
Ascending aorta
Inferior cardiac branch
Superior vena cava
Aortic arch
Left subclavian artery
Pulmonary trunk

lar branches. This is the largest and the longest of the cervical plexus. It moves the diaphragm [63] and its sensory fibers innervate the pleura, the pericardium, and the posterior part of the abdomen and the inferior surface of the diaphragm. Some of its fibers reach the celiac plexus through the diaphragm.

-Brachial plexus: is made up of the 5th, 6th, 7th, and 8th anterior branches of the cervical nerves and of the 1st thoracic nerve, to which fibers of the 4th cervical nerve and of the 2nd thoracic nerve are attached. The plexus rests in the axillary fossa and in the subclavicular groove; once in contact, the nerves exit in the shape of trunks, continuing on to the upper limbs [64-120].

Many parasympathetic and sympathetic nerves are also found inside the cranial-cervical area: while the first ones follow the sensory spinal nerves, the second ones make up other plexi (complex and branched networks). Among these, the main ones are the carotid plexus, the cavernous plexus, the intercarotid plexus and the subclavian plexus. They are made up of the perivascular branches of the cervical and sympathetic segments. The also accompany the carotid artery and its ramifications along their entire course.

① lateral branch
① medial branch

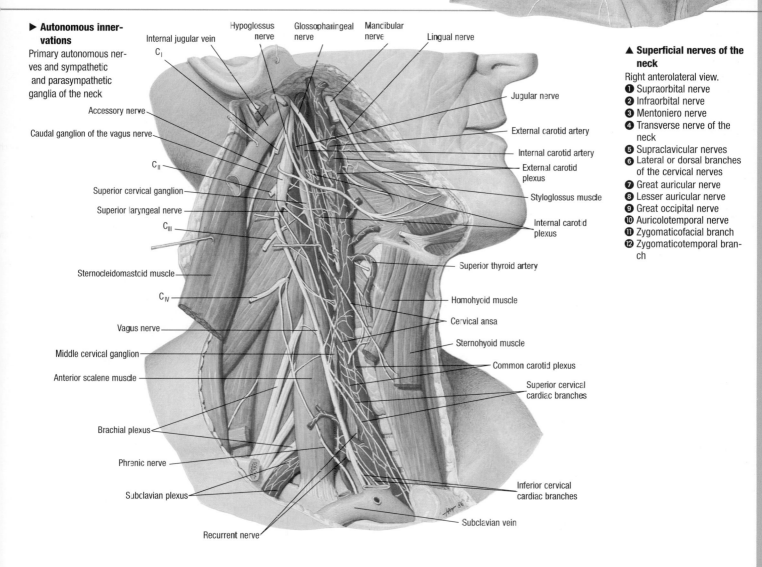

▶ **Autonomous innervations**

Primary autonomous nerves and sympathetic and parasympathetic ganglia of the neck

Hypoglossus nerve
Glossopharingeal nerve
Mancibular nerve
Lingual nerve
Internal jugular vein
C I
Accessory nerve
Caudal ganglion of the vagus nerve
C II
Superior cervical ganglion
Superior laryngeal nerve
C III
Sternocleidomastoid muscle
C IV
Vagus nerve
Middle cervical ganglion
Anterior scalene muscle
Brachial plexus
Phrenic nerve
Subclavian plexus
Recurrent nerve

Jugular nerve
External carotid artery
Internal carotid artery
External carotid plexus
Styloglossus muscle
Internal carotid plexus
Superior thyroid artery
Homohyoid muscle
Cervical ansa
Sternohyoid muscle
Common carotid plexus
Superior cervical cardiac branches
Inferior cervical cardiac branches
Subclavian vein

▲ **Superficial nerves of the neck**
Right anterolateral view.
❶ Supraorbital nerve
❷ Infraorbital nerve
❸ Mentoniero nerve
❹ Transverse nerve of the neck
❺ Supraclavicular nerves
❻ Lateral or dorsal branches of the cervical nerves
❼ Great auricular nerve
❽ Lesser auricular nerve
❾ Great occipital nerve
❿ Auricolotemporal nerve
⓫ Zygomaticofacial branch
⓬ Zygomaticotemporal branch

THORACIC-ABDOMINAL NERVES

They include the nerve ending of the thoracic and lumbar segments of the sympathetic, and the spinal nerves of the thoracic, lumbar, sacral, and the coccygeal tract of the spinal column. Because of the need to regulate the activity of each internal organ, the sympathetic nerves are found here in large quantities. While short visceral branches of the sympathetic innervate the esophagus and the alimentary tube (esophageal branches, great and lesser splanchnic nerves), the others are organized into plexi, named after the territory or the innervated organ: aortic plexus, cardiac plexus, pulmonary plexus, and preaortic plexus.

Cardiac plexus: It is made up of 3 cardiac nerves (superior, middle, and inferior) that descend from their cervical ganglion, from the fibers of the first 4 thoracic ganglia and from 3 cardiac branches of the vagus nerve (parasympathetic). Other parasympathetic cervical endings innervate the aorta, the coronary walls, the cardiac cavities and the pericardium.

Preaortic plexus: it is a single interlacement of sympathetic fibers, which extends from the abdominal;

aorta all the way to the ganglia. It is divided into:

-a celiac plexus: the secondary even plexi (phrenic, renal, surrenal, spermatic, pelvic) and the odd ones (lineal, empathic, superior gastric, superior mesenteric) are connected to it.

-an aorticoabdominal plexus: located under the superior mesenteric artery.

-a hypogastric plexus: located in front of the middle sacral artery. The spinal nerves are divided into:

-12 pairs of thoracic nerves: the 1st pair emerges between the 1st and the 2nd thoracic vertebra, the last one between the 12th and the 1st lumbar. The anterior bran-

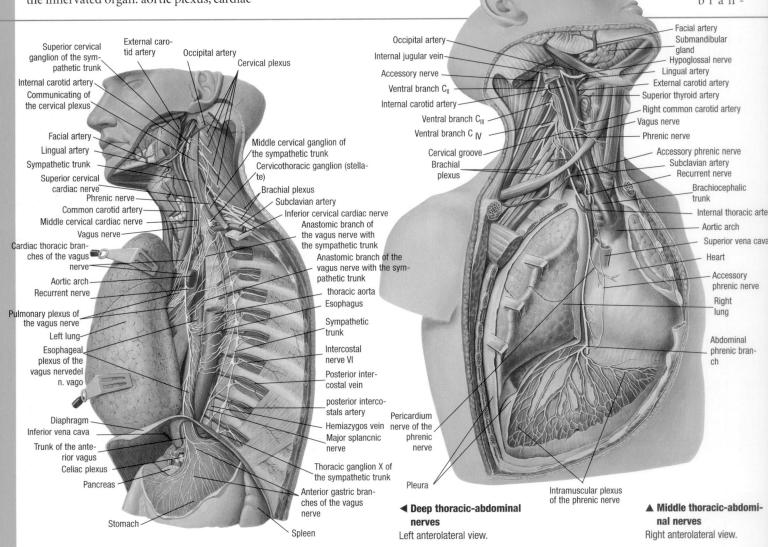

◄ Deep thoracic-abdominal nerves
Left anterolateral view.

▲ Middle thoracic-abdominal nerves
Right anterolateral view.

▶ **Deep innervations of the costal tract**
Frontal view.

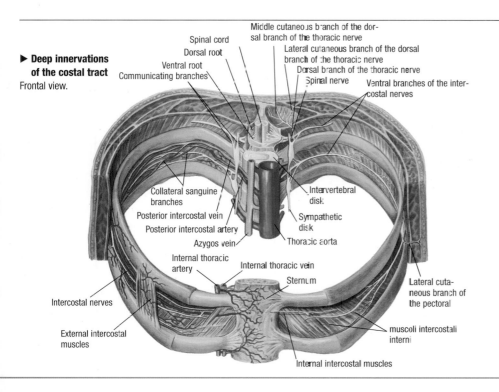

- Spinal cord
- Dorsal root
- Ventral root
- Communicating branches
- Middle cutaneous branch of the dorsal branch of the thoracic nerve
- Lateral cutaneous branch of the dorsal branch of the thoracic nerve
- Dorsal branch of the thoracic nerve
- Spinal nerve
- Ventral branches of the intercostal nerves
- Collateral sanguine branches
- Posterior intercostal vein
- Posterior intercostal artery
- Azygos vein
- Intervertebral disk
- Sympathetic disk
- Thoracic aorta
- Internal thoracic artery
- Internal thoracic vein
- Sternum
- Lateral cutaneous branch of the pectoral
- Intercostal nerves
- External intercostal muscles
- muscoli intercostali interni
- Internal intercostal muscles

ches are named after the intercostals nerves, since each one of them runs through the corresponding intercostal space: they innervate the intrinsic muscular structure of the thorax, the thoracic-abdominal wall and the cutis. Once they arrive at the sternum, the first 6 will outwardly cross the thoracic wall, and will distribute in the anterolateral region of the thorax. The 2nd, 3rd and 4th anterior cutaneous branches are named after the medial mammary branches, because they innervate the medial part of the mammilla.

The last 6 intercostal nerves insert in the abdominal wall, crossing the internal oblique and transverse muscles of the abdomen, up to the rectus muscle of the abdomen. They end with cutaneous branches as well. The 12th nerve runs underneath the

▶ **Superficial thoracic-abdominal nerves of a man's body**
1. Frontal view
2. Dorsal view

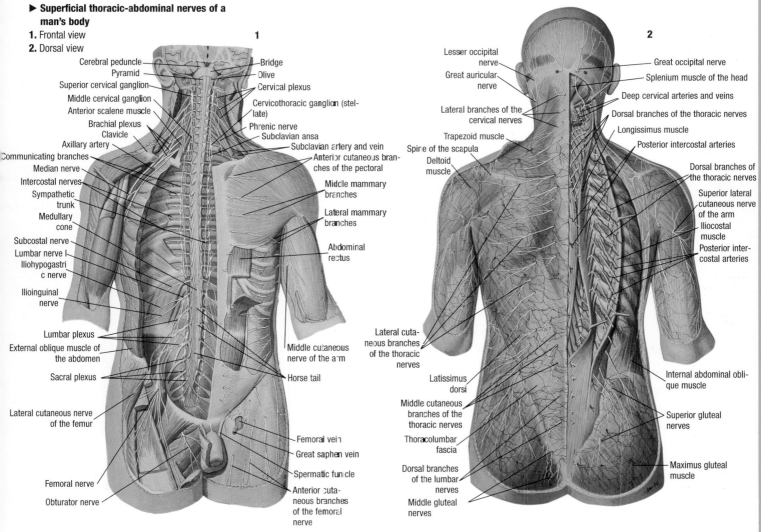

1

- Cerebral peduncle
- Pyramid
- Superior cervical ganglion
- Middle cervical ganglion
- Anterior scalene muscle
- Brachial plexus
- Clavicle
- Axillary artery
- Communicating branches
- Median nerve
- Intercostal nerves
- Sympathetic trunk
- Medullary cone
- Subcostal nerve
- Lumbar nerve I
- Iliohypogastric nerve
- Ilioinguinal nerve
- Lumbar plexus
- External oblique muscle of the abdomen
- Sacral plexus
- Lateral cutaneous nerve of the femur
- Femoral nerve
- Obturator nerve
- Bridge
- Olive
- Cervical plexus
- Cervicothoracic ganglion (stellate)
- Phrenic nerve
- Subclavian ansa
- Subclavian artery and vein
- Anterior cutaneous branches of the pectoral
- Middle mammary branches
- Lateral mammary branches
- Abdominal rectus
- Middle cutaneous nerve of the arm
- Horse tail
- Femoral vein
- Great saphen vein
- Spermatic funicle
- Anterior cutaneous branches of the femoral nerve

2

- Lesser occipital nerve
- Great auricular nerve
- Lateral branches of the cervical nerves
- Trapezoid muscle
- Spine of the scapula
- Deltoid muscle
- Great occipital nerve
- Splenium muscle of the head
- Deep cervical arteries and veins
- Dorsal branches of the thoracic nerves
- Longissimus muscle
- Posterior intercostal arteries
- Dorsal branches of the thoracic nerves
- Superior lateral cutaneous nerve of the arm
- Iliocostal muscle
- Posterior intercostal arteries
- Lateral cutaneous branches of the thoracic nerves
- Latissimus dorsi
- Middle cutaneous branches of the thoracic nerves
- Thoracolumbar fascia
- Dorsal branches of the lumbar nerves
- Middle gluteal nerves
- Internal abdominal oblique muscle
- Superior gluteal nerves
- Maximus gluteal muscle

12th rib (costa) and it is therefore called subcostal. The thoracic nerves emit connecting anastomotic branches and muscular and nervous collateral branches.

-5 pairs of lumbar nerves: the 1st pair originates between the 1st and 2nd lumbar vertebra; the last one originated between the 5th and the sacral one. The anterior branch of the 1st lumbar nerve is thinner and it gives origin to the iliohypogastric and the ilioinguinal nerves; the anterior branch of the 2nd lumbar nerve is divided into a lateral cutaneous femoral nerve and a genito-femoral nerve. From the 2nd anastomotic ansa originate the superior roots of the obturator and the femoral nerves.

The anterior branch of the 3rd lumbar nerve originates the middle roots of the obturator and the femoral nerves, while the anterior branch of the 4th lumbar nerve originates the inferior roots of those same nerves.

The anastomotic ansa, which originates from this lumbar nerve, connects to the anterior branch of the 5th lumbar nerve originates the lumbosacral trunk which participate in the development of the sacral plexus. In turn, the anterior branches of the 1st, 2nd, 3rd, and 4th lumbar nerve and a branch of the 12th thoracic nerve make up the lumbar plexus. Triangular in shape, its base is at the level of the spinal column and the vertex is located where the roots of the femoral nerve converge.

The following emerge from this plexus:
- Connecting anastomic branches;
Short collateral branches. They have a motor function. These are the nerves of the lateral transverse muscles of the trunk, of the great psoas muscle, of the lesser psoas muscle and of the lumbar quadrate muscle. Long collateral branches ▶122-123. They are generally mixed, and reach the inferior part of the abdomi-

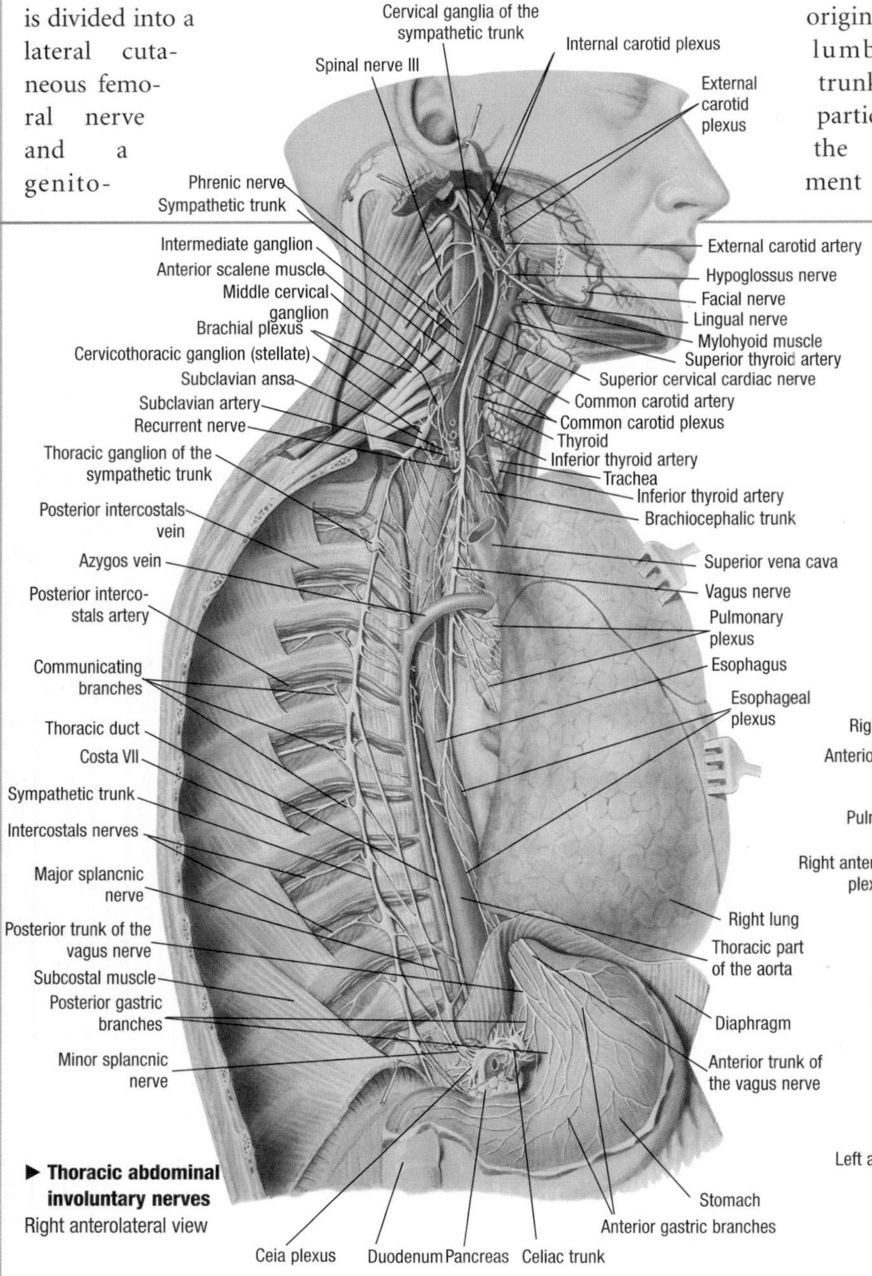

▶ **Thoracic abdominal involuntary nerves**
Right anterolateral view

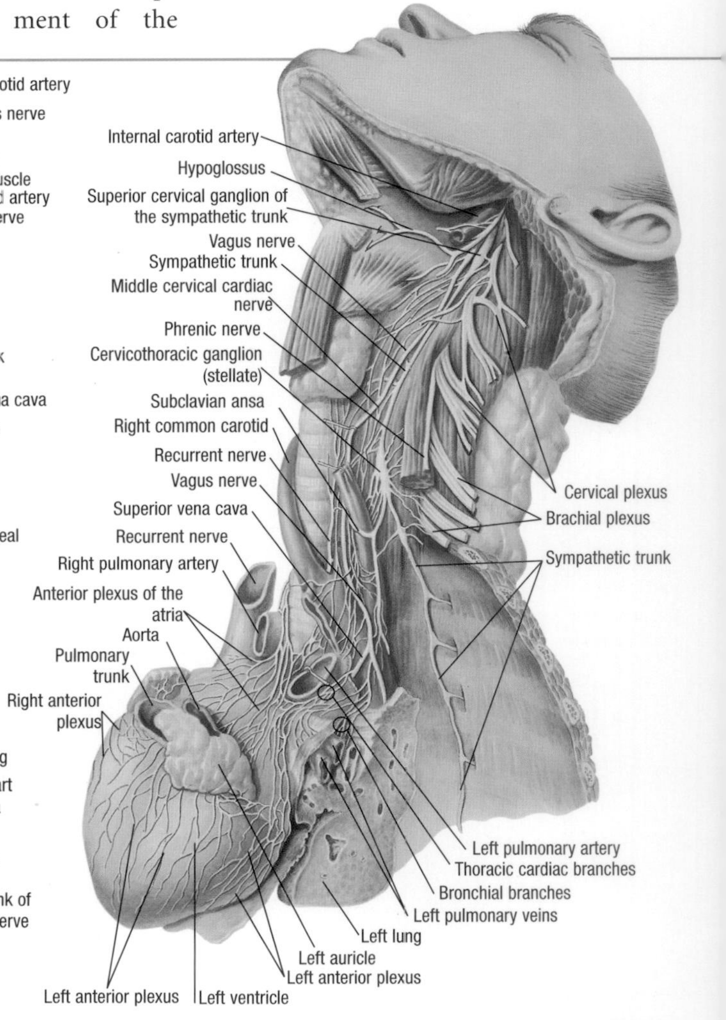

▲ **Involuntary cardiac innervations**
Left lateral view.

nal wall, the genitalia and the lower limbs: these are the iliohypogastric, ilioinguinal, genitofemoral nerves and the lateral cutaneous of the femur.

-Terminal branches [122-123]. They branch out from the pelvis and are arranged on the cutis and the muscle of the lower limb: this is the obturator and the femoral nerve.

5 pairs of sacral nerves. They branch out of the sacral nerves.

The union between the anterior branches of the 1st, 2nd and 3rd sacral nerve and the lumbosacral trunk, which collects the fibers of the anterior branch of the 4th lumbar nerve and the entire 5th lumbar nerve, originates at the sacral plexus, a flattened triangular group of nerves, whose base is located in the sacral bone, and the vertex in the inferior outline of the ischia-tic foramen. A large terminal branch (the ischiatic nerve, better known as the sciatic nerve), and the posterior and anterior collateral branches.

All of the anterior collateral branches are motor, and they innervate the superior and inferior gemellus muscles, the internal obturator and the quadratus femoris. The posterior collateral branches include 3 motor nerves, which innervate the gluteal muscles, tensor fasciae latae, and pyriform and a sensory nerve: the posterior cutaneous nerve of the femur, which reaches the posterior surface of the thigh and the leg, and the perineal region.

The 2nd, 4th and mostly the 3rd sacral nerves make op the prostate plexus, which innervates the genital organs, part of the intestine, part of the urinary tract, of the muscles and of the cutis of the perineum. The prostate plexus send out visceral branches (parasympathetic), muscular and cutaneous somatic branches.

On the sides of the coccyx, the anterior branches of the 1st coccygeal nerve and of the 5th sacral nerve along with some fibers from the 4th sacral nerve make up the coccygeal plexus.

The anterior visceral branches are parasympathetic fibers, which reach the hypogastric plexus [116]; the posterior somatic branches reach the coccygeal muscle and the cutis.

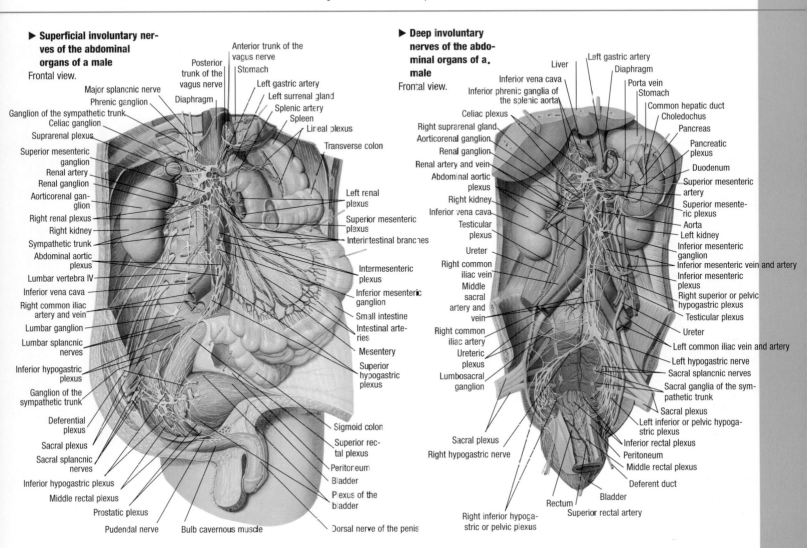

▶ **Superficial involuntary nerves of the abdominal organs of a male**
Frontal view.

Anterior trunk of the vagus nerve
Posterior trunk of the vagus nerve
Stomach
Left gastric artery
Left surrenal gland
Major splancnic nerve
Phrenic ganglion
Diaphragm
Splenic artery
Spleen
Ganglion of the sympathetic trunk
Lineal plexus
Celiac ganglion
Suprarenal plexus
Transverse colon
Superior mesenteric ganglion
Renal artery
Renal ganglion
Aorticorenal ganglion
Right renal plexus
Left renal plexus
Right kidney
Superior mesenteric plexus
Sympathetic trunk
Interintestinal branches
Abdominal aortic plexus
Lumbar vertebra IV
Intermesenteric plexus
Inferior vena cava
Inferior mesenteric ganglion
Right common iliac artery and vein
Small intestine
Lumbar ganglion
Intestinal arteries
Lumbar splancnic nerves
Mesentery
Inferior hypogastric plexus
Superior hypogastric plexus
Ganglion of the sympathetic trunk
Deferential plexus
Sigmoid colon
Superior rectal plexus
Sacral plexus
Sacral splancnic nerves
Peritoneum
Bladder
Inferior hypogastric plexus
Plexus of the bladder
Middle rectal plexus
Prostatic plexus
Pudendal nerve
Bulb cavernous muscle
Dorsal nerve of the penis

▶ **Deep involuntary nerves of the abdominal organs of a. male**
Frontal view.

Liver
Left gastric artery
Diaphragm
Inferior vena cava
Porta vein
Inferior phrenic ganglia of the splenic aorta
Stomach
Celiac plexus
Common hepatic duct
Choledochus
Right suprarenal gland
Aorticorenal ganglion
Pancreas
Renal ganglion
Pancreatic plexus
Renal artery and vein
Duodenum
Abdominal aortic plexus
Superior mesenteric artery
Right kidney
Superior mesenteric plexus
Inferior vena cava
Testicular plexus
Aorta
Ureter
Left kidney
Right common iliac vein
Inferior mesenteric ganglion
Middle sacral artery and vein
Inferior mesenteric vein and artery
Inferior mesenteric plexus
Right common iliac artery
Right superior or pelvic hypogastric plexus
Ureteric plexus
Testicular plexus
Lumbosacral ganglion
Ureter
Left common iliac vein and artery
Left hypogastric nerve
Sacral splancnic nerves
Sacral ganglia of the sympathetic trunk
Sacral plexus
Left inferior or pelvic hypogastric plexus
Sacral plexus
Inferior rectal plexus
Right hypogastric nerve
Peritoneum
Middle rectal plexus
Deferent duct
Rectum
Bladder
Right inferior hypogastric or pelvic plexus
Superior rectal artery

NERVES OF THE UPPER LIMBS

These nerves leave the brachial plexus ▶115, and run through the arm and the forearm, until they reach the hand. As they leave the shoulder, they split into 3 *primary trunks*:

-The superior primary trunk is made up of the anterior branches of the 4th, 5th and 6th cervical nerve.

-The middle primary trunk is made up of the 7th cervical nerve and is an independent nerve.

The inferior primary trunk is made up of the anterior branches of the 1st and 2nd thoracic nerves, which unite with the anterior branch of the 8th cervical nerve. Each of the three primary trunks split near the axillary fossa into an anterior and a posterior branch. In turns, the anterior and the posterior branches make up 3 *secondary trunks*:

-The posterior secondary trunk is made of the posterior branches of the 3 primary trunks. From here the radial and the axillary nerves originate.

-The lateral secondary trunk is made of the anterior branches of the 3 middle primary trunks. The muscolocutaneous nerve and the lateral root of the median nerve originate from here.

-The medial secondary trunk, made up of the sole independent anterior branch of the inferior primary trunk; the medial root of the median nerve, the ulnar nerve, the medial cutaneous nerve of the arm and the medial cutaneous nerve of the forearm originate from here. All of these

Lateral supraclavicle nerves
Lateral root of the median nerve
Muscolocutaneous nerve
Axillary nerve
Superior lateral cutaneous nerve of the arm
Deep branch of the radial nerve
Superior branch of the radial nerve
Antebrachial lateral cutaneous nerve
Common palmar digital nerves
Superficial branch of the ulnar nerve
Deep branch of the ulnar nerve
Dorsal digital nerves
Proper palmar digital nerves

Brachial plexus
- Lateral fascicle
- Posterior fascicle
- Middle fascicle

Middle root of the median nerve
Radial nerve
Median nerve

Middle cutaneous nerve of the arm
Ulnar nerve

Antebrachial middle cutaneous nerve

▲ **Main nerves of the right arm**
Frontal view.

▶ **Nerves of the right shoulder**
Frontal view.

Posterior auricular artery
Parotid gland
Facial artery
Hypoglossus nerve
Internal jugular vein
Internal carotid artery
Accessory nerve
Minor occipital nerve
Vagus nerve
Cervical ansa
Cervical plexus
Middle scalene muscle
Phrenic nerve
Anterior scalene muscle
Inferior thyroid artery
Internal jugular vein
Lateral supraclavicular nerve
Subclavian vein
Subclavian artery
Brachial plexus
Cephalic vein
Axillary artery
Lateral fascicle
Lateral thoracic artery
Middle fascicle

Submandibular gland
Lingual artery
External carotid
Superior thyroid artery
Superior laryngeal artery
Thyroid

Superior trunk
Middle trunk
Inferior trunk

Anterior pectoral cutaneous bran-
Lateral and middle ches pectoral nerves

Intercostobrachial nerves (intercostals II, III)
Pectoral lateral cutaneous branch
Lateral mammary branches
Long thoracic nerve
Middle mammary branches

nerves are terminal branches of the brachial plexus, long nerves anastomotic branches for the interconnection of different nerves, and collateral branches, which distribute to the various dorsal and thoracic muscles. The main nerve endings of the hand originate from the median and the ulnar nerve; they branch out into progressively thinner nerves, until they reach the finger tips.

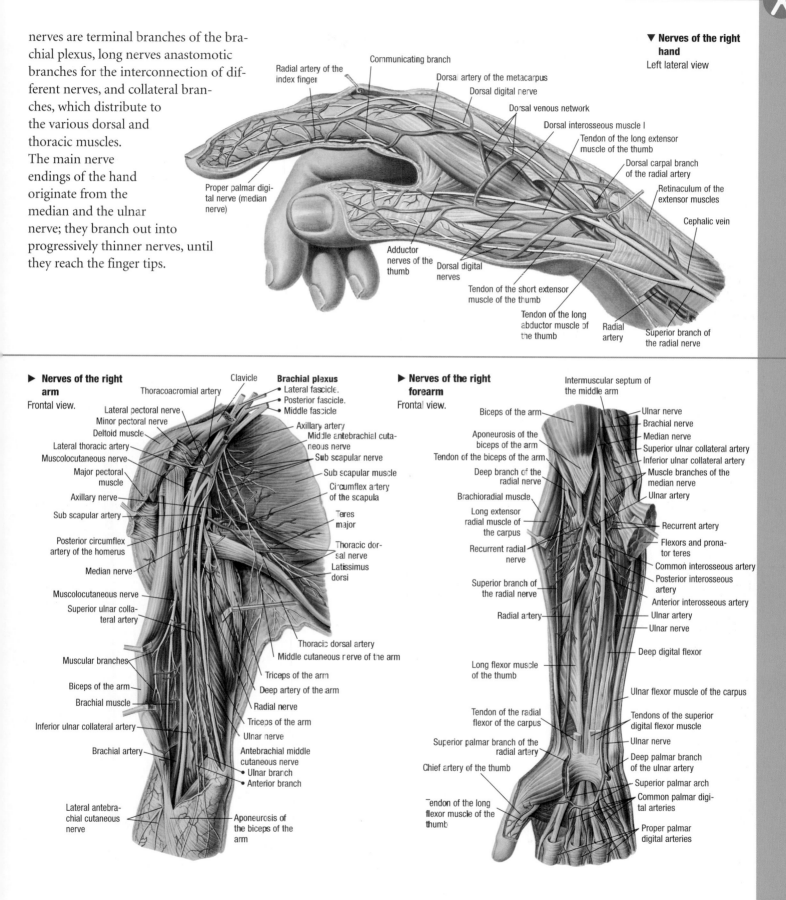

▼ **Nerves of the right hand**
Left lateral view

▶ **Nerves of the right arm**
Frontal view.

▶ **Nerves of the right forearm**
Frontal view.

121

NERVES OF THE LOWER LIMBS

The nerves that distribute to the lower limbs originate in the sacral plexus ▶119 and in the lumbar plexus ▶118: these are the ischiatic nerve (the only extension of the sacral plexus), and **the long collateral branches** and **the terminal branches** of the lumbar plexus. Except of the lateral cutaneous nerve of the thigh and of the internal saphen, all the others are mixed nerves: they have sensory and motor functions, often subdivided into cutaneous branches (sensory) and muscular branches (motor) of each nerve.

ISCHIATIC NERVE OR SCIATIC

It is the largest and the longest in the body; it originates from the anterior branches of the 4th and 5th lumbar nerve and of the 1st, 2nd and 3rd sacral nerve. Its collateral and terminal branches innervate the posterior muscles of the thigh, the leg, the foot, the articulations of the hip and of the knee: particularly the muscular collateral branches made up of the nerves of the long head of the biceps muscle of the femur, of the semitendineus muscle, of the semimembranous of the adductor muscle. Originating from the side of the sacral bone, the ischiatic nerve departs form the pelvis, crosses the buttock, runs parallel to the ischiatic artery behind the superior and inferior gemellus muscles and the quadratus femuris, until it reaches the upper corner of the popliteus, where it becomes superficial and divides into two terminal branches:

-The tibial nerve is the largest. It moves toward the malleolus and then divides into two branches: the medial plantar nerve and the lateral plantar nerve. With its muscular

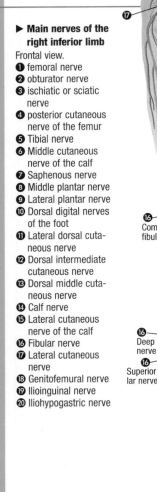

▶ **Main nerves of the right inferior limb**
Frontal view.
❶ femoral nerve
❷ obturator nerve
❸ ischiatic or sciatic nerve
❹ posterior cutaneous nerve of the femur
❺ Tibial nerve
❻ Middle cutaneous nerve of the calf
❼ Saphenous nerve
❽ Middle plantar nerve
❾ Lateral plantar nerve
❿ Dorsal digital nerves of the foot
⓫ Lateral dorsal cutaneous nerve
⓬ Dorsal intermediate cutaneous nerve
⓭ Dorsal middle cutaneous nerve
⓮ Calf nerve
⓯ Lateral cutaneous nerve of the calf
⓰ Fibular nerve
⓱ Lateral cutaneous nerve
⓲ Genitofemural nerve
⓳ Ilioinguinal nerve
⓴ Iliohypogastric nerve

Common fibular nerve

Deep fibular nerve

Superior fibular nerve

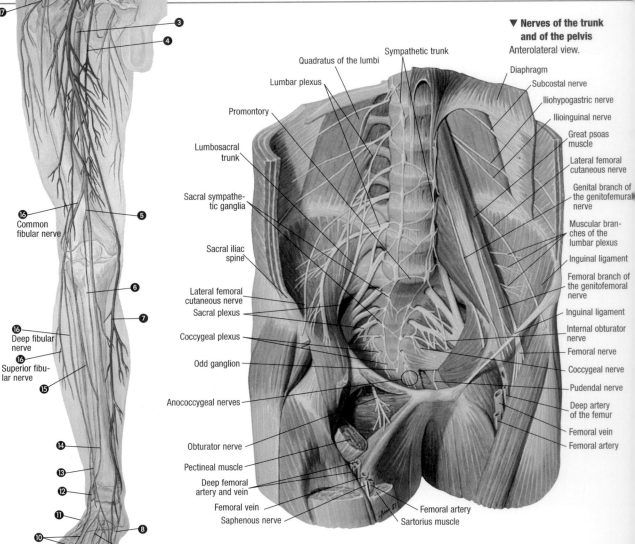

▼ **Nerves of the trunk and of the pelvis**
Anterolateral view.

Quadratus of the lumbi
Sympathetic trunk
Lumbar plexus
Diaphragm
Subcostal nerve
Promontory
Iliohypogastric nerve
Ilioinguinal nerve
Lumbosacral trunk
Great psoas muscle
Lateral femoral cutaneous nerve
Sacral sympathetic ganglia
Genital branch of the genitofemural nerve
Muscular branches of the lumbar plexus
Sacral iliac spine
Inguinal ligament
Lateral femoral cutaneous nerve
Femoral branch of the genitofemural nerve
Sacral plexus
Inguinal ligament
Coccygeal plexus
Internal obturator nerve
Odd ganglion
Femoral nerve
Coccygeal nerve
Anococcygeal nerves
Pudendal nerve
Deep artery of the femur
Obturator nerve
Femoral vein
Pectineal muscle
Femoral artery
Deep femoral artery and vein
Femoral vein
Femoral artery
Saphenous nerve
Sartorius muscle

and sensory collateral branches, it innervates the popliteal muscles, the posterior superficial of the leg (gemellus, soleus, plantar, gracilis), the deep posterior of the leg (posterior tibial, longus flexor of the hallux), the foot, the knee joint and the talocrural joint (ankle), the cutis of the heel and the foot. The medial cutaneous nerve of the calf (or external saphenous) is a sensory branch, which originates from the calf nerves, the dorsal lateral of the foot and dorsal lateral cutaneous of the foot, which gives sensitivity to the fingers and the post lateral cutis of the inferior part of the limb;
-The common peroneal nerve runs along the internal margin of the biceps of the femur all the way to the external surface of the leg, where it divides into 2 terminal branches: the superficial peroneal nerve, which innervates the peroneal muscles, and the inferolateral cutis of the leg (intermediate cutaneous nerves and medial dorsal cutaneous), and the deep peroneal nerve, which innervates the deep muscles of the leg, and after passing through the entire foot, it reaches the first two fingers. Its muscular and cutaneous collateral branches contribute to innervate the superficial and deep muscles of the leg and the knee joint. Along with the cutaneous branches of the tibial nerve, they make up the calf nerve.

LONG COLLATERAL BRANCHES
-Genitofemoral: it originates from the 2nd lumbar nerve, and it arrives at the inguinal ligament through the great psoas muscle.

Here it divides into 2 terminal branches: the genital branch and the femoral branch, which depart from the pelvis, near the external iliac artery, become subcutaneous and innervate the anterosuperior part of the thigh.
-Lateral cutaneous nerve of the femur: it is solely a sensory nerve, and it originates from the anterior branch of the 2nd lumbar nerve. It runs through the great psoas, the iliac fossa, and exits the pelvis as a subcutaneous nerve. Here it branches out into the gluteal and the femoral branch, which arrives at the knee on the anterolateral surface of the thigh.

TERMINAL BRANCHES
-*Obturator nerve*: it originates from the 2nd, 3rd and 4th lumbar nerve, and the 3

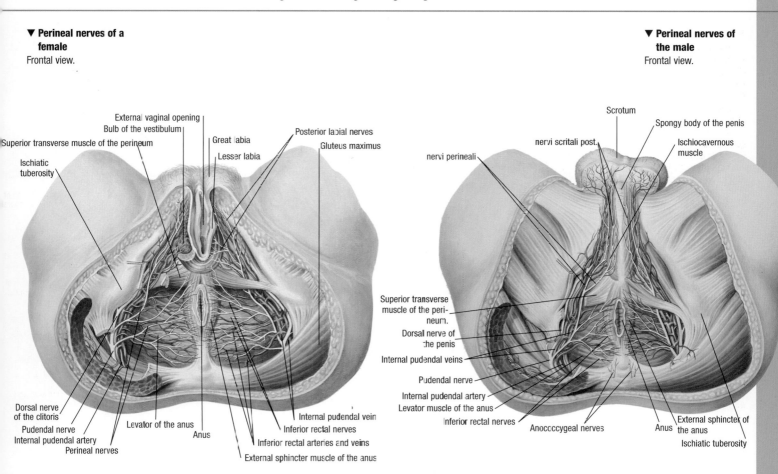

▼ **Perineal nerves of a female**
Frontal view.

▼ **Perineal nerves of the male**
Frontal view.

roots connect downward inside a sole trunk. This trunk crosses the great psoas toward the pelvis, and passing above the sacroiliac joint, it reaches the obturator canal where it produces the only collateral branch: the nerve for the external obturator muscle. As it exits the obturator canal along with the main blood vessels, it divides into 2 branches:

-The anterior branch: this is the biggest; it runs down along the external obturator muscle, it innervates the long and short adductor muscles of the thigh, the gracilis muscle, and it ends as a cutaneous branch in the inferomedial region of the thigh;

-The posterior branch: this innervates the external obturator muscle and the great adductor, and it sends out articular branches into the hip and the knee.

-The femoral nerve: it is the largest of the lumbar plexus; it has 3 roots in the anterior branches of the 2nd, 3rd and 4th lumbar nerves, and of some anterior fibers of the 1st lumbar nerve. By the 5th lumbar vertebra the roots fuse into a single trunk, and the nerve comes out the great psoas and moves toward the pelvis. Once it arrives at the inguinal ligament, it continues on with the iliopsoas and then divides into its terminal branches that reach the extremities of the limb.

In the abdominal and pelvic tract, the femoral nerve has many collateral nerves, which are primarily motorial:
-The nerves for the great psoas muscle
-The nerves for the iliac muscle
-The nerves for the pectineal muscle
-The nerves for the femoral artery, which

follow the vessel to half the thigh.
Among the terminal nerves there are the
-Lateral muscolocutaneous nerve, whose muscular branches are made up of motor fibers of the sartorius muscle and sensory cutaneous branches of the anterior surface of the thigh, divided into anterior and middle perforating nerves and in internal accessory of the saphenous nerve;
-The medial muscolocutaneous nerve, with muscular and cutaneous branches, which innervate the pectineus muscle, the longus adductor and the superomedial surface of the thigh;
-The quadriceps muscle of the femur: it is the deepest, is almost completely motor, and is divided into 4 muscular branches, which innervate the anterior rectus muscle, the vastus lateralis, the middle and inter-

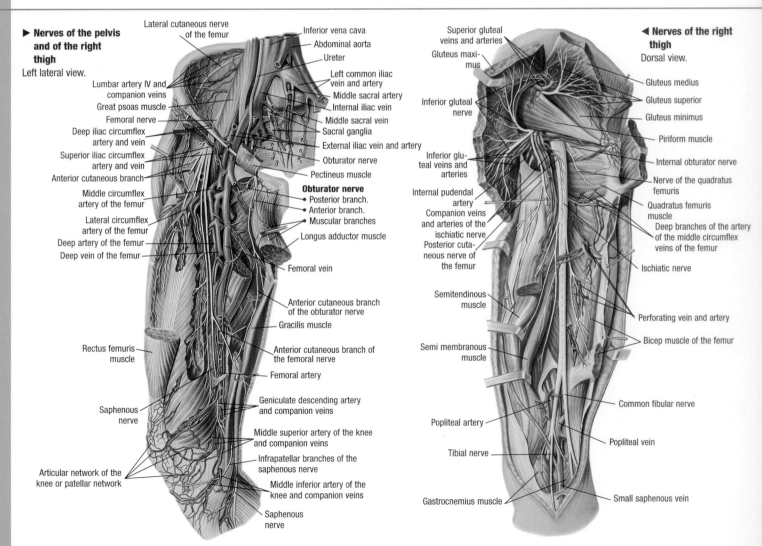

▶ **Nerves of the pelvis and of the right thigh**
Left lateral view.

Lateral cutaneous nerve of the femur
Inferior vena cava
Abdominal aorta
Ureter
Lumbar artery IV and companion veins
Great psoas muscle
Left common iliac vein and artery
Middle sacral artery
Femoral nerve
Internal iliac vein
Deep iliac circumflex artery and vein
Middle sacral vein
Sacral ganglia
Superior iliac circumflex artery and vein
External iliac vein and artery
Anterior cutaneous branch
Obturator nerve
Middle circumflex artery of the femur
Pectineus muscle
Obturator nerve
• Posterior branch.
• Anterior branch.
Lateral circumflex artery of the femur
• Muscular branches
Deep artery of the femur
Longus adductor muscle
Deep vein of the femur
Femoral vein
Anterior cutaneous branch of the obturator nerve
Rectus femuris muscle
Gracilis muscle
Anterior cutaneous branch of the femoral nerve
Femoral artery
Saphenous nerve
Geniculate descending artery and companion veins
Middle superior artery of the knee and companion veins
Articular network of the knee or patellar network
Infrapatellar branches of the saphenous nerve
Middle inferior artery of the knee and companion veins
Saphenous nerve

◀ **Nerves of the right thigh**
Dorsal view.

Superior gluteal veins and arteries
Gluteus maximus
Gluteus medius
Gluteus superior
Gluteus minimus
Inferior gluteal nerve
Piriform muscle
Internal obturator nerve
Inferior gluteal veins and arteries
Nerve of the quadratus femuris
Internal pudendal artery
Quadratus femuris muscle
Companion veins and arteries of the ischiatic nerve
Deep branches of the artery of the middle circumflex veins of the femur
Posterior cutaneous nerve of the femur
Ischiatic nerve
Semitendinous muscle
Semi membranous muscle
Perforating vein and artery
Bicep muscle of the femur
Common fibular nerve
Popliteal artery
Popliteal vein
Tibial nerve
Gastrocnemius muscle
Small saphenous vein

mediate muscle of the quadriceps. The few sensory branches innervate the periostium of the femur, the patella and the knee joint; -The internal saphenous nerve, is a sensory nerve and it goes to the foot. Some collateral branches innervate the cutis of the internal fascia of the thigh, of the knee and of the leg. Through the sartorius muscle, the saphenous nerve divides into 2 terminal branches:

-The patellar branch (or infrapatellar), which innervates the cutis of the knee

-The tibial nerve, which, thanks to its collateral branches, innervates the medial and posteromedial surface of the leg, along with the malleolus and the medial margin of the foot.

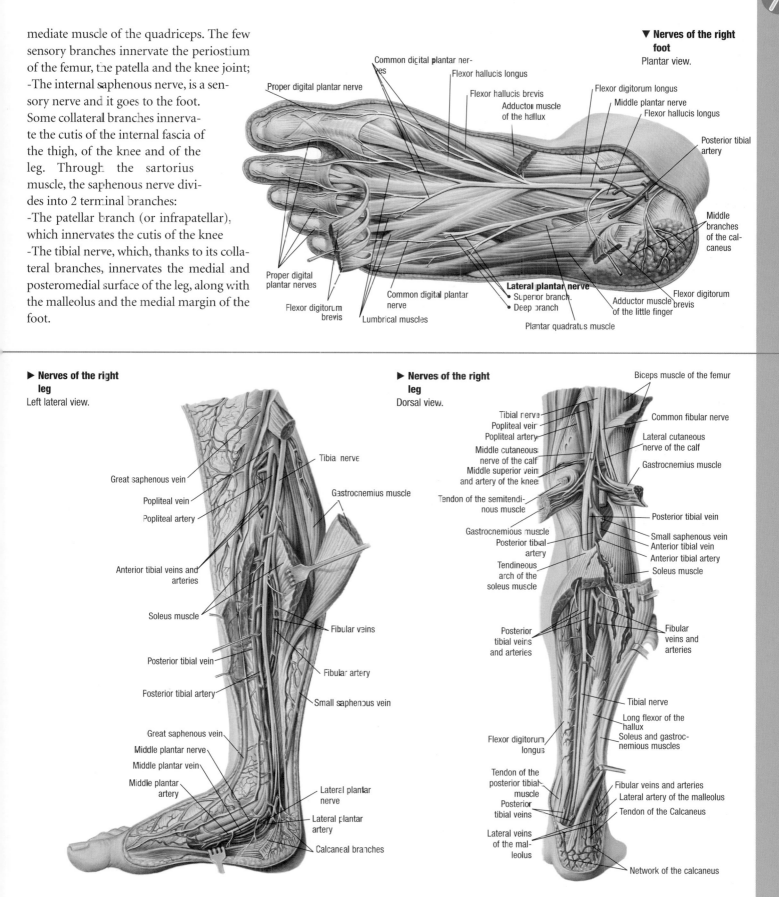

▼ **Nerves of the right foot**
Plantar view.

Common digital plantar nerves
Proper digital plantar nerve
Flexor hallucis longus
Flexor hallucis brevis
Adductor muscle of the hallux
Flexor digitorum longus
Middle plantar nerve
Flexor hallucis longus
Posterior tibial artery
Middle branches of the calcaneus
Proper digital plantar nerves
Common digital plantar nerve
Flexor digitorum brevis
Lumbrical muscles
Lateral plantar nerve
• Superior branch.
• Deep branch
Adductor muscle of the little finger
Flexor digitorum brevis
Plantar quadratus muscle

▶ **Nerves of the right leg**
Left lateral view.

Great saphenous vein
Popliteal vein
Popliteal artery
Tibia nerve
Gastrocnemius muscle
Anterior tibial veins and arteries
Soleus muscle
Fibular veins
Posterior tibial vein
Fibular artery
Posterior tibial artery
Small saphenous vein
Great saphenous vein
Middle plantar nerve
Middle plantar vein
Middle plantar artery
Lateral plantar nerve
Lateral plantar artery
Calcaneal branches

▶ **Nerves of the right leg**
Dorsal view.

Biceps muscle of the femur
Tibial nerve
Popliteal vein
Popliteal artery
Middle cutaneous nerve of the calf
Middle superior vein and artery of the knee
Tendon of the semitendinous muscle
Gastrocnemius muscle
Posterior tibial artery
Tendineous arch of the soleus muscle
Posterior tibial veins and arteries
Flexor digitorum longus
Tendon of the posterior tibial muscle
Posterior tibial veins
Lateral veins of the malleolus
Common fibular nerve
Lateral cutaneous nerve of the calf
Gastrocnemius muscle
Posterior tibial vein
Small saphenous vein
Anterior tibial vein
Anterior tibial artery
Soleus muscle
Fibular veins and arteries
Tibial nerve
Long flexor of the hallux
Soleus and gastrocnemius muscles
Fibular veins and arteries
Lateral artery of the malleolus
Tendon of the Calcaneus
Network of the calcaneus

125

THE MOST HETEROGENOUS SYSTEM OF ALL. IT IS MADE UP OF NUMEROUS GLANDULAR
ORGANS DISTRIBUTED IN VARIOUS AREAS OF THE BODY. THESE GLANDS CHEMICALLY
CONTROL EACH PART OF THE BODY IN A DIFFERENT WAY.

THE ENDOCRINE SYSTEM

As we can see from the image on the right, the organs which make up this anatomical system are very different from each other, not only because of their shape: in fact, even their structure, the substances they produce and the tissues they are made of is different. Why then are they considered as part of a single system? What they have in common is the fact that they are glands without excretory ducts, they are fed by a thick network of blood capillaries and their basic activity is to produce substances (hormones) which go directly to the blood stream ▶176.

Beside their embryonic origin, besides a very different morphological and functional characteristics one from the next, and besides being anatomically separated, these glands are in fact functionally linked to each other and make up a chemical network of signals, which keep the "basic" functioning of the entire body, under control, by constantly balancing actions and retroactions. Just as it happens at a macroscopic level, where the neuronal network, due to antagonistic nervous signals, regulates the activity of organs and muscles by rapidly intervening where it's necessary, at a cellular level, the "network" of chemical messages produced by the endocrine organs, balances and regulates all of the vital metabolic activities.

Unlike nervous action, hormonal action is slower (there are exceptions of course: the adrenalin produced by the surrenal

glands, for example, has an immediate effect on the targeted organs!) but it can offer continuous stimulus and a long lasting effect. In fact, the body doesn't only need stimuli and rapid and efficient reactions: it involves also delicate equilibriums, a constant and low growth, the regeneration and the breakdown of tissues, the assimilation of substances and the elimination of waste products….in other words, it is made up of a series of interrupted processes, which are in constant equilibrium and must be regulated and kept under control.

Also, just as the nervous control of the muscular activity constantly varies in response to the information that reaches the central nervous system from the periphery, even the activity of the endocrine glands is constantly modified based on the nervous, hormonal, and chemical information received by the hypothalamus, by other endocrine glands and by the organs of the body.

But the nervous system and the endocrine system are not just very similar: they also constantly collaborate in controlling the various activities of the body. An example of it is digestion: while the autonomous nervous system manages the muscular aspect of the involuntary movement of the digestive and circulatory apparatuses, the centralized nervous system coordinates their activities and it partially regulates the gastric secretion. Such activity is assisted by the endocrine system, which directly organizes the absorption of nutrients, establishes the secretion and the assimilating activities of the cells and the different organs, and influences the various exchanges between tissues. Such "collabora-

◀ **The endocrine system of a baby boy**

It is made up of very heterogeneous elements (in color) even if all made of endocrine epithelial tissue; it helps regulating the vital activities.

❶ Pituitary gland
❷ Thyroid
◼ Surrenal gland
  ❸ Marrow
  ❹ Cortex
❺ Testicle
❻ Aortic glomus
❼ Pancreas
❽ Surrenal gland
❾ Liver
❿ Para aortic bodies
⓫ Thymus
◼ Parotid
  ⓬ Inferior.
  ⓭ Superior.
⓮ Carotid body
⓯ Pineal body

tion" occurs in all the vital processes: at any time, any activity of the body is kept under control by the central nervous system, autonomously managed by the peripheral nervous system, and constantly stimulated and regulated by the endocrine glands.

A SYSTEM THAT INVOLVES THE ENTIRE BODY

In a way, all of the cells of our body could be considered as being part of the endocrine system: their metabolic products end up in the blood, and often they modulate the activity of other cells (for example, an increasing concentration of carbon dioxide stimulates the neurons of the respiratory centers, the chemioreceptors of the aortic arch and of the carotid body, and it has a direct effect on the smooth muscles of the blood vessels). Furthermore, all of the cells of the body also produce 'chemical messages" (such as the interferon, or the interleukine) which induce specific reactions in the surrounding cells, that not only have an important role in different cellular activities, but also modulate the immune response ▶180-181.

However, most of these substances manifest their actions in a relatively restricted area, right next to the cell that has produced them this is the reason why they are also called "local hormones". Finally, there are some cells that produce specific substances through extensive physiological activity: these are parahormones such as the histamine, which produces a dilation of the blood capillaries, an increased permeability of the tissues and of the

▼ **Insulin**
Until new biomedical technologies have invented a safe and precise method of dosing insulin, other than injections, diabetics will continue to depend on syringes for their survival.

✛ ENDOCRINE PROBLEMS: DIABETES AND PRE-MENSTRUAL SYNDROME

Most of the endocrine glands remain active for one's life: they are the so-called perennial endocrine glands, such as the pituitary gland, the thyroid and the surrenal glands. Others, such as the female glands, function in a limited span, and they are called transient endocrine glands. A faulty functioning of the endocrine glands (the excessive or insufficient production of hormones), along with a natural variation of the endocrine activity, causes visible psychological and physical changes, and often diseases and malfunctions.

Problems caused by a lack of a hormone are much more solvable than others: by in taking an adequate dose of the missing hormone, complications and symptoms are drastically limited. An example of this is diabetes (or diabetes mellitus). This is a disease which is becoming more and more diffused in industrialized countries, and is caused by a faulty functioning of the pancreas ▶154, which produces an amount of insulin insufficient for metabolizing the glucose present in the blood (the sugar found in food). As this substance collects in the blood, it can cause many problems: besides coma (only in extreme cases) malaise, nausea, loss of appetite, dehydration of the skin,

shrinking of the pupils, and acetone breath are very frequent.

With time, many organs can suffer: loss of vision, the kidneys, the heart, but in particular the blood vessels and the nervous system, can alter their functionality: problems such as nephropathies, coronary diseases, gangrene of the lower limbs and the alteration of the tactile sensitivity and of the motor ability are quite common. Complications can be avoided or limited with the proper cure, specific for every type of diabetes (there are 5, all with different origins and symptoms ▶154-155).

On the other hand, the premenstrual syndrome, which affects million of women, is an example of how physiological changes of the endocrine activity, drastically influence both the body and the mind. Even if it is part of the normal physiological cycle of a

woman, the intensity of the symptoms can vary a lot, becoming in some cases a true problem. Anxiety and emotional swings appear half way through ovulation, and their intensity increases as they approach the menstrual cycle. Besides personality changes (intense irritability, emotional instability, crying fits, and aggression) physical changes can also occur: headaches, abdominal cramps and swellings of the breasts and the legs. Fortunately these problems disappear as the menstrual cycle begins.

ENDOCRINE ACTIVITY OF THE MOST IMPORTANT GLANDS

GLAND	HORMONES	EFFECTS
anterior pituitary gland or adenohypophysis	thyreotropic hormone (**THS**) or thyrotropin	it stimulates the thyroid hormonal activity
	adenocorticotropic hormone (**ACTH**)	it activates the hormonal secretion of the surrenal cortex.
	prolan a or follicle-stimulant hormone (**LH**)	stimulates the maturation of the ovarian follicles and controls spermatogenesis.
	prolan b or luteinizing gonadotropin	it stimulates ovulation, the formation of the luteal body, the production of gametes, the synthesis and the secretion of testosterone.
	prolactin (luteotropic hormone, **lsh** or **lth**)	it stimulates the lactation throng
	somatotropic hormone (**sth** or **gh**), growth hormone, or somatotropin	affect the growth of muscle tissues and bone tissues.
	melanocyte stimulating hormone (**MSH**)	it stimulates the melanocytes activated by the ultraviolet or by the sexual hormones to produce melanin.
posterior pituitary gland or neurohypophysis	oxitocina	it stimulates muscular concentrations, particularly the uterine ones, and the production of milk.
	anti diuretic hormone (adh) or vasopressin	induces kidneys to retain water
	prolactin inhibiting factor (**PIF**)	neurohormone (maybe dopamine) that inhibits the secretion of prolactin
epiphysis or pineal gland	melatonin	acts on the hypothalamus, and it regulates the sleeping and waking cycles, the resting and active cycles, along with the ovarian cycle
thyroid	thyroxin (**T4**)	it affects energy, thermoregulation, and growth, accelerating the metabolism.
	triiodothyronin (**T3**)	it regulates the basal metabolism, maintains homoeothermic balance, and it affects growth.
	calcitonin or hypocalcemic hormone	it reduces calcium in the bloodstream and its action is opposite than the parathormone (**PTH**)
parathyroid	parathormone (**PTH**)	it increases the level of calcium in the blood
pancreas	insulin	it lowers the blood glucose level
	glucagon	it increases the blood glucose level
surrenal gland, cortex corticosteroid	glucocorticoids (hydrocortisone, cortisol, cortisone, deidrocortisone)	regulates proteins, and carbohydrates
	mineralocorticoids (aldosterone and desossicoticosterone)	they regulate the metabolism of electrolytes and water (concentration of fluids inside tissues)
surrenal gland, marrow	adrenaline or epinephrine	it acts on the muscles of the blood vessels.
	non adrenaline	it increases the cardiac and the respiratory rhythm, and mobilizes the energy reserves (fugue reactions)
male gonads	androgenous steroid hormones (testosterone, androsterone, androstenedione, deidroepiandrosterone)	regulates the development of male gonads, they induce the secondary male characteristics
female gonads	progesterone or lutein	it encourages the implantation of the zygote, modifying the structure of the endometrium of the uterus.
	steroid estrogenic hormones or follicular hormones, such as estradiol, estrone, estriol and folliculine	they produce phenomena, which precede, accompany and follow the ovulating process, prepare the uterus for pregnancy, and induce the secondary female characteristics.

ORGANS WITH AN ENDOCRINE ACTIVITY

ORGANS	HORMONES	EFFECTS
kidney	erythropoietin	it stimulates the maturation of red blood cells inside the spinal cord
stomach	gastrin	it stimulates the production of the gastric juices of the gastric glands
hypothalamus peptic hormones	*releasing hormone* (**RH**)	it stimulates the activity of the hypophysial gland
	releasing factor (**RF**)	stimulates the adenohypophysis to produce and /or lets certain hormones into the bloodstream.
	tireotropin releasing hormone (**TRH**)	controls the release of the tireotropin (**tsh**) hormone by the adenohypophysis.
	releasing inhibiting hormones (**RIH**)	it inhibits the activity of the hypophysial gland (such as somatostatine)
	growth hormone relasing factor (**GHRF**) or somatotropic releasing factor (**SRF**)	it regulates the adenohyophyseal activity
	luteinizing hormone releasing factor (**LRF**, o **LHRH** o **Gn-RH**)	it induces the synthesis and the excretion of the adenohypophysis of the luteinizing gonadotopin (**LH**)
lesser cul de sac	gastrina	stimola soprattutto la secrezione dell'acido cloridrico dello stomaco
duodeno	secretina	stimola la secrezione nel pancreas di bicarbonati e acqua, inbisce la produzione di gastrina
	pancreozimina	stimola la secrezione degli enzimi pancreatici
	colecistochinina	stimola: secrezione di enzimi pancreatici, contrazione della coleciti e motilità intestinale
duodeno e digiuno	*gastric inhibitory polypeptide* (**GIP**)	iit inhibits: the gastric motility and the chloropeptic excretions; it stimulates intestinal and endocrine pancreatic secretion
	vasoactive intestinal polypeptide (**VIP**)	powerful vasodilator, it facilitates water and electrolytes secretion in the intestine
placenta	human chorionic somatotropic hormone (**HCS**)	growth hormone
	human chorionic gonadotropin (**HCG**)	it stimulates the development of the placenta
	human placental lactogen (**HPL**)	it stimulates the growth of the fetus, and has a galctogenic and luteotropic function

gastric, salivary, sudoriferous secretions, or the serotonin, which has numerous antagonistic actions, both active in allergic responses [181].

Often, the muscular electrical signals transform into chemical signals (such as the synapses) and vice versa: the neurotransmitters can have parahormonal functions, tightly connecting the two control systems.

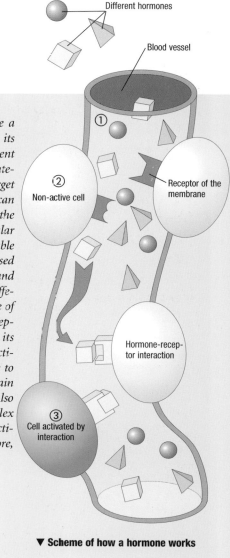

▼ Scheme of how a hormone works

HOW DO HORMONES WORK

A hormone is a molecule (a protein, an amino acids chain or a cholesterol compound), which is produced by specialized secretory cells, which are usually collected inside endocrine glands, but often also distributed into various organs (brain, kidney, digestive tract.). Once it is released by the secretory cell, the hormone reaches, through direct diffusion or through the bloodstream, specific cells called target cells, which are part of the target organs where the hormonal stimulus is destined: in fact, only on the membrane of these cells we can find specific receptors that are able to connect with that particular hormone.

The interaction between a hormone and a membrane creates a reaction inside the target cell, which can be energetic, productive or plastic: in other words, this interaction is able to modify the metabolism of the cell, its activity of protein synthesis or its structure.

This occurs even if the concentration level of the hormone is infinitesimal (oligodynamic action). In this way, the hormone acts as a "chemical message" which unleashes specific reactions.

①.Different types of hormones circulate inside the bloodstream. Glands with different regulatory functions produce such hormones.

Each hormone is "destined" for only one specific type of cell. In fact it is only able to interact with specific receptors (indicated, in each cell, by an indentation in its own shape).

②.Until the receptors of the membrane of the target cell interact with the hormone, the cell remains inactive.

③.Usually, as soon as this interaction is established," a chain reaction" occurs inside the cell, causing the production of AMP from ATP: this process makes a large amount of chemical energy readily available. The cell, in order to carry on metabolic functions, utilizes such energy.

④.The hormone is then deactivated by the cell or by the liver, in case it remains in the bloodstream. The byproducts of its breakdown are excreted or recycled during the synthesis of vital substances. How can a hormone

induce a chain reaction inside a cell, by simply connecting to its surface? It can carry on different activities based on the type of interaction established with the target cell. In the bottom figure, we can observe the reproduction of the simplified scheme of a cellular plasmatic membrane. The double layer of phospholipids is crossed and interrupted by proteins and protein complexes that have different roles: by connecting to one of these proteins (the specific receptor), the hormone changes its shape, and consequently its activity. So, for example, it is able to change the permeability of certain substances or their structure. Also the hormone-receptor complex could "collapse into" the cell, activating (or inhibiting), therefore, precise functions or processes.

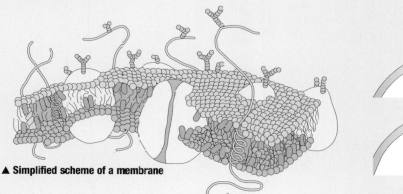

▲ Simplified scheme of a membrane

THE ENDOCRINE ACTIVITY OF THE ENCEPHALON: THE HYPOTHALAMUS, THE PITUITARY GLAND AND THE EPIPHYSIS

The cerebral endocrine activity is limited to the three bodies of the hypothalamus, of the pituitary gland and of the pineal body. All these have different roles and they secrete different hormones.

THE HYPOTHALAMUS

This part of the encephalon acts as a connection between the central nervous system and the endocrine system ▶126: in particular, its neursecretory activity is limited to the supraoptic and the paraventricular nuclei, and to the so-called medial eminence. The neurosecreting cells that are located in this area produce respectively vasopressin, oxytocin and RF (releasing factors, which are molecules that activate the specific release of hormones from the pituitary glands and the thyroid). These substances are transported through the axons of the hypothalamus-pituitary fasciae, all the way to the neurohypophysis. Here they accumulate and then released into the blood stream, stimulating or inhibiting the thyroid.

ACTIVITY OF SPECIFIC AREAS OF THE HYPOTHALAMUS	
thalamic region	
anterior nucleus of the hypothalamus and preoptical zone	Regulation of body temperature, perspiration, ,and of breathing, Production of **RF-TSH** e **RF-LH**
supraoptical nucleus	Production of vasopressin
paraventricular nucleus	Production of oxytocin, center for thirst
infibulotuberal region	
ventromedial nucleus	Center for thirst and satiety
dorsomedial nucleus	Stimulates the gastrointestinal functions
posterior nucleus	Increasing blood pressure, pupil dilation, "thrills" center, production of **RF-ACHT**
perifornical nucleus	Center for hunger, increasing blood pressure, center for anger
infibular center	Production of **RF-FSH**
nucleo ipotalamico laterale	centro della fame
nuclei tuberali	produzione di **RF**
mamillary region	
mamillary nucleus	Reflex vegetative activity, and emotional/instinctive activities

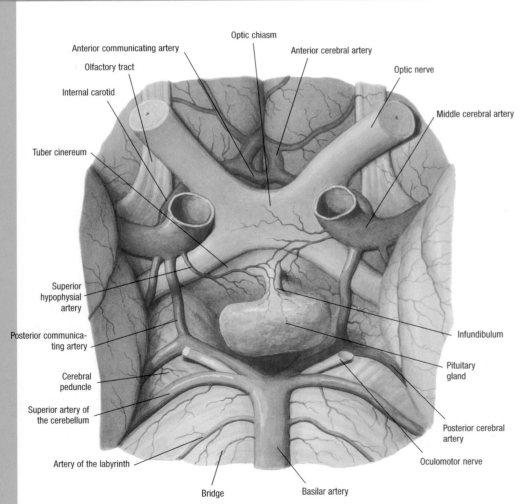

Optic chiasm
Anterior communicating artery
Anterior cerebral artery
Olfactory tract
Optic nerve
Internal carotid
Middle cerebral artery
Tuber cinereum
Superior hypophysial artery
Posterior communicating artery
Infundibulum
Cerebral peduncle
Pituitary gland
Superior artery of the cerebellum
Posterior cerebral artery
Artery of the labyrinth
Oculomotor nerve
Bridge
Basilar artery

◀ **Pituitary gland**
External view, from the bottom.

▶ **Formation of the pituitary gland**
It occurs during embryonal development, and can be divided into 3 main phases. The two parts (adenohypophysis and neurohypophysis) unite starting from different tissues: the first one originates in the ectoderm, which makes up the oral cavity of the embryo; the second one originates from the diacephalic vesicle. Such vesicle will create the brain and it remains in contact with the pituitary gland.
1 Diacephalic vesicle and oral cavity are just sketched.
2 An extroflexion of the oral ectoderm detaches and migrates to the top; **3** The two types of tissue create the pituitary gland.

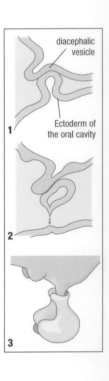

diacephalic vesicle
Ectoderm of the oral cavity
1
2
3

THE HYPOPHYSIS OR PITUITARY GLAND

It is a gland whose diameter is a little over 1cm long. It is located at the base of the brain inside the sella turcica of the skull. It is divided into two parts or lobes. They have an embryonal origin, and a different endocrine structure and function: the posterior pituitary gland, or **neurohypophysis** and the **anterior pituitary gland** or adenohypophysis. The neurohypophysis or posterior pituitary gland, originates from the same tissues that form the encephalon inside the embryo. It is made up of pituicytes, a particular type of glia whose extensions create a thick network of nerve fibers. Many of them come from the hypothalamus, and have numerous uneven enlargements along their entire course: they are the secretion granules that form inside the hypothalamic nuclei, and that move toward the pituitary gland until they are released near the numerous blood vessels which wet it. Therefore, the role of this part of the pituitary gland is to deposit and to distribute oxytocin and the antidiuretic hormone (**ADH**), or vasopressin, both polypeptides produced by the hypothalamus. The **adenohypophysis** or **anterior pituitary gland**, originates from tissues that make up the oral epithelium inside the mouth: it doesn't receive any nervous fiber from the hypothalamus: it is connected to it through a thick blood network (the pituitary gland portal system) which guarantees the afflux of the release factors (**RF**) produced by the hypothalamus, into the pituitary gland. For each hormone secreted by the adenohypophysis there is a RF produced by the hypothalamus, and just as many neurohormones with an antagonistic function. The role of the adenohypophysis is the result of a constant equilibrium between hypothalamic neurohormones. It is made up of different cells (acidophil, basophil, chromophobe), which produce 7 proteic hormones:

-The *somatotropic hormone* (**STH**) or *somatotropin* or *growth hormone* (**GH**), stimulates the activity of the thyroid, the synthesis of proteins, it slows down the cellular consumption of glucose, and it directly influences the growth and the development of osseous and muscular tissues;

-The *thyrotropic hormone* (**TSH**) or *thyrotropin* stimulates the elaboration and the secretion of the thyroid hormones. If this hormone is missing, the thyroid

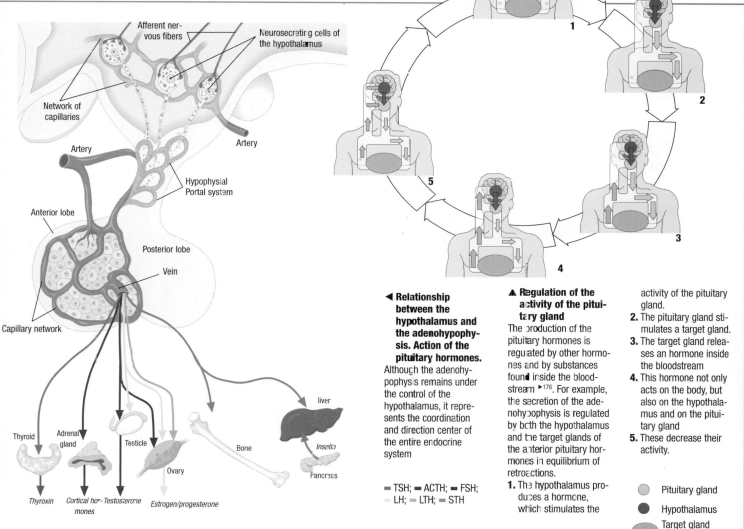

◄ **Relationship between the hypothalamus and the adenohypophysis. Action of the pituitary hormones.**
Although the adenohypophysis remains under the control of the hypothalamus, it represents the coordination and direction center of the entire endocrine system

■ TSH; ■ ACTH; ■ FSH;
■ LH; ■ LTH; ■ STH

▲ **Regulation of the activity of the pituitary gland**
The production of the pituitary hormones is regulated by other hormones and by substances found inside the bloodstream ►176. For example, the secretion of the adenohypophysis is regulated by both the hypothalamus and the target glands of the anterior pituitary hormones in equilibrium of retroactions.

1. The hypothalamus produces a hormone, which stimulates the activity of the pituitary gland.
2. The pituitary gland stimulates a target gland.
3. The target gland releases an hormone inside the bloodstream
4. This hormone not only acts on the body, but also on the hypothalamus and on the pituitary gland
5. These decrease their activity.

Labels on left figure: Afferent nervous fibers; Neurosecreting cells of the hypothalamus; Network of capillaries; Artery; Artery; Hypophysial Portal system; Anterior lobe; Posterior lobe; Vein; Capillary network; Thyroid; Adrenal gland; Testicle; Bone; liver; Insulin; Ovary; Pancreas; Thyroxin; Cortical hormones; Testosterone; Estrogen/progesterone

○ Pituitary gland
● Hypothalamus
◯ Target gland
→ Secreting activity
⇒ Hormonal

131

► **Structure of the pituitary gland**
Left lateral section.

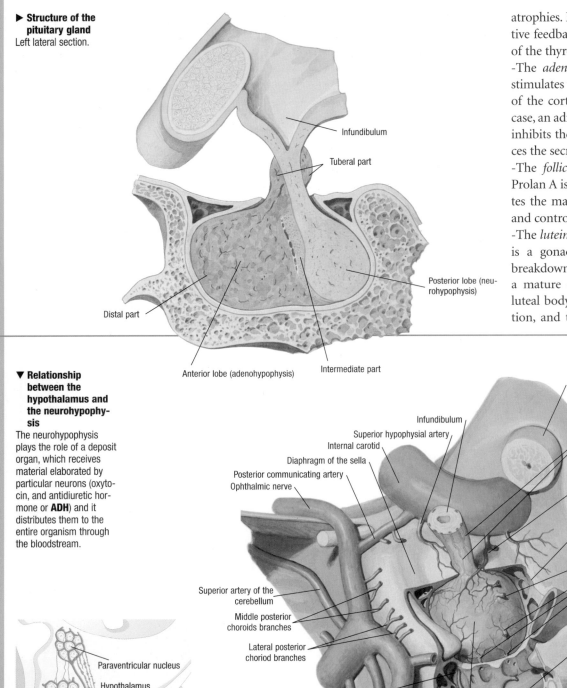

Infundibulum

Tuberal part

Posterior lobe (neurohypophysis)

Distal part

Anterior lobe (adenohypophysis)

Intermediate part

atrophies. Its level is regulated by the negative feedback of the thyroxin (a byproduct of the thyroid activity ►134);

-The *adenocorticotropic hormone* (**ACTH**) stimulates the development and the growth of the cortex of the adrenal gland; in this case, an adrenal gland product (the *cortisol*) inhibits the hypothalamic RF, which induces the secretion of the pituitary hormone.

-The *follicle-stimulant hormone* (**FSH**) or Prolan A is a gonadotropin, which stimulates the maturation of the ovarian follicles and controls for the spermatogenesis;

-The *luteinizing hormone* (**LH**) or prolan B, is a gonadotropin which stimulates the breakdown of the follicle and the release of a mature egg, the transformation of the luteal body of the follicle during fecundation, and the secretion of testosterone by

▼ **Relationship between the hypothalamus and the neurohypophysis**

The neurohypophysis plays the role of a deposit organ, which receives material elaborated by particular neurons (oxytocin, and antidiuretic hormone or **ADH**) and it distributes them to the entire organism through the bloodstream.

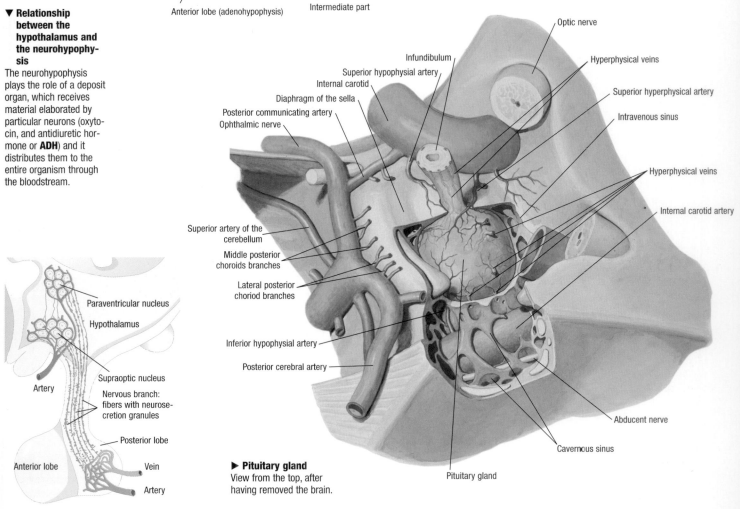

Optic nerve

Infundibulum

Superior hypophysial artery

Hyperphysical veins

Internal carotid

Superior hyperphysical artery

Diaphragm of the sella

Posterior communicating artery

Intravenous sinus

Ophthalmic nerve

Hyperphysical veins

Internal carotid artery

Superior artery of the cerebellum

Middle posterior choroids branches

Lateral posterior choroid branches

Inferior hypophysial artery

Posterior cerebral artery

Abducent nerve

Cavernous sinus

Pituitary gland

► **Pituitary gland**
View from the top, after having removed the brain.

Paraventricular nucleus

Hypothalamus

Artery

Supraoptic nucleus

Nervous branch: fibers with neurosecretion granules

Posterior lobe

Anterior lobe

Vein

Artery

the interstitial cells of the testicle. Used in connection with prolan A, it stimulates the secretion of estrogens;

-*Prolactin* or *luteotropic hormone* (**LSH** or **LTH**) stimulates the lactation throng;

-*Melanocyte stimulating hormone* (**MSH**) stimulates the melanocytes that are activated by ultraviolet rays or by sexual hormones in order to produce melanin, the skin pigment. From the intermediate zone, located between the anterior and posterior pituitary gland, also called "intermediate lobe" of the pituitary gland, the prolactin-inhibiting factor (**PIF**) is secreted. It is probably dopamine, which inhibits the secretion of prolactin.

EPIPHYSIS, ALSO CALLED PINEAL BODY OR PINEAL GLAND

It is a small gland, shaped like a small pine nut (from here the name "pineal").

We do not know much about its endocrine activity: we are sure that it produces melatonin, which acts on the hypothalamus. The production of this hormone is stimulated by noradrenalin (produced by the adrenal glands ▶[138]) and is controlled by the sympathetic nervous system, activated, in turn, by visual stimuli. The activity of this hormone helps regulating the sleep-awake cycle, the rest-activity cycle and the ovarian cycle. A defective functioning of this gland has been related to certain forms of depression and to seasonal affective disorder (**SAD**).

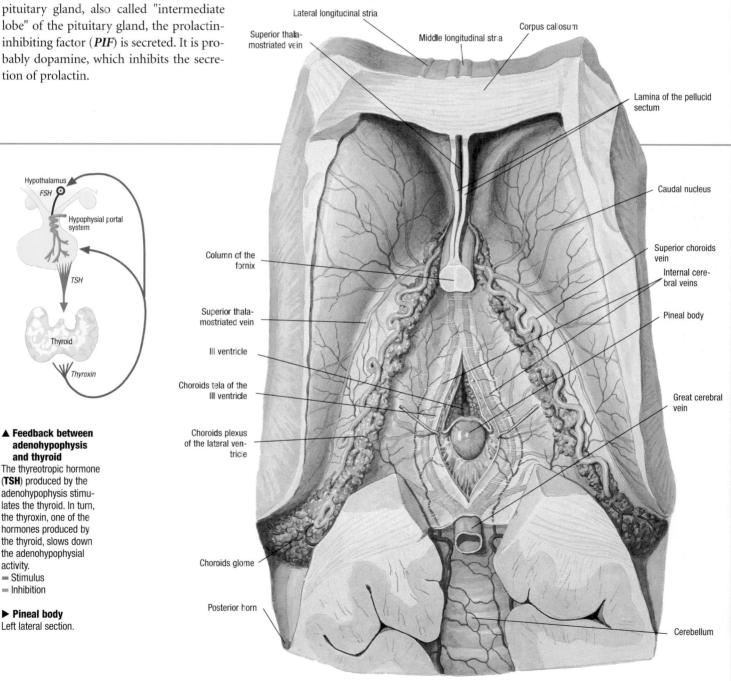

▲ **Feedback between adenohypophysis and thyroid**
The thyreotropic hormone (**TSH**) produced by the adenohypophysis stimulates the thyroid. In turn, the thyroxin, one of the hormones produced by the thyroid, slows down the adenohypophysial activity.
━ Stimulus
━ Inhibition

▶ **Pineal body**
Left lateral section.

THYROID AND PARATHYROID

These glands are all found in the neck: the first is divided into 2 lobes (about 5cm x 2,5cm and about 15g each) connected by a transversally extended isthmus, located in front of the trachea; the second ones are 4 little glands, the size of a pea, located on the posterior surface of the lobes of the thyroid: 2 higher ones (*superior parathyroid*), 2 lower ones (*inferior parathyroid*). They are embedded at different depths inside the thyroid tissue.

THE THYROID

It is one of the most important endocrine glands, because thanks to the hormones it produces, it regulates all of the metabolic processes of the organism. The thyroid is made up of cubic epithelial cells that make up a group of cavities called follicles, which are filled with colloid ▶27, an iodate protein, which originates, through hydrolysis, the three hormones produced by this gland:

-The *thyroloodstream*, acting in the opposite way to parathormone (**PTH**), which is produced by the parathyroid glands; it prevents the absorption of calcium at an intestinal level, it blocks its re-absorption in the bones and it facilitates the excretion at a renal level.

THE PARATHYROIDS

They produce only one hormone: the parathormone (**PTH**) or parathyroid hormone, which essentially controls the metabolic processes, which regulates the level of calcium and phosphorus inside

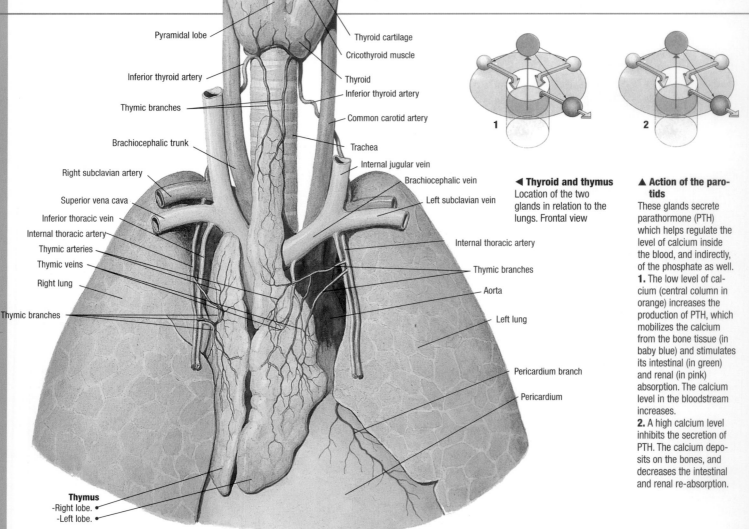

Pyramidal lobe

Inferior thyroid artery

Thymic branches

Brachiocephalic trunk

Right subclavian artery

Superior vena cava

Inferior thoracic vein

Internal thoracic artery

Thymic arteries

Thymic veins

Right lung

Thymic branches

Thymus
-Right lobe.
-Left lobe.

Thyroid cartilage

Cricothyroid muscle

Thyroid

Inferior thyroid artery

Common carotid artery

Trachea

Internal jugular vein

Brachiocephalic vein

Left subclavian vein

Internal thoracic artery

Thymic branches

Aorta

Left lung

Pericardium branch

Pericardium

1   2

◀ Thyroid and thymus
Location of the two glands in relation to the lungs. Frontal view

▲ Action of the parotids
These glands secrete parathormone (PTH) which helps regulate the level of calcium inside the blood, and indirectly, of the phosphate as well.
1. The low level of calcium (central column in orange) increases the production of PTH, which mobilizes the calcium from the bone tissue (in baby blue) and stimulates its intestinal (in green) and renal (in pink) absorption. The calcium level in the bloodstream increases.
2. A high calcium level inhibits the secretion of PTH. The calcium deposits on the bones, and decreases the intestinal and renal re-absorption.

the body.

In particular, parathormone increases the level of calcium inside the blood, fostering its intestinal and bone absorption, and preventing the renal excretion. Its action is the opposite to that of calcitonin. It is secreted every time that the level of calcium inside the blood lowers. A lack of it causes serious neuromuscular and bone problems.

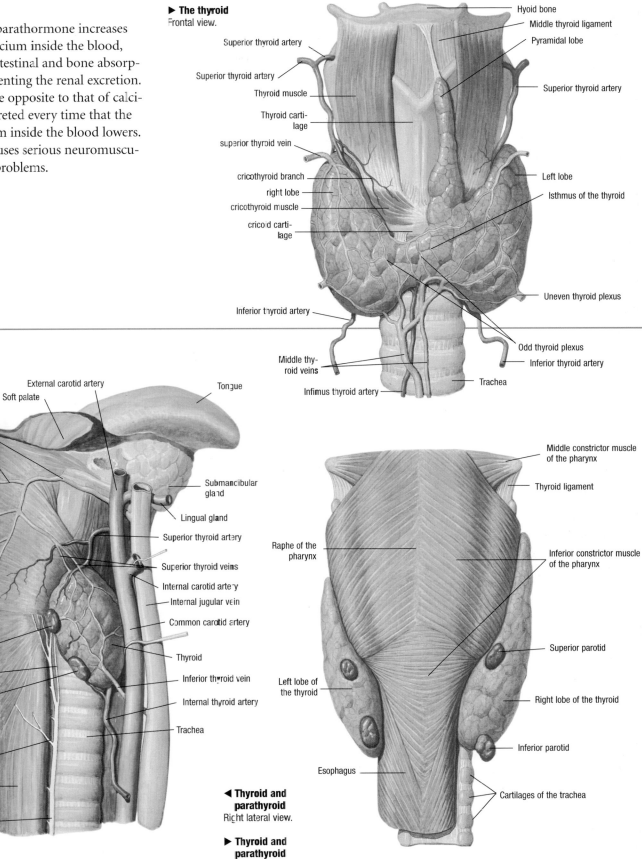

► **The thyroid**
Frontal view.

Superior thyroid artery
Superior thyroid artery
Thyroid muscle
Thyroid cartilage
superior thyroid vein
cricothyroid branch
right lobe
cricothyroid muscle
cricoid cartilage
Inferior thyroid artery
Middle thyroid veins
Infimus thyroid artery

Hyoid bone
Middle thyroid ligament
Pyramidal lobe
Superior thyroid artery
Left lobe
Isthmus of the thyroid
Uneven thyroid plexus
Odd thyroid plexus
Inferior thyroid artery
Trachea

External carotid artery
Soft palate
Pharyngeal veins
Pharynx
Pharyngeal plexus
Superior parathyroid
Inferior laryngeal nerve
Inferior parathyroid
Esophagus branch
Esophagus
Recurrent laryngeal nerve

Tongue
Submandibular gland
Lingual gland
Superior thyroid artery
Superior thyroid veins
Internal carotid artery
Internal jugular vein
Common carotid artery
Thyroid
Inferior thyroid vein
Internal thyroid artery
Trachea

◄ **Thyroid and parathyroid**
Right lateral view.

Raphe of the pharynx
Left lobe of the thyroid
Esophagus

Middle constrictor muscle of the pharynx
Thyroid ligament
Inferior constrictor muscle of the pharynx
Superior parotid
Right lobe of the thyroid
Inferior parotid
Cartilages of the trachea

► **Thyroid and parathyroid**
Frontal view

THE PANCREAS

Irregular cellular clusters are found inside the exocrine tissue of the pancreas, which secretes the pancreatic juice necessary for the digestion of fats [155]. These clusters vary in size and have an endocrine function. They get their name from the scientist who discovered it first: called "Langerhans' islands", they are surrounded by blood capillaries into which they secrete their hormones.

The Langerhans Islands are made up of 3 types of "producing" cells:
-The *alpha cells*: they secrete glucagone, a proteic- insuline's antagonist hormone, which increases the concentration of blood glucose, stimulating the enzymatic system of the liver, which transforms glycogen in glucose, and slows down the cellular consumption of glucose;
-The *beta cells*: they secrete insulin, a proteic hormone that plays an essential role in the metabolism of sugars. It decreases the blood glucose level (*glycemia* or *glucose level*) allowing this sugar to "enter" all the cells of the body, stimulating their use (oxidation) and assisting in the transformation into glycogen. Under normal conditions, the amount of circulating glucose is kept constant mainly by the action of the insulin, along with that of other mechanisms involving the liver, the peripheral nervous system and other endocrine gland;
-The *delta cells*: they secrete somatotatin, a hormone that regulates the production of the

ampulla   pancreatic   acini   splenic arteries   Langerhans' islands   beta cells   alpha cells   mesenteric veins

Mesenteric arteries

▲ Structures and functions
99% of the pancreatic tissue is made up of exocrine glandular epithelial tissue, which secretes an alkaline digestive juice that reaches the small intestine [156-157] though the pancreatic duct. The remaining endocrine areas, about 1 million cellular aggregates called Islands of Langerhans (from the name of the scientist who discovered them) are made up of two types of cells: alpha cells (they secrete glucagon) and beta cells (secrete insulin). These two hormones play an antagonist role in the regulation of the level of blood glucose (a sugar). Insulin reduces the liver production of glucose and promotes the absorption and the use of glucose by the adipose tissue, and the skeletal muscle tissue. The glucagon acts in the opposite way, and it favors the secretion of glucose in the liver.

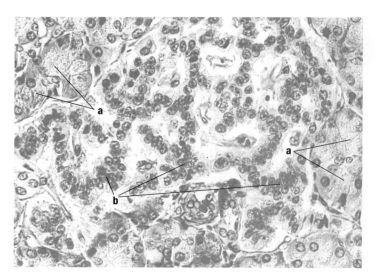

◄ Endocrine pancreas
Inside the exocrine glandular epithelial tissue (**a**) there is an endocrine Langerhans Island made up of cells, which are arranged in cords (**b**)..

other two, inhibiting their production. The production of somatostatin is not the only mechanism, which regulates the pancreatic endocrine activity. In fact, the production of insulin is always kept under control, thanks to a positive *feedback*: an increase in the blood glucose level stimulates its production, while a decrease in its level inhibits it. Also, the production of insulin must constantly balance the action of other hyperglycemizing hormones (such as the **STH**, the *epinephrine* or *adrenalin*). This imbalanced production of insulin can lead to diabetes mellitus that can originate for different reasons: -The *insulin-dependent* diabetes (Type I), affects primarily young people, and is caused by the total absence of beta cells; -The *non insulin-dependent* diabetes (Type II) affects primarily adults, and it is brought on by an insufficient production of insulin by the beta cells; -The *secondary diabetes* is produced by any agent, which may damage the pancreas, reducing its endocrine capability. Any level of insulin deficiency prevents the glucides (sugars) from feeding the cells, from being oxidized and transformed into glycogen; the reserves of this substance (inside the liver and the muscles) rapidly run out. Even the metabolism of proteins and lipids is altered.

Just as it is for other endocrine glands, even an excessive pancreatic hormonal production can cause serious complications: the hyperinsulinism, or hypoglycemic syndrome occurs when beta cells produce and excessive amount of insulin.

▼ **Pancreas**
Frontal view.

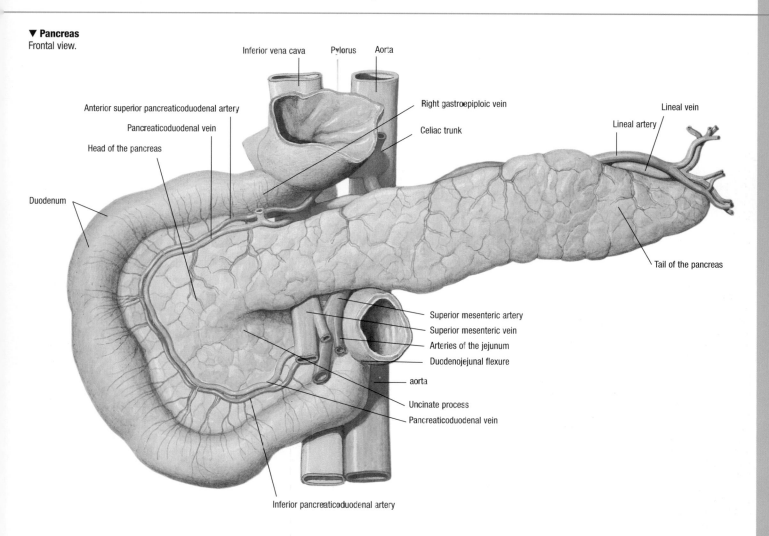

Inferior vena cava
Pylorus
Aorta
Right gastroepiploic vein
Lineal vein
Anterior superior pancreaticoduodenal artery
Lineal artery
Pancreaticoduodenal vein
Celiac trunk
Head of the pancreas
Duodenum
Tail of the pancreas
Superior mesenteric artery
Superior mesenteric vein
Arteries of the jejunum
Duodenojejunal flexure
aorta
Uncinate process
Pancreaticoduodenal vein
Inferior pancreaticoduodenal artery

THE SURRENAL (ADRENAL) CAPSULES

They are also called adrenal glands due to their position in relation to such organs (they are located above the kidneys). They are about 5cm x 2,5 cm in size. Each adrenal gland is divided into two parts easily distinguishable for both structure, and endocrine activity: the adrenal cortex or cortical portion on the outside, and the portion or medullary substance, on the inside.

THE ADRENAL CORTEX

This part of the adrenal gland is further divided into 3 concentric layers, different in appearance, in their cellular grouping and endocrine activity:

-The more external layer is the *glomerular zone*, where the aldosterone is secreted. *Aldosterone* is a mineralcorticoid, which regulates the metabolism of sodium and of potassium, stimulating the re-absorption of sodium by the renal tubules, and incrementing the secretion of potassium in the urine. The secretion of aldosterone is regulated mainly by the renin-angiotensine ►128 system, by the hypophysial hormone ACTH, and by a positive feedback of potassium inside the bloodstream;

-The second layer is the *fascicular zone*: here the cells are arranged in large radial cords. In this layer, influenced by the hypophysial hormone **ATCH** ►130, the *glycocorticoids cortisone* and *cortisol* are produced: these hormones are very important in the metabolism of carbohydrates, by speeding up the synthesis of glucose; also, they trigger the proteic metabolism and mobilize the lipids from their deposits. The cortisol, in particular, is able to prevent inflammatory reactions;

-The deepest part of the adrenal gland is called *reticular zone*; it represents only a transit area between the cortex and the medullary portion: in fact it borders with the latter, through a trabecular system and it doesn't have an endocrine activity.

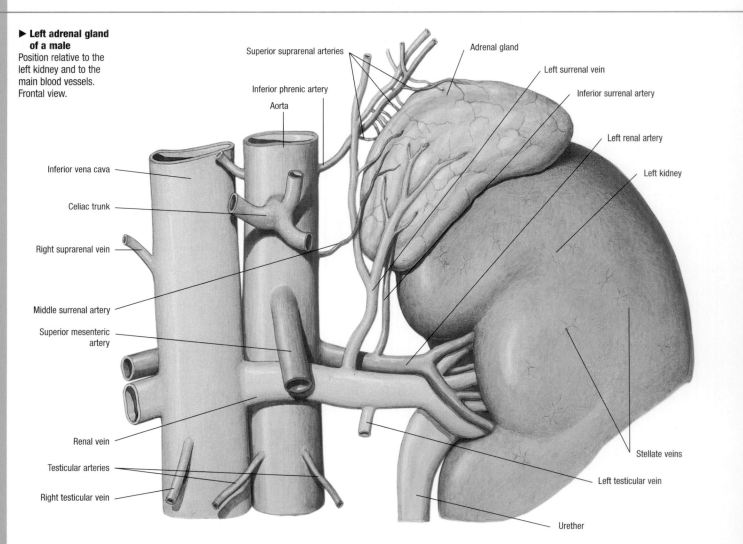

► Left adrenal gland of a male
Position relative to the left kidney and to the main blood vessels. Frontal view.

Superior suprarenal arteries

Adrenal gland

Inferior phrenic artery

Aorta

Left surrenal vein

Inferior surrenal artery

Left renal artery

Inferior vena cava

Left kidney

Celiac trunk

Right suprarenal vein

Middle surrenal artery

Superior mesenteric artery

Renal vein

Stellate veins

Testicular arteries

Left testicular vein

Right testicular vein

Urether

THE MEDULLARY PORTION

It is made of cells, which contain granules that easily absorb oxidizing substances, containing chrome, which color them: this is the reason why the tissue that they form is called chromophil. This part of the adrenal

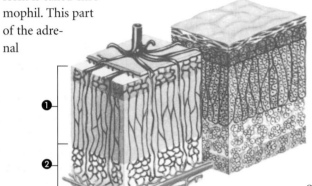

gland produces two different hormones with similar characteristics: the adrenaline and the noradrenalin. Both of them are released in response to stimuli arriving from pregangliar fibers of the sympathetic [111]: in fact, the sympathetic nervous fibers, which innervate the adrenal glands, are numerous. Adrenaline and noradrenaline have a very powerful vasoconstricting action on the peripheral arterioles of the circulatory system [182], causing an increase in the systolic pressure [187]; they influence the metabolism of the saccharides, by stimula-

ting the degradation of the glycogen and increasing the glycemia; they also do so, by modifying the action of most visceral organs: stimulating heart rate [185], inhibiting intestinal peristalsis [156-157], relaxing the bronchial muscles [168] and by accelerating the respiratory movements [163]. All of these changes are globally defined as "escape reactions": in fact, they are often stimulated by fear or by a sudden fright, predisposing the body to a quick burst of energy and to a reduced peripheral sensitivity.

▲ **Cortex and medullary portion**
Inside the adrenal glands, which are richly vascularized and innervated by ending of the sympathetic, two different areas have been distinguished: the cortex ❶), sensitive to the ACTH hormone, and the medullary portion ❷ which acts as a gland on its own, under the stimuli of the orthosympathetic autonomous nervous system.

▶ **Adrenal glands**
1. Anterior surface.
2. Posterior surface.

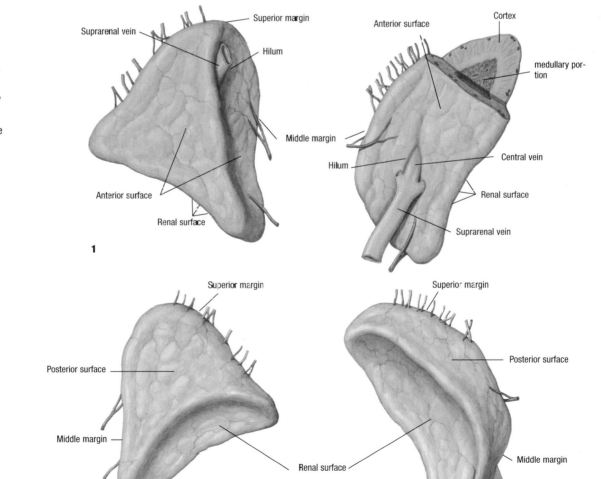

THE GONADS

Inside those organs that produce reproductive cells (male and female), there are some cells specialized in elaborating those hormones that act on certain organs that collaborate in the reproductive process.

IN THE MALE

The interstitial cells scattered in the connective tissue of the testicle (Leydig cells) do not make up a real gland, but they secrete androsterone, the androsteinedione and the deidroepiandrone. This occurs under the direct control of the huypophyseal hormone **LH**, and various *androgen steroid hormones*, such as the testosterone, which is the one with the highest androgen activity. While testosterone also affects metabolism (mainly the anabolism of proteins), this entire group of hormones affect the functionality and the development of the male genital organs, and the development of male secondary characteristics ▸222.

IN THE FEMALE

The granulose cells of the growing ovarian follicles secrete estrogen steroid hormones, or follicular hormo-

1
0 4 8 12 16 20 24 28 32 36 40
Weeks

2
0 2 4 6 8 10 12 14 16 18 20 22 24 26 28
Days

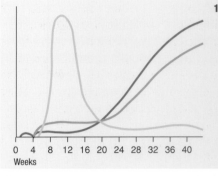

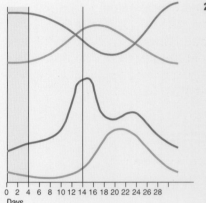

▲ Hormonal course
Reciprocal variations of the average blood concentrations of some sexual hormones during pregnancy (**1**) and ovulation (**2**).
In this last graphic, the pink fascia indicates the menstrual cycle.

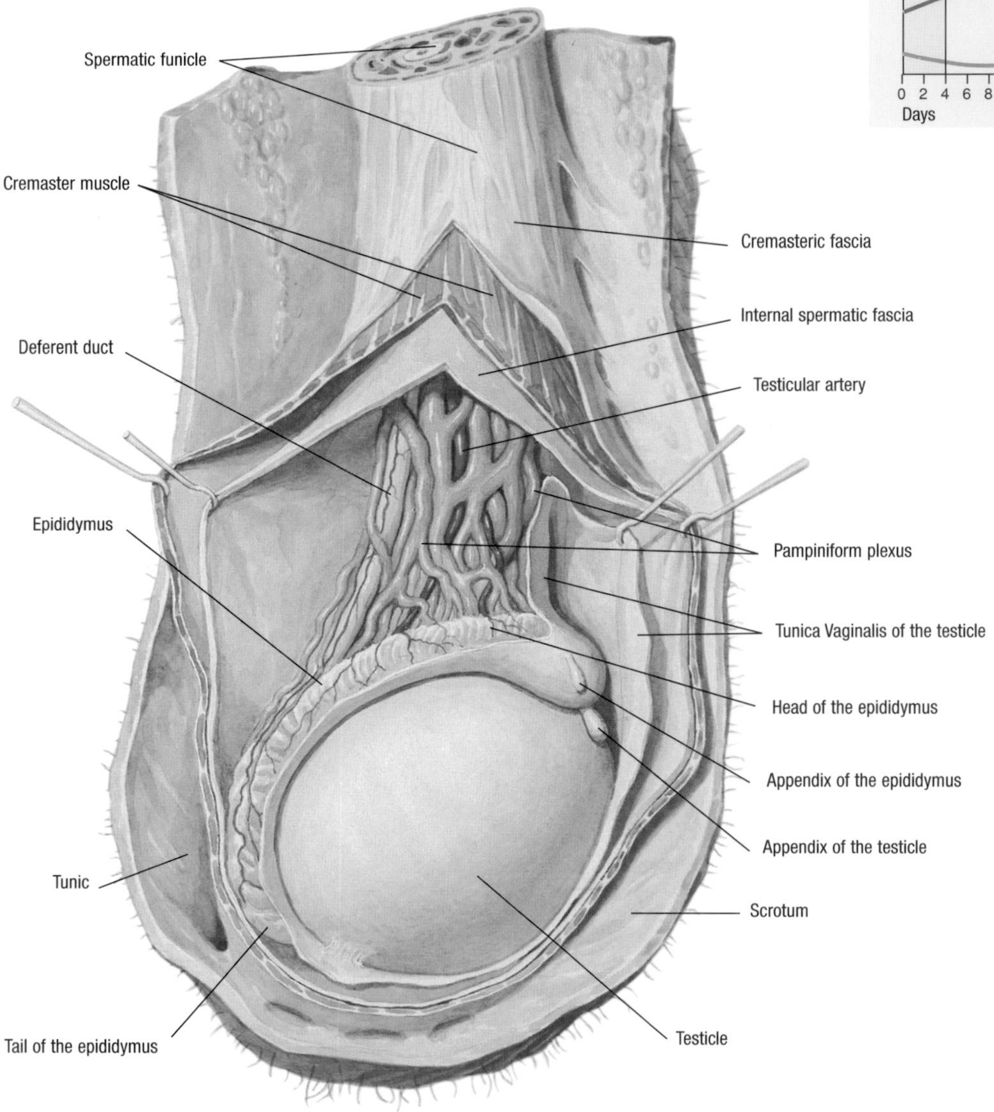

Spermatic funicle
Cremaster muscle
Cremasteric fascia
Internal spermatic fascia
Deferent duct
Testicular artery
Epididymus
Pampiniform plexus
Tunica Vaginalis of the testicle
Head of the epididymus
Appendix of the epididymus
Appendix of the testicle
Tunic
Scrotum
Tail of the epididymus
Testicle

◀ Right male gonad
Frontal view.

nes, such as estradiol, estrone, estriol and folliculine Under the direction of the pituitary gland, they induce all of the phenomena of the ovarian cycle: also they prepare and maintain the characteristics of the uterus during pregnancy, and they induce secondary female characteristics (mainly estradiol). Right after ovulation, the follicle transforms into a temporary endocrine tissue, which secretes progesterone, or lutein, another hormone, which favors the implant of the zygote, by modifying the structure of the uterine endometrium, and changes the structure of the mammary glands, essential for lactation.

◄ Hypophysial activity and menstrual cycle
① Fluctuation of the hypophysial production of hormones
② Development and dehiscence of the ovarian follicle
③ Fluctuation of the ovarian hormonal production
④ Development and degeneration of the uterine endometrium
⑤ Course of the average body temperature

▼ Left female gonads
frontal view.

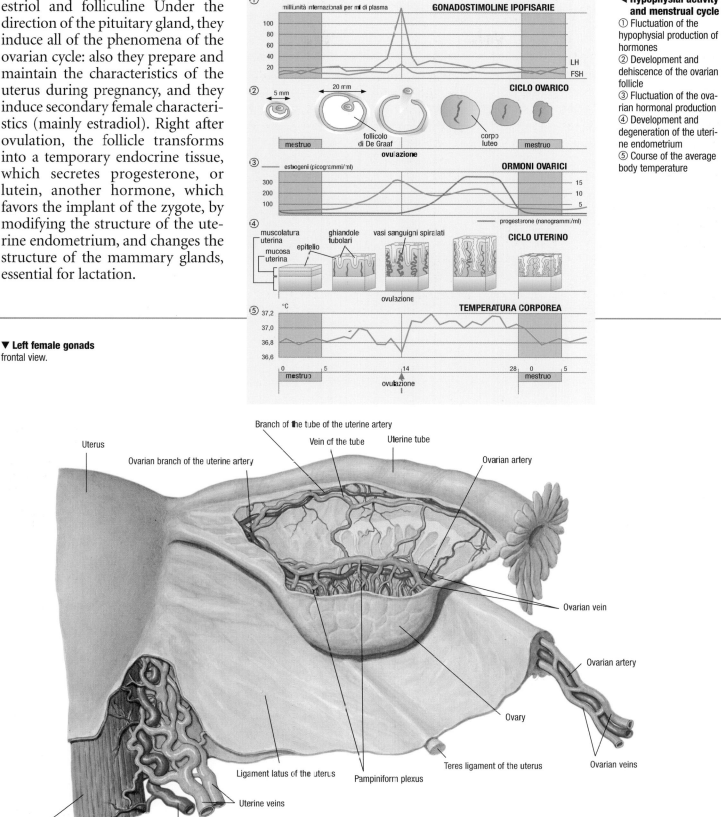

① milliunità internazionali per ml di plasma — GONADOSTIMOLINE IPOFISARIE
100 80 60 40 20 — LH FSH

② 5 mm — 20 mm — CICLO OVARICO
mestruo — follicolo di De Graaf — corpo luteo — mestruo
ovulazione

③ estrogeni (picogrammi/ml) — ORMONI OVARICI
300 200 100 — 15 10 5
progesterone (nanogrammi/ml)

④ muscolatura uterina — ghiandole tubolari — vasi sanguigni spiralati — CICLO UTERINO
mucosa uterina — epitelio
ovulazione

⑤ °C — TEMPERATURA CORPOREA
37,2 37,0 36,8 36,6
0 5 14 28 0 5
mestruo — ovulazione — mestruo

Branch of the tube of the uterine artery
Uterus
Ovarian branch of the uterine artery
Vein of the tube
Uterine tube
Ovarian artery
Ovarian vein
Ovarian artery
Ovary
Ovarian veins
Teres ligament of the uterus
Pampiniform plexus
Ligament latus of the uterus
Uterine veins
Vagina
Uterine artery

✚ Biopsychology

The relationship between our psyche and our physical well-being has turned from a simple suspicion into a confirmed scientific reality. Statistics have aroused interest in **biopsychology***. Although this new discipline maintains a medical investigative system, it attempts to identify and quantify the influence that the mind has on organic functions (and vice versa), paying particular attention to the immune components ▶180. For example, research has shown that people who have recently lost someone dear, tend to see the doctor more often, and develop a higher incidence of tumors and circulatory diseases, almost as if they were depressed not only psychologically but also in their immune system. Extensive laboratory research has shown that the number of lymphocytes ▶178-179 in this group of people is drastically lower than the normal number: this is probably why they are more susceptible to bacterial, viral and tumoral attacks. The connection between "mind" and "body" is very complex, and it involves a group of factors so heterogeneous that research becomes very complex. Nevertheless, some data is already available. For example, it was discovered that some hormones produced by the hypothalamus (the* **CRF** *▶128,130-32) can also directly influence the immune system: by attaching to the receptors of white blood cells membrane, they could activate them, and prepare them to react as soon as they get in contact with an antigen. However, besides being influenced by the hormones directly produced by the hypothalamic and hypophysial nervous cells, the body's defense can also be directly influenced by the central nervous system: The lymphatic organs ▶202) are directly connected to the brain and the spinal cord. Through the nervous fibers, active substances released in locu directly stimulate them. This happens for example inside the intestine, where the neurons release a peptide, which bonds with the white globules of the lymphocytes, changing their distribution and their activity.*

On the other hand, the activity of our immune system can also clearly influence our psyche. We have observed, in fact, that the production of interleukin-1 (a chemical message that coordinates the protective activity of the white globules) stimulates a series of functional and biochemical changes in the brain, causing a loss in appetite, drowsiness and irritability. If the psychological factors can depress or stimulate the immune system of our organism, such immune activities can also affect our psychological well-being.

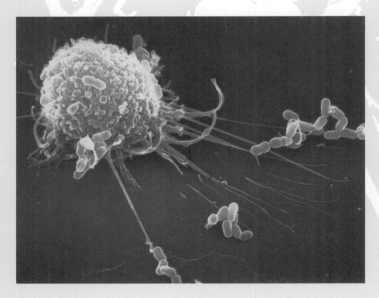

◀ Macrophage attacking a group of bacteria

These immune cells are less selective, and they swallow "extraneous" cells, "degenerated" body cells, and undifferentiated waste material. In this image we can see different types of pseudopods (the "tentacles"), which attract captured bacteria inside the cell (photograph taken with an electron microscope).

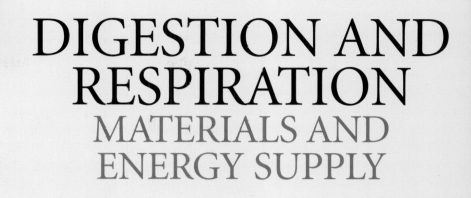

DIGESTION AND RESPIRATION
MATERIALS AND ENERGY SUPPLY

*Vitally important substances
such as "building material",
energy substances, oxygen and water
enter our mouth and thanks
to two specialized systems
that flow into our body.*

THE DIGESTIVE SYSTEM IS A 12 METER LONG MUSCULAR TUBE IN MOTION, WETTED BY MUCOUS, ACIDIC AND ENZIMATIC SECRETIONS OF GLANDS AS BIG AS THE LIVER, OR AS SMALL AS CELLS. THROUGH THIS SYSTEM, THE BODY RECEIVES ENERGY AND ALL THE NECESSARY RAW MATERIAL.

THE DIGESTIVE SYSTEM

In order to survive, our body must constantly renew its cells, reintegrate the energy spent by the organs, maintain constant the level of its internal substances, and it must intake new material for its growth and development (even hair, and nail growth need extra "materials"….)

What we eat and drink is used to fulfill all these needs: *nutrition* is integrated with *digestion*, a group of chemical-physical processes that transforms food into "raw material" that can be absorbed easily by our body. Digestion takes place inside the digestive system, which is a muscular tube, in which many glands and glandular organs pour all of the substances

they produce, in order to "breakdown" the complex chemical structures of food, reducing them into simple structures easily absorbable at a cellular level.

Food enters the mouth, crosses the isthmus of the fauces and the pharynx, reaches the esophagus and then the stomach, the most dilated part of the digestive system. Here the most "laborious" part of digestion takes place. At this point, the already modified food, goes through the *pyloric sphincter,* and arrives into the *intestine.* The intestine is divided into 2 parts, with different lengths and functions The *small intestine* (about 6,80 m) and the *large intestine* (1,80 m). In turn, the small

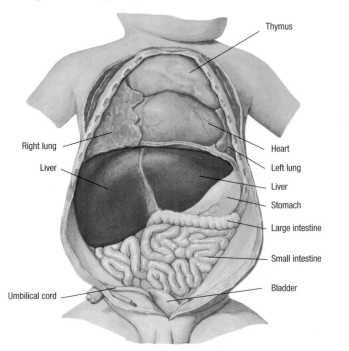

Thymus

Right lung

Liver

Heart

Left lung

Liver

Stomach

Large intestine

Small intestine

Bladder

Umbilical cord

DIGESTIVE ENZYMES			
SECRETION	ENZYMES	SUBSTRTTE	PRODUCTS
saliva	ptyalin	starch	dextrin
			maltose
gastric juices	pepsin	proteins	protease
			peptones
	curd	Caseinogenus	casein
pancreatic juices	trypsin	protein	amino acids
	lipase	lipids	fatty acids
			glycerin
	amylase	starch	maltose
	disaccharides	disaccharides	monosaccharide
succo enterico	enterokinase	typsinogen	trypsin
	peptidase	polypeptides	amino acids
	amylase	starch	maltose
	lipase	lipids	fatty acids
			glycerin

◄ **Digestive system of a newborn**
Unlike the adult body, the fetal digestive system does not have a digestive function from birth. The stomach and the intestine are reduced, while the liver (with its hemopoietic functions) is proportionally much more developed.

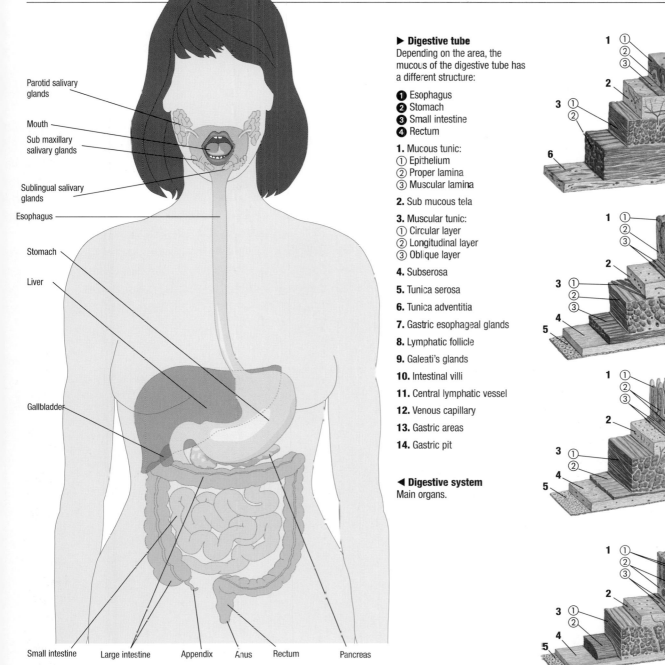

Parotid salivary glands

Mouth

Sub maxillary salivary glands

Sublingual salivary glands

Esophagus

Stomach

Liver

Gallbladder

Small intestine

Large intestine

Appendix

Anus

Rectum

Pancreas

▶ **Digestive tube**
Depending on the area, the mucous of the digestive tube has a different structure:

❶ Esophagus
❷ Stomach
❸ Small intestine
❹ Rectum

1. Mucous tunic:
① Epithelium
② Proper lamina
③ Muscular lamina

2. Sub mucous tela

3. Muscular tunic:
① Circular layer
② Longitudinal layer
③ Oblique layer

4. Subserosa

5. Tunica serosa

6. Tunica adventitia

7. Gastric esophageal glands

8. Lymphatic follicle

9. Galeati's glands

10. Intestinal villi

11. Central lymphatic vessel

12. Venous capillary

13. Gastric areas

14. Gastric pit

◀ **Digestive system**
Main organs.

intestine is further divided into 3 parts: the duodenum, where we find the pancreatic juices and the bile, which are produced by the liver; the jejunum and the ileum, which communicate with the large intestine through the ileocecal valve.

Even the large intestine is further divided into 3 parts: the *ascending colon*, the *descending colon*, and the rectum that ends with the anal sphincter. Although the colon might not be very important in the digestive process, it is however important in the production of vitamins, thanks to its large amount of bacterial flora, and to its re-absorption of fluids.

THE MOUTH

Here we find the tongue, located between the dental arches and the cheeks. The mouth has bone and muscle walls, and it opens on the outside with the lip fissure. The isthmus of the fauces is the limit that separates it from the pharynx, on the back. In front of the isthmus there are the palatine tonsils and in the middle there is the uvula. The mouth has many different functions: it grinds (masticates) food; it mixes it with the saliva produced by the salivary glands (the parotids, located on the top part of the neck, under the ear; submaxillary, on the internal surface of the maxillary bone; sublingual, located a little

higher) and it swallows it, sending it to the pharynx. Once chewed and mixed in, the food looks like a smooth mixture, called the alimentary bolus. If mastication is a voluntary act, managed by the centers of the temporal cerebral cortex and of the bulb, the deglutition is "piloted" by a series of reflected signals: they

stimulate the contraction of the muscles of the pharynx, the lifting of the velum palatinum, the lowering of the epiglottis and the contemporaneous lifting of the hyoid bone; at the same time respiration is automatically interrupted (swallowing apnea). This way, fluids and the food directed to the stomach, are blocked from entering the larynx, blocking or even preventing respiration.

◀ The mouth
1 Frenulum of the upper lip.
2 Superior gum
3 Anterior lingual gland
4 Margin of the tongue
5 Lingual nerve
6 Inferior longitudinal muscle
7 Frenulum of the tongue
8 Sublingual salivary gland
9 Submandibular duct
10 Inferior gum.
11 Frenulum of the inferior lip.
12 Sublingual caruncle
13 Sublingual fold
14 Inferior surface of the tongue
15 Commisure of the lips
16 Fimbriated fold
17 Back of the tongue

The mouth — internal anatomical elements

- Transverse palatine folds
- Upper lip
- Mucous tunic of the hard palate
- Palatine gland
- Gingival margin
- Parotid papilla
- Interdental gingival papilla
- Mucous membrane
- Buccinator muscle
- Greater palatine artery
- Greater palatine foramen
- Parotid duct
- Tendon of the tensor muscle of the velum palatinum
- Elevator muscle of the velum palatinum
- Buccopharyngeal part of the superior constrictor muscle of the pharynx
- Palatopharingeal muscle
- Palatoglossal muscle
- Palatoglossal arch
- Palatine tonsil
- Isthmus of the fauces
- Superior longitudinal muscle
- Inferior longitudinal muscle
- Gingival sulcus
- Buccinator muscle
- Pterygomandibular velum
- Palatopharingeal muscle
- Palatoglossal muscle
- Styloglossal muscle
- Uvular muscle
- Lingual aponeurosis
- Superior longitudinal muscle
- Vertical muscle of the tongue
- Transversal muscle of the tongue

▶ The mouth
Internal anatomical elements

Tooth — anatomical elements

- Crown
- Enamel
- Ivory
- Dentin
- Gum
- Collar
- Pulp cavity
- Root
- **Pulp**
  - Of the crown
  - Of the root
- Alveolar periosteum
- Cement
- Spongy substance of the maxillary or mandibular bone
- Nerve ending
- Venous capillary
- Arterial capillary

▶▲ Tooth
Anatomical elements.

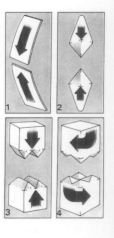

▶ Dentition
For each form, a specific function:
1. Incisor tooth, to cut;
2. Canine tooth, to tear;
3. Premolar tooth, to break;
4. Molar tooth, to grind.

THE DIGESTION OF COMPLEX CARBOHYDRATES

The first tract of the digestive system

Hard palate
Velum palatinum
Superior nasal concha
Supreme nasal concha
Opening of the sphenoid sinus
Sphenoid sinus
Fornix of the pharynx
Frontal sinus
Middle nasal concha
Inferior nasal concha
Salpingopalatine fold
Pharyngeal opening of the auditory tubes
Pharyngeal tonsil
Proper oral cavity
levator Torus
Torus tubarius
Arch of the atlas
Oral vestibule
Nasopharynx
Crescent fold
Salpingopharingeal fold
Uvula
Palatoglossal arch
Supratonsil fossa
Triangular fold
Geniglossal muscle
Palatine tonsil
Palatopharyngeal arch
Genohyoid muscle
Mylohyoid muscle
Oropharynx
Epiglottis
Hyoid bone
Laringopharynx
Cavity of the larynx
Thyroid cartilage
Cricoid cartilage
Esophagus
Trachea

▲ **The first tract of the digestive system**
Anatomical elements of the mouth and of the pharynx.

▼ **The movements of the deglutition**

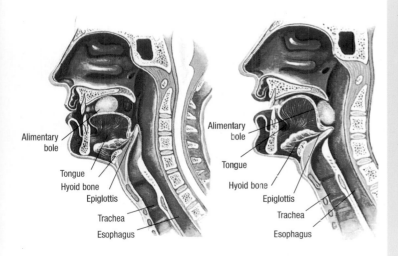

Alimentary bole
Tongue
Hyoid bone
Epiglottis
Trachea
Esophagus

Alimentary bole
Tongue
Hyoid bone
Epiglottis
Trachea
Esophagus

The saliva is used in the first important enzymatic elaboration of food inside the mouth. Saliva is produced by the salivary glands, whose location appears in the next door picture. Such glands do not produce the same type of saliva: the parotids secrete a lot of ptyalin, the most active enzyme in the breakdown process of starches; the sublingual have a very viscous saliva, rich in mucine but without ptyalin; the saliva of the submaxillary glands is a mixture of the two. However, these differences are limited to less than 0,5% of the weight of the saliva itself: in fact, saliva is mostly made of water (98,7%). The organic substances (mucin and amylase, the salivary enzymes that attack starch) represent only 0,5%; while the 0,8% is made up of mineral salts such as bicarbonates, phosphates and mostly chlorides, needed in the activation of the amylases and to keep the pH of saliva constant.

Starch, a polymer of glucose, is the complex carbohydrate in which plants "deposit" chemical energy, and it is common in all food. The enzymatic activity of the amylase breaks the bonds of many molecules of carbohydrates. The process continues until each starch sub-molecule is reduced to dextrin and maltose; in fact, with these molecules the enzyme is not able to break other bonds, and ceases to be active. This process of breaking down starch is carried out by all amylases; the ptyalin, in particular, keeps acting until it reaches the stomach, where the high acidity level de-activates it. The process of hydrolysis of the starch, conducted by the salivary enzymes, is more effective if the starch is cooked: in fact, when cooked, the granules of starch are no longer protected by a natural cellulose coating (impossible to breakdown), and no longer an obstacle for an enzymatic attack.

①

②

① Anatomical location of the main groups of salivary gland
② Nervous mechanisms for the production of saliva. Saliva is constantly produced, but its quantity varies depending on the presence and type of food. Salivary secretion is not only regulated by a reflex produced by the mechanical stimuli of mastication (chewing) and by the chemical stimuli of taste and smell, but the sight and the memory of a certain food also regulate it.
In fact, glands are innervated by fibers of the parasympathetic and sympathetic system, in connection with the cerebral bulb.

THE ESOPHAGUS

25-26 cm long, being the muscular canal that, despite not being very elastic, runs almost vertically from the pharynx (at the level of the 6th cervical vertebra) all the way to the stomach, after passing through the diaphragm. It is divided into 4 regions that take their name from the part of the body in which they are located: cervical (4-5 cm), thoracic (16 cm), diaphragmatic (1-2 cm) and abdominal (3 cm). It also has 4 shrinkable areas: the corticoid, which corresponds to the beginning of the esophagus; the aortic and the bronchial, near the aortic arch and arch of the left bronchus; and the diaphragmatic, at the same level as the diaphragm. The esophagus, located between shrinkable areas, is slightly dilated, and shaped like a fuse. The esophagus is full of muciparous glands that have a lubricating role. Innervated by the vagus nerve and by the sympathetic nerve ▶110-111, the esophagus pushes the food all the way into the stomach, thanks to a rhythmic contraction of a muscular tunic called peristalsis. While the tract moving the food along remains contracted, the one in front of it relaxes, allowing for the food to rapidly move forward. Once peristalsis has started, it propagates along the entire digestive tract.

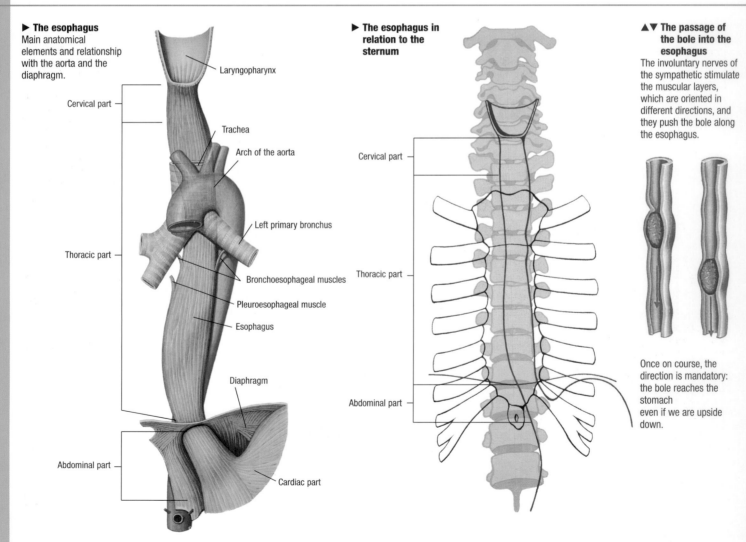

▶ **The esophagus**
Main anatomical elements and relationship with the aorta and the diaphragm.

Cervical part

Laryngopharynx

Trachea

Arch of the aorta

Left primary bronchus

Thoracic part

Bronchoesophageal muscles

Pleuroesophageal muscle

Esophagus

Diaphragm

Abdominal part

Cardiac part

▶ **The esophagus in relation to the sternum**

Cervical part

Thoracic part

Abdominal part

▲▼ **The passage of the bole into the esophagus**
The involuntary nerves of the sympathetic stimulate the muscular layers, which are oriented in different directions, and they push the bole along the esophagus.

Once on course, the direction is mandatory: the bole reaches the stomach even if we are upside down.

THE STOMACH

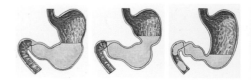

Located in the abdominal cavity right under the diaphragm. The stomach of an adult has an average capacity of 1300 cm_, even if it is able to dilate depending on the eating habits of each subject (for example, vegetarians have a bigger stomach). Here, in fact, an entire meal is collected and subjected to the digestive action of gastric juices: it takes 3-4 hours to digest an average meal; if the meal is high in fats, digestion becomes a much slower process.

The walls of the stomach are made up of a number of overlapping layers: the muscular tunic, which follows the peristaltic movements of the esophagus, and ensures that the food is remixed with the gastric juices and that the food reaches the intestine. The muscular tunic is made up of 3 layers: an outer layer made of longitudinal fibers; a middle layer, made of circular fibers concentric to the longitudinal axis of the stomach, and a deeper layer, made up of oblique fibers. In this way, the movements of the stomach are different, and the food is well mixed. Also, as digestion proceeds, the stomach changes its shape and pushes the food toward the intestine, where different digestive juices will further break it down. The *gastric mucosa*, covers the internal surface of the stomach and it is

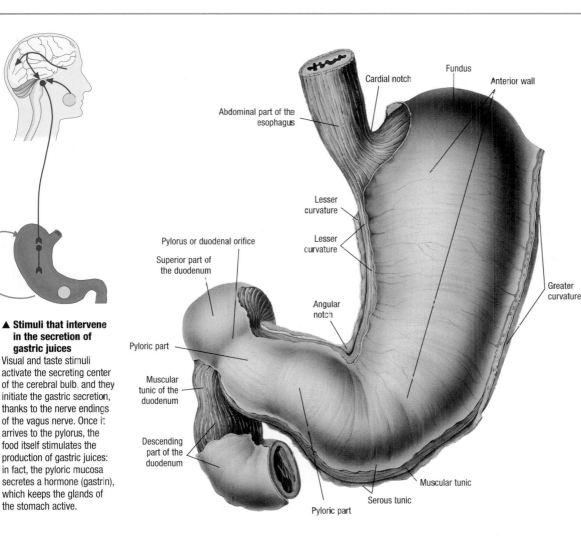

▲ Stimuli that intervene in the secretion of gastric juices
Visual and taste stimuli activate the secreting center of the cerebral bulb, and they initiate the gastric secretion, thanks to the nerve endings of the vagus nerve. Once it arrives to the pylorus, the food itself stimulates the production of gastric juices: in fact, the pyloric mucosa secretes a hormone (gastrin), which keeps the glands of the stomach active.

Abdominal part of the esophagus

Cardial notch

Fundus

Anterior wall

Lesser curvature

Pylorus or duodenal orifice

Superior part of the duodenum

Lesser curvature

Angular notch

Greater curvature

Pyloric part

Muscular tunic of the duodenum

Descending part of the duodenum

Muscular tunic

Serous tunic

Pyloric part

▲▲ The active stomach
The stomach holds about one and a half liters of food, which is mixed with the gastric juices for about 3 hours, becoming therefore a semi liquid mass called chyme. As soon as the stomach fills up, its wall begins to contract downward in a swaying motion. When the chime has been sufficiently digested, the sphincter, which usually closes the stomach, opens up and lets the chyme move to the duodenum.

▲ The muscles of the stomach
3 layers of muscles coat the stomach walls:
1. Longitudinal external layer;
2. Circular intermediate layer;
3. Oblique internal layer.

◄ The stomach
Superficial anatomical elements.

✚ GASTRIC ULCERS

Commonly defined as an interruption of the continuity of the stomach mucosa: the walls of the affected area, are exposed to the erosive action of the hydrochloric acid and of the digestive enzymes, and therefore they are deeply altered. This results in the formation of an inflamed area of the unprotected mucosa. At times the area can be a few cm in diameter. Nausea, vomiting, anorexia, satiety, gastric tension, cramps, and spasms are all symptoms of this well-known disease.

Until few years ago it was believed that ulcer was brought on by psychosomatic issues; today we believe that it is caused by an inflammation of a digestive strain of the bacterium Helicobacter pylori, commonly present inside the bacterial flora, made possible by a concomitant weakening of the natural defenses of the body.

For example, a temporary decrease of gastric mucous, of hydrochloric acid or of the gastric enzymes, could increase the habitat of these microorganisms. Other irritating factors can favor the onset of an ulcer: among these are the backflow of pancreatic enzymes and of bile from the duodenum to the stomach, the delayed emptying of the stomach, the abuse of alcohol and coffee, smoking and the use of drugs (especially anti inflammatory drugs, taken at empty stomach, or steroidal preparations taken for a long time).

Stress often contributes to lowering the natural defenses of the body, and it remains a very important cause for the onset of ulcers. Once diagnosed, the ulcer can be cured with specific antibiotics, a proper diet, the elimination of irritating substances (smoke, alcohol, spicy food, coffee, drugs…) and an adequate and specific drug therapy, which is able to increase the defenses of the mucosa, lowering an aggressive gastric secretion.

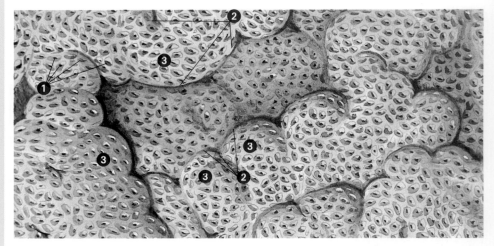

rich with *gastric glands*: these are simple or branched tubular glands that emerge on the deeper part of the *fovea* or *gastric foveola* ❶, inside the gastric areas ❷ and outlined by the folds ❸ of the mucosa. The glands are made up of muciparous calceiform cells of chief cells or alelomorph that produce *pepsinogen* (an inactive

▶ **The stomach**
Internal muscles.

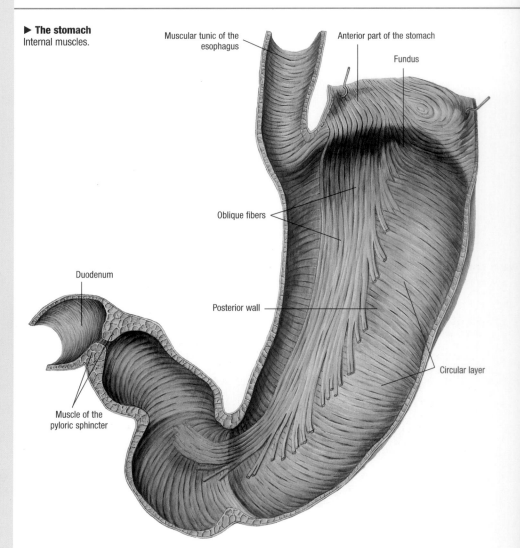

Muscular tunic of the esophagus

Anterior part of the stomach

Fundus

Oblique fibers

Duodenum

Posterior wall

Circular layer

Muscle of the pyloric sphincter

precursor of the pepsin enzyme) and of parietal or delomorph cells, which secrete hydrochloric acid when stimulated by *gastrin*. This hormone, produced by the *G cells*, which are distributed inside the gastric mucous, is not the only one secreted by the stomach: the *A cells* produce *glucagon*, which mobilizes the hepatic glycogen, and the enterochromoaffine cells that produce *serotonin*, which stimulates the contraction of the smooth muscles.

The 2 liters of gastric juice produced every day by the gastric glands, are a mixture of different substances, in different proportions: those enzymes which are able to breakdown proteins (for example the *pepsin*) and lipids (*lipase*), that function perfectly at the pH guaranteed by the hydrochloric acid, which also denatures the proteins and transforms pepsinogen into pepsin.

The thin layer of gastric mucosa, which covers the entire internal surface of the stomach, prevents the mucosa from digesting itself. The stomach is richly vascularized and surrounded by the fibers of the vagus nerve and of the 5th-8th thoracic segment of the sympathetic nerve. It is surrounded by parasympathetic and orthosympathetic plexuses (anterior gastric, posterior, superior and inferior, celiac and myenteric), which regulate both the motor activity of the muscles and the secretory activity. In particular, the parasympathetic is a stimulant, while the orthosympathetic is an inhibitor.

DIGESTION INSIDE THE STOMACH

Preliminary digestion of proteins occurs inside the stomach, thanks to the activity of **pepsin**, *an enzyme that catalyzes the hydrolysis of the peptic bonds between the aromatic amino acids. The protein molecules, large and insoluble, are transformed into much smaller soluble molecules called* **peptones***. The chief cells of the gastric glands release the pepsin as pepsinogen and it is rapidly activated by hydrochloric acid; once formed, it activates other pepsinogens itself.*

This enzyme is particularly efficient: under natural conditions (37° C and pH 1,5) it can digest in one hour an amount of proteins 1000 times higher than its weight.

Hydrochloric acid is essential in the entire gastric digestion process. It is essential in the activation of the first pepsin molecules; it is essential in maintaining the pH of the gastric juice within the optimal limits of the enzymatic action. It has antiseptic properties and it prevents pathogen agents from orally invading the body. It also denatures proteins by "unwinding" the ball of proteins, finally, it facilitates the action of the pepsin.

Inside the stomach of newborns, the digestion of proteins is assisted by another enzyme- **rennet***, or* **chymosis***, which causes the coagulation of milk. Therefore, it remains longer inside the gastric juice, allowing for the pepsin to breakdown the proteins in the milk more efficiently*

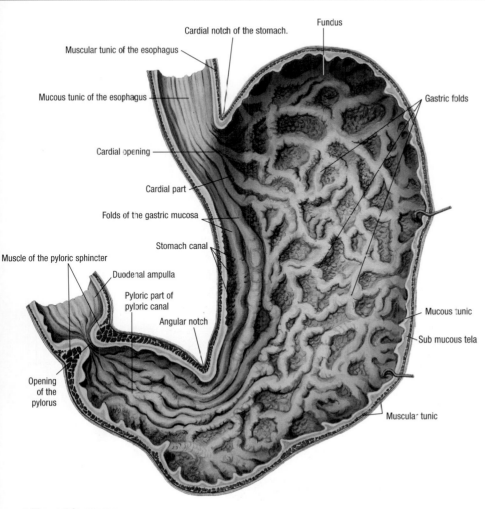

Cardial notch of the stomach.
Fundus
Muscular tunic of the esophagus
Mucous tunic of the esophagus
Gastric folds
Cardial opening
Cardial part
Folds of the gastric mucosa
Stomach canal
Muscle of the pyloric sphincter
Duodenal ampulla
Pyloric part of pyloric canal
Angular notch
Mucous tunic
Sub mucous tela
Opening of the pylorus
Muscular tunic

▲ **The gastric mucosa**

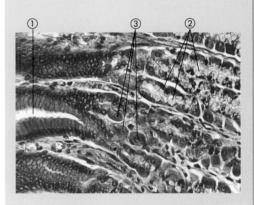

Gastric mucosa under the microscope: ① gastric pit, ② chief cells; ③) parietal cells.

151

THE LIVER

It is the biggest gland in the body: brownish-red in color, and is located in the upper part of the abdominal cavity, underneath the diaphragm, on the right side.

It weighs about 1,5 Kg, and it is divided into two lobes (right and left), one about 3 times bigger than the other. Impressed on it are the marks of its surrounding organs. It is richly vascularized, and it is connected to the digestive tract by the duodenum with its excretory ducts: the biliary extra hepatic pathways.

It is a vital organ, because it is involved in regulating metabolism, and in many other functions, some of which are complex.

Besides secreting the bile, a very important digestive juice in the absorption of food, it also produces proteins, and regulates and controls the formation of the majority of the byproducts of the metabolism of proteins. It produces urea, stores and utilizes fats and glucose (in the form of glycogen), by regulating the glycemic level. It "filters" the blood of any toxic substance and it helps keep the body clean from forei-

gn substances. It also produces prothrombine along with other factors that regulate the coagulation of blood, as well as synthesizing and storing many substances that are necessary in the formation of red blood cells and of other blood components. The microstructure of the liver is based on hepatic lobules: they are delimited by little layers of connective tissue and are made up of many cellular laminae, which show a system of many canals with a thick network of biliary and blood capillaries. The first ones follow a tortuous course (sinusoid) and are arranged in a radial

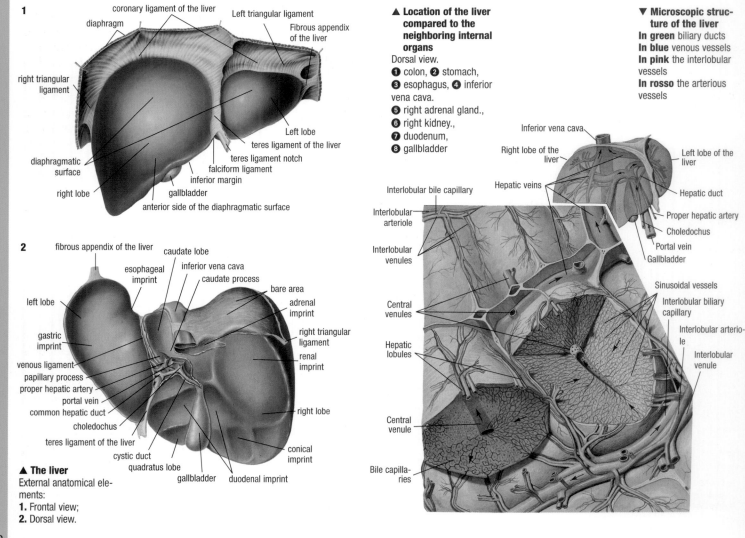

1

coronary ligament of the liver
diaphragm
Left triangular ligament
Fibrous appendix of the liver
right triangular ligament
diaphragmatic surface
right lobe
Left lobe
teres ligament of the liver
teres ligament notch
falciform ligament
inferior margin
gallbladder
anterior side of the diaphragmatic surface

▲ **Location of the liver compared to the neighboring internal organs**
Dorsal view.
❶ colon, ❷ stomach,
❸ esophagus, ❹ inferior vena cava.
❺ right adrenal gland.,
❻ right kidney.,
❼ duodenum,
❽ gallbladder

▼ **Microscopic structure of the liver**
In green biliary ducts
In blue venous vessels
In pink the interlobular vessels
In rosso the arterious vessels

2

fibrous appendix of the liver
caudate lobe
esophageal imprint
inferior vena cava
caudate process
bare area
adrenal imprint
left lobe
gastric imprint
right triangular ligament
renal imprint
venous ligament
papillary process
proper hepatic artery
portal vein
common hepatic duct
right lobe
teres ligament of the liver
cystic duct
quadratus lobe
gallbladder
duodenal imprint
conical imprint
choledochus

▲ **The liver**
External anatomical elements:
1. Frontal view;
2. Dorsal view.

Inferior vena cava
Right lobe of the liver
Left lobe of the liver
Hepatic veins
Hepatic duct
Interlobular bile capillary
Proper hepatic artery
Interlobular arteriole
Choledochus
Interlobular venules
Portal vein
Gallbladder
Central venules
Sinusoidal vessels
Interlobular biliary capillary
Hepatic lobules
Interlobular arteriole
Interlobular venule
Central venule
Bile capillaries

fashion, from the periphery toward the center of the lobe. Here, in the center, there is a vein. Branches of the portal vein, of the hepatic artery, the interlobular biliary canalicula (bile capillaries), the lymphatic vessels and the hepatic nerves are located at the confluence of more lobules (portal spaces or portobiliary). Bile is secreted by the hepatic cell inside the biliary canals, and these make up the three-dimensional labyrinth network inside each lobule. It ends up inside the biliary canalicula that converge inside increasingly larger ducts, until

they become two large intra hepatic ducts of the right and the left lobes. Both of them overflow into the common hepatic duct, which feeds the choledochus, along with the cystic duct that originates in the gallbladder. The common hepatic duct opens up into the duodenum through a minuscule muscular cleft and a circular opening inside the mucosa through which the pancreatic duct goes as well.

✠ GALLSTONES

The bile secreted by the liver is essential for digestion: it activates the pancreatic lipase intensifying the digestive activity of the pancreas, neutralizing the hydrochloric acid of the stomach, facilitating the intestinal absorption of lipids, exciting intestinal peristalsis and acting as an antiseptic against the intestinal flora. About 1 liter of bile is collected every day inside the gallbladder, or **cholecystis** *or bile vesicles: here it becomes enriched with a* **cholecistic** *mucosa, while* **bilirubine** *(a biliary pigment that originates from the transformation of hemoglobin ▶[180]), the biliary salts (sodic salts of the glycolic and taurocholic acids), the enzymes and the fatty substances that make up the bile, are concentrated for the absorption of water and mineral salts. Biliary dysfunctions cause various digestion problems: inflammations, bacterial infections, problems with the gallbladder, or malfunctioning of the choledochal sphincter (its inability to contract prevents the bile from relaxing) are very frequent, and cholecistectomy is the most common abdominal surgical intervention. However, the most common problem remains that of the gallstones: their origin is either chemical or they are made of cholesterol clusters, which arise spontaneously or due to an infection. They affect about 25% of women and about 12% of men over the age of 60. Their presence doesn't always manifest itself: in fact many people are affected by it without showing any symptoms. But sometimes, a gallstone can obstruct a cystic duct, or it can block the choledochal sphincter, preventing bile and pancreatic juice from reaching the duodenum. In such cases, due to nausea, jaundice, acute inflammation and other symptoms, surgery or* **lithotripsy** *(the breaking of the calculus through ultrasounds) is the best solution.*

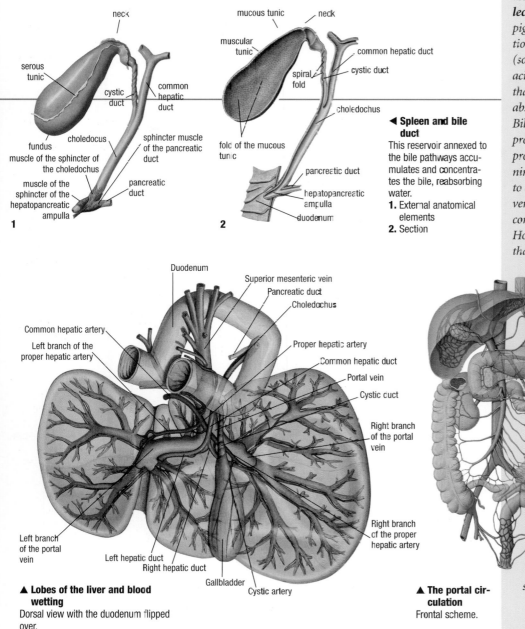

1

neck
serous tunic
cystic duct
common hepatic duct
choledocus
fundus
muscle of the sphincter of the choledochus
muscle of the sphincter of the hepatopancreatic ampulla
sphincter muscle of the pancreatic duct
pancreatic duct

2

mucous tunic
neck
muscular tunic
common hepatic duct
spiral fold
cystic duct
choledochus
fold of the mucous tunic
pancreatic duct
hepatopancreatic ampulla
duodenum

◀ **Spleen and bile duct**
This reservoir annexed to the bile pathways accumulates and concentrates the bile, reabsorbing water.
1. External anatomical elements
2. Section

Duodenum
Superior mesenteric vein
Pancreatic duct
Choledochus
Common hepatic artery
Left branch of the proper hepatic artery
Proper hepatic artery
Common hepatic duct
Portal vein
Cystic duct
Right branch of the portal vein
Right branch of the proper hepatic artery
Left branch of the portal vein
Left hepatic duct
Right hepatic duct
Gallbladder
Cystic artery

▲ **Lobes of the liver and blood wetting**
Dorsal view with the duodenum flipped over.

▲ **The portal circulation**
Frontal scheme.

THE PANCREAS

It is a mixed gland: exocrine (complex tubuloacinar consisting of a serous secretion) and endocrine. The pancreas is long and flat, and it is transversally arranged in the upper part of the abdomen, and is divided into:

-a *head*: more voluminous, and in contact with the duodenal ansa. It is "separated" from the body of the pancreas by an isthmus, which is a restricted area demarcated by two notches;

-a *body*: slightly oblique from the bottom to the top, placed in front of the aorta and the inferior vena cava;

-a *tail*: in contact with the spleen, and covered by the parietal peritoneum.

It is innervated by fibers that originate from the celiac plexus, and its structure is different depending on the role played by each part. Endocrine secretion is conducted by the Langerhans Islands, which produce hormones that regulate the metabolism of sugars ▶147; they are not connected to the excretory ducts, to which, on the other hand, the pancreatic acini and the areas of exocrine production are afferent to.

Stimulated by nervous mechanisms and by two hormones (the secretin and the pancrozinine-cholecistochinin) produced by the duodenal mucosa, the pancreatic acini pour their secretions into small canals, which flow into two ducts:

-The *chief duct* or *Wirsung duct*, which runs through the pancreas for its entire length and flows into the major papilla of the duodenum (papilla of Vater) along with the choledochus, to which it connects distally;

-The *accessory duct* or *Santorini's canal*, which flow into the minor papilla of the duodenum;

The *pancreatic juice* constantly produced by the pancreas in low and high quantities, following neuroendocrine stimuli (vagus, duodenal hormones), penetrates the duodenum and plays a very important role in the digestive process. Besides being full of bicarbonate ions, that along with the bile, can tampon the acidity of the food arriving from the stomach, the pancreatic

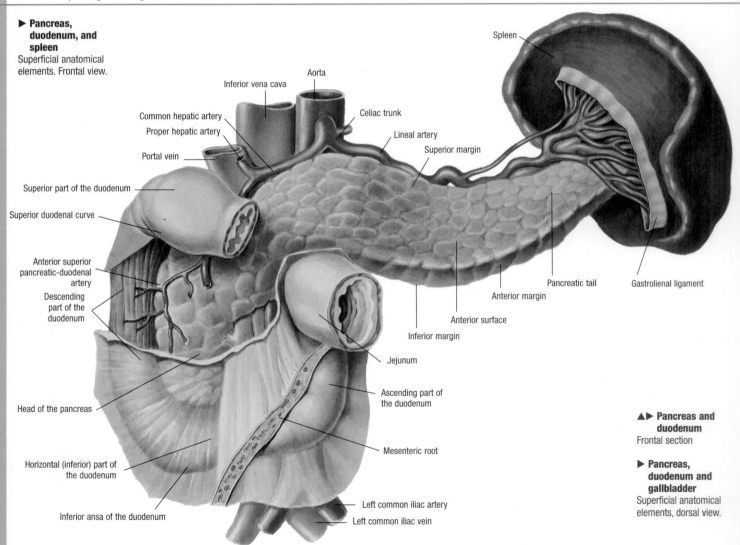

▶ **Pancreas, duodenum, and spleen**
Superficial anatomical elements. Frontal view.

- Spleen
- Inferior vena cava
- Aorta
- Common hepatic artery
- Celiac trunk
- Proper hepatic artery
- Lineal artery
- Portal vein
- Superior margin
- Superior part of the duodenum
- Superior duodenal curve
- Anterior superior pancreatic-duodenal artery
- Descending part of the duodenum
- Pancreatic tail
- Gastrolienal ligament
- Anterior margin
- Anterior surface
- Inferior margin
- Jejunum
- Head of the pancreas
- Ascending part of the duodenum
- Horizontal (inferior) part of the duodenum
- Mesenteric root
- Inferior ansa of the duodenum
- Left common iliac artery
- Left common iliac vein

▲▶ **Pancreas and duodenum**
Frontal section

▶ **Pancreas, duodenum and gallbladder**
Superficial anatomical elements, dorsal view.

juice contains different enzymes. The most important ones are the *trypsinogen, the amylase and the lipase*: the first one is activated as trypsin, by the intestinal enterokinase, and it acts on the proteins and on the peptones, reducing them into amino acids; the second one attacks those carbohydrates that have not yet been transformed by the ptyalin of the saliva ▶147, and transforms them into disaccharides; the third one, assisted by the bile, acts on the neutral fats and breaks them down into their components (fatty acids and glycerin). The role of bile is essential in the lipase: in fact, bile salts bind to the fats, creating the so called micelle, that have a gaseous-fatty interface, in which the pancreatic is able to act. The action of these enzymes is facilitated by the movements of the small intestine, which constantly re-mixes it, without moving the food forward.

THE ENZYMES OF THE PANCREATIC JUICE

*The acinous cells of the pancreas contain numerous **zymogen granules**, made up of proenzymes, which are only activated when they reach the alimentary canal (digestive tract). After being excreted inside the lumen of the pancreatic ducts, they reach the duodenum where the proenzymes have arrived from other enzymes and from duodenal substances, starting their specific chemical activity.*

Amylase: it catalyzes the hydrolysis of the alpha bonds inside the entire amylose chain; it transforms all of the carbohydrates entering the duodenum into a mixture of simple sugars (glucose and maltose) that can easily cross the intestinal membrane and enter the bloodstream.

Lipase: it catalyzes the hydrolysis of those fats that make up the micelle, transforming them into free fatty acids and glycerol, easily absorbable by the cell.

***Ribo-** and **deoxyribonuclease**: these are two types of enzymes (alpha and beta), which demolish the deoxyribonucleic (DNA) and the ribonucleic (RNA) acids, catalyzing the hydrolysis of the phosphodiesteric bridges found inside these molecules.*

***Trypsin**: it catalyzes the hydrolysis of the peptic bonds in which the carbonyl function has a residue of lysine and arginine: the higher the number of these amino acids inside the structure of a protein, the higher the degree of fragmentation of the protein.*

***Chimotrypsin**: it catalyzes the hydrolysis of the peptic bonds in which the carbonyl function has a residue of phenylalanine, tyrosine or tryptophan. Again, the higher the number of amino acids is, the higher the degree of protein fragmentation.*

***Carboxypeptidase**: it catalyzes the hydrolysis of the peptic bonds of COOH-terminal. Unlike trypsin and chimotrypsin, the carboxypeptidase continues to "digest" the protein detaching one amino acid after the other*

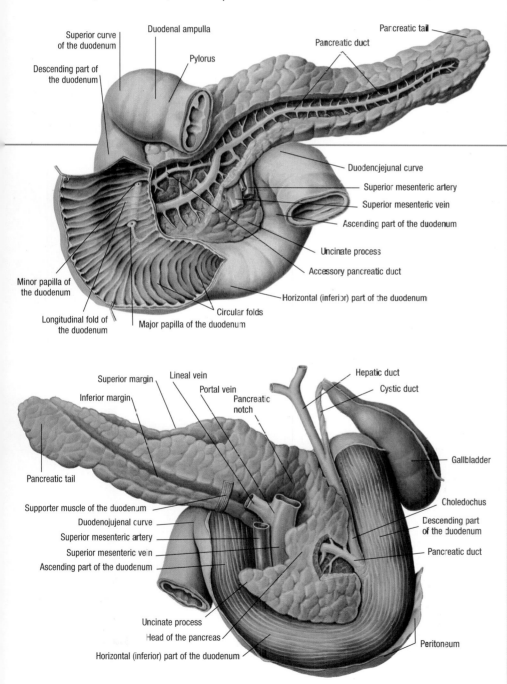

Superior curve of the duodenum
Duodenal ampulla
Pylorus
Pancreatic tail
Pancreatic duct
Descending part of the duodenum
Duodencjejunal curve
Superior mesenteric artery
Superior mesenteric vein
Ascending part of the duodenum
Uncinate process
Accessory pancreatic duct
Minor papilla of the duodenum
Horizontal (inferior) part of the duodenum
Circular folds
Longitudinal fold of the duodenum
Major papilla of the duodenum

Superior margin
Lineal vein
Portal vein
Pancreatic notch
Hepatic duct
Cystic duct
Inferior margin
Gallbladder
Pancreatic tail
Choledochus
Supporter muscle of the duodenum
Duodenojujenal curve
Superior mesenteric artery
Superior mesenteric vein
Ascending part of the duodenum
Descending part of the duodenum
Pancreatic duct
Uncinate process
Head of the pancreas
Horizontal (inferior) part of the duodenum
Peritoneum

THE INTESTINE

It is divided into 6 parts, which all have a different structure and function. These parts are grouped into two main camps, 3 on 3:

-the *small intestine* (about 6,80 m long) is characterized by an extremely folded internal surface (folds) filled with intestinal villi, whose density progressively increases as we get closer to the large intestine (where there are no villi). It includes:

-The *duodenum* (25-30 cm long), which is the first part of the intestine, right after the opening of the pylorus. It looks like an incomplete ring that surrounds the head of the pancreas and which is divided into 4 parts: superior, descending, horizontal, ascending. The pancreatic juices ▶155 and the bile ▶152-153 feed into the descending part. Along with the duodenal mucous secretion (in particular of the Brunner's Glands, and of Galeazzi-Lieberkuhn) they conclude the digestion of food. The enteric juice secreted by the duodenal mucosa is alkaline, and it contains many enzymes such as the enterochinase, essential in the activation of the pancreatic proenzymes, and also a lot of mucous. The duodenal mucosa, stimulated by the acidic chyme that arrives from the stomach, also secretes two hormones: the secretin, which stimulates the pancreatic production of bicarbonates and water, inhibiting the action of the gastrin (gastric), and the production of pancreozimine-cholecystochinin, which stimulates the production of pancreatic enzymes, the contraction of the gallbladder and the intestinal peristalsis;

-The *jejunum*, full of intestinal villi

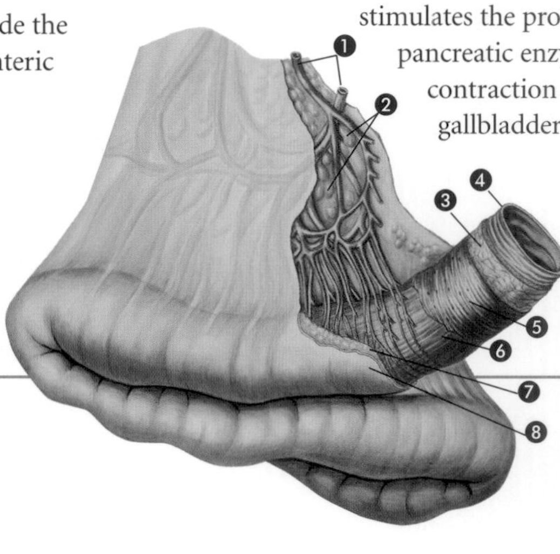

▼ **Intestine**
1. Superficial anatomical elements
2. Deep anatomical elements

▶ **Small intestine**
Superficial anatomical elements.
❶ Jejunal artery and vein
❷ Lymphatic nodules
❸ Submucouse tela

❹ Mucous tunic
❺ circular layer of the muscular tunic
❻ Longitudinal layer of the muscular tunic
❼ sub serous tela
❽ serous tunic

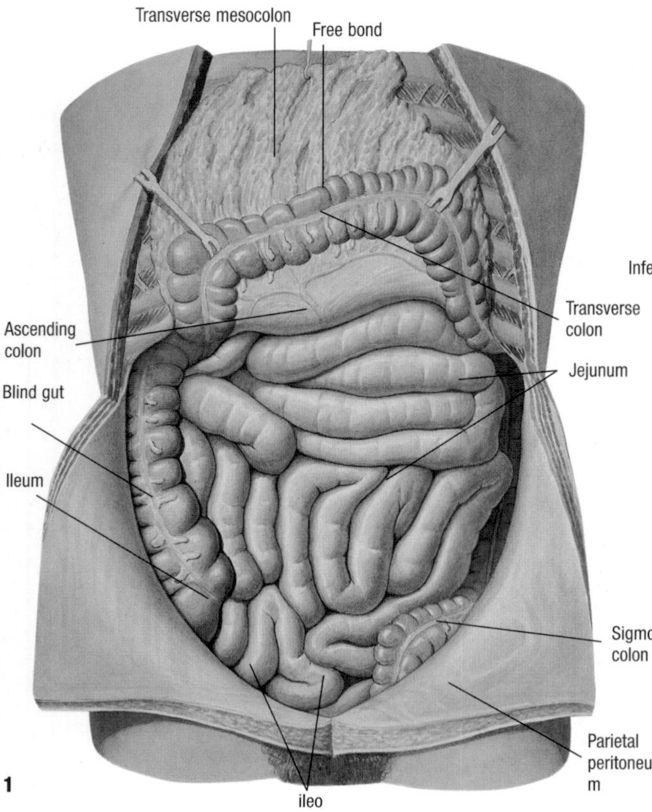

1 (figure labels):
- Transverse mesocolon
- Free bond
- Ascending colon
- Blind gut
- Ileum
- Transverse colon
- Jejunum
- Sigmoid colon
- Parietal peritoneum
- ileo

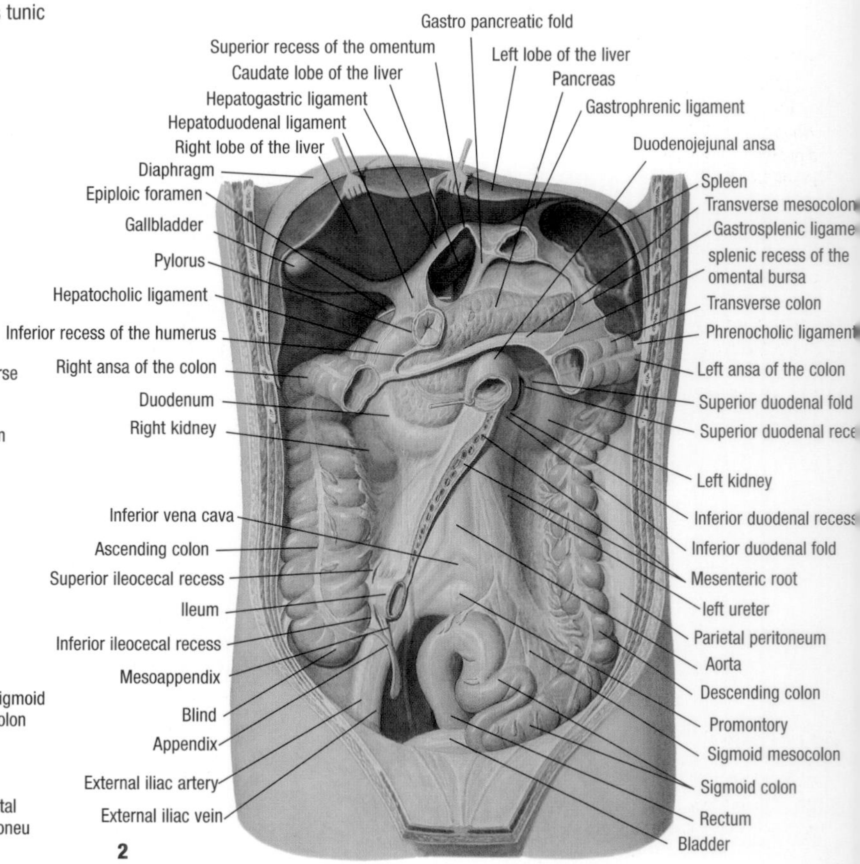

2 (figure labels):
- Superior recess of the omentum
- Caudate lobe of the liver
- Hepatogastric ligament
- Hepatoduodenal ligament
- Right lobe of the liver
- Diaphragm
- Epiploic foramen
- Gallbladder
- Pylorus
- Hepatocholic ligament
- Inferior recess of the humerus
- Right ansa of the colon
- Duodenum
- Right kidney
- Inferior vena cava
- Ascending colon
- Superior ileocecal recess
- Ileum
- Inferior ileocecal recess
- Mesoappendix
- Blind
- Appendix
- External iliac artery
- External iliac vein
- Gastro pancreatic fold
- Left lobe of the liver
- Pancreas
- Gastrophrenic ligament
- Duodenojejunal ansa
- Spleen
- Transverse mesocolon
- Gastrosplenic ligame
- splenic recess of the omental bursa
- Transverse colon
- Phrenocholic ligament
- Left ansa of the colon
- Superior duodenal fold
- Superior duodenal rece
- Left kidney
- Inferior duodenal recess
- Inferior duodenal fold
- Mesenteric root
- left ureter
- Parietal peritoneum
- Aorta
- Descending colon
- Promontory
- Sigmoid mesocolon
- Sigmoid colon
- Rectum
- Bladder

which absorb nutrients;
-The *ileum*, much richer with assimilating villi: we can count up to 1000/cm_, and physics-chemical laws regulate their action.

Their action is also regulated by the selective activity of the specific epithelium that covers them: it is made up of cells with microvilli, and has its own movements (the intestinal villi stretch and

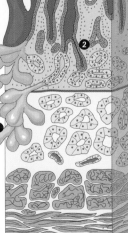

▲ Secretion of the peptic juice and of the bile
When the partially digested food arrives inside the duodenum, the duodenal mucous produces secretin, a hormone, which stimulates the pancreas and the liver to secrete gastric juices: respectively the pancreatic juice and the bile. Secretin was one of the first hormones to be discovered.

► Duodenum
Scheme of a three dimensional section.
❶ Brunner's glands
❷ Lieberkuhn glands

THE ABSORPTION OF NUTRIENTS

Just as plants absorb water and mineral salts through their roots, we absorb through the small intestine all the simple sugars, fatty acids, amino acids, water, mineral salts, vitamins and many other substances necessary for our survival and for the growth of our cells and our organism. In order to increase and maximize absorption, the intestinal walls are richly folded in order to increase the interface surface between the body and the food ingested. As a result, besides refolding into so-called plicae (big folds of the mucosa), the inside of the intestine is covered by villi, which, in turn, are lined with a cylindrical epithelium of cells provided with microvilli. A surface of about 1 cm_ can be covered by an average of 1000 villi and by more than one million and a half of microvilli: this creates a surface of absorption of over 300 m_, which is 200 times larger than the surface of the skin! Barely visible to the naked eye (they are about 1mm high) the intestinal villi have a particular structure that permits the immediate absorption of nutrients into the bloodstream and the lymph. Each one of them contains a network of blood capillaries and a lymphatic vessel (chyle vessel): amino acids, glucose, vitamins, minerals, and all the other

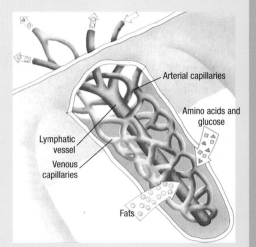

Arterial capillaries
Amino acids and glucose
Lymphatic vessel
Venous capillaries
Fats

hygrophilous substances that go directly into the bloodstream, and from here they reach the mesenteric veins, the portal vein and finally the liver. On the other hand, most of the fatty acids, some vitamins and the glycerol, all water insoluble, must first go through the lymphatic vessel and then they are let in the bloodstream The process of absorption of nutrients by the villi is a complex one, and it involves cellular membranes in both the active transport of specific nutrients, and the passive transport (osmotic) of others. The rhythmic shortening and stretching movement of the villi, contributes to this process. Such movement is stimulated by the villikinin, a hormone, which originates inside the intestinal mucosa during digestion, which stimulates the intestinal lymphatic circulation of each vessel

Intestinal villi

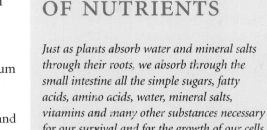

Microvilli

Mitochondria

Nucleus

①

②

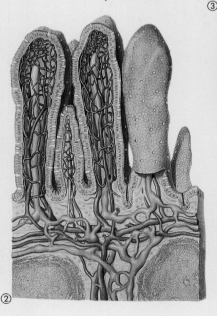

③

① Scheme of the structure of an epithelial cell of an intestinal villus, whose microvilli surface is facing the intestinal lumen.

② Scheme of an intestinal mucosa: the section shows the network of blood and lymphatic capillaries

③ Microphotography of villi: enlarged 300 times

contract in rhythm), which contribute in the passage of the nutrients inside the vessels underneath. The ileum communicates with the large intestine through the ileo-cecal valve;

-The *large intestine* (about 1,80 m long) reabsorbs water from the liquid mass (chyle) arriving from the small intestine, and it also pushes out waste products through peristalsis. Here fermentation and putrefaction take place thanks to bacterial flora. It lacks villi and it includes:

-The *blind gut*, which represents the part that follows the ileo-cecal valve. The appendix is here.

-The ascending colon, that from the blind gut goes toward the inferior surface of the liver;

-The *transverse colon*, which runs through the inferior surface of the liver until it reaches the inferior margin of the spleen;

-The *descending colon*, that descends all the way to the iliac crest, from the inferior margin of the spleen;

-The *ileo-pelvic* or *sigmoid colon*, that with a small curve connects the descending colon to the rectum;

-The *rectum*, located deep at the hypogastric level, in a retro-subperineal position, it ends up with the anal sphincter. Inside the rectum the absorption of water terminates and feces are accumulated. Its peristalsis, regulated by both sympathetic and the orthosympatetic stimuli, contributes to defecation.

The peritoneum, covers almost the entire intestine, and all the other internal organs: it facilitates movement, even if it keeps the organs in their original place; it protects the abdominal cavity from pathogenic agents, thanks to the secretive and absorbent characteristics of its epithelium. It is highly vascularized and it allows amino acids, monosaccharides, glycerin, vitamins, water and mineral salts to enter the mesenteric blood circulation. On the other hand, the fatty acids enter the lymphatic vessels.

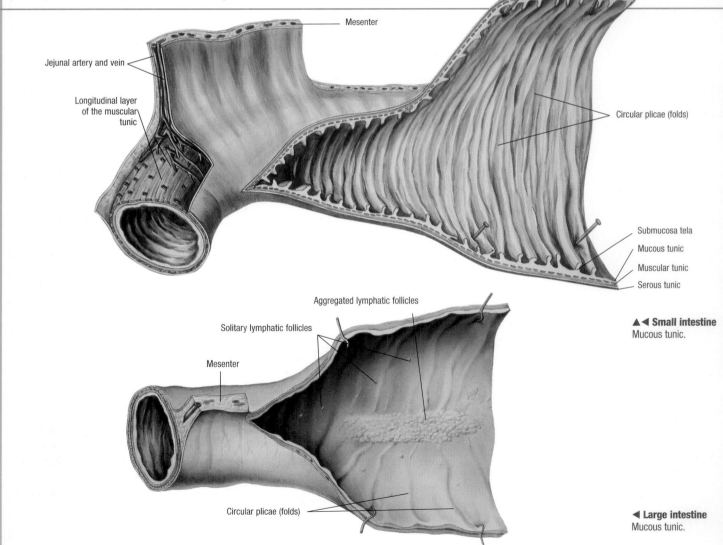

Mesenter

Jejunal artery and vein

Longitudinal layer of the muscular tunic

Circular plicae (folds)

Submucosa tela

Mucous tunic

Muscular tunic

Serous tunic

▲◀ **Small intestine**
Mucous tunic.

Aggregated lymphatic follicles

Solitary lymphatic follicles

Mesenter

Circular plicae (folds)

◀ **Large intestine**
Mucous tunic.

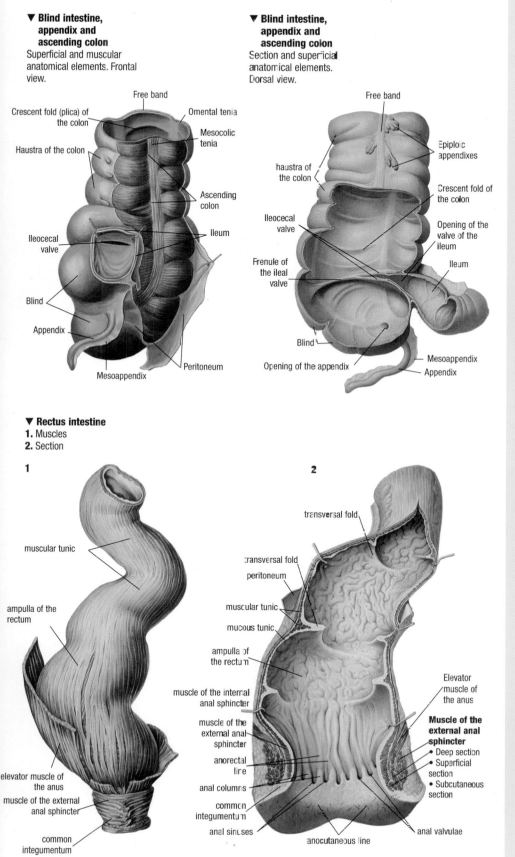

▼ Blind intestine, appendix and ascending colon
Superficial and muscular anatomical elements. Frontal view.

Crescent fold (plica) of the colon
Free band
Omental tenia
Mesocolic tenia
Haustra of the colon
Ascending colon
Ileocecal valve
Ileum
Blind
Appendix
Mesoappendix
Peritoneum

▼ Blind intestine, appendix and ascending colon
Section and superficial anatomical elements. Dorsal view.

Free band
Epiploic appendixes
haustra of the colon
Crescent fold of the colon
Ileocecal valve
Opening of the valve of the ileum
Frenule of the ileal valve
Ileum
Blind
Mesoappendix
Opening of the appendix
Appendix

▼ Rectus intestine
1. Muscles
2. Section

1

muscular tunic
ampulla of the rectum
elevator muscle of the anus
muscle of the external anal sphincter
common integumentum

2

transversal fold
transversal fold
peritoneum
muscular tunic
mucous tunic
ampulla of the rectum
muscle of the internal anal sphincter
muscle of the external anal sphincter
anorectal line
anal columns
common integumentum
anal sinuses
anocutaneous line
Elevator muscle of the anus
Muscle of the external anal sphincter
• Deep section
• Superficial section
• Subcutaneous section
anal valvulae

✚ DUODENAL ULCER AND COLITIS

Even though therapy is very similar and recently we have discovered the important role of the Helicobacter Pylori, the **duodenal ulcer,** *unlike the gastric ulcer* ▶ *150, is primarily caused by a hyper secretion of hydrochloric acid by the stomach. Unlike the gastric ulcer, the duodenal ulcer is 10 times more popular and it affects more men than women, generally around the ages of 30. The psychosomatic components of this disease are very clear and scientifically proven: energetic, dynamic, highly emotional, unsatisfied, frustrated and stressed people are at higher risk to develop such type of ulcer. On the other hand* **colitis** *is an inflammation of the colon. It can be caused by a viral, bacterial or parasitic infection (***dysenteric colitis***), by a scarce blood supply to the intestine (***ischemic colitis***), by an excessive exposure to ionizing radiations (***radiotherapy colitis***), by the use of pharmaceutical products (mostly antibiotics) which, by altering the bacterial flora of the intestine, allow for an abnormal development of the Clostridium difficile, a bacteria that produces a toxin which is able to cause a serious local necrobiosis (* **pseudo membranous enterocolitis** *).*

However, there are times when the origin of the inflammation is unknown: this happens in the case of an **irritable colon,** *probably psychosomatic, and common among depressed, insecure and stressed individuals. Something similar occurs with the* **ulcerative colitis.** *This is chronic inflammation of the mucosa and the submucosa of the descending colon, of the sigmoid and of the rectum, which causes hemorrhages, diarrhea, abdominal cramps, weight loss and fever. Often such ulcer affects other parts of the body, such as skin, the sacral bone, other peripheral joints, the kidneys, the liver and even the eyes (conjunctivitis).*

Therapy varies depending on the type and the severity of the colitis

ABOUT 400 ALVEOLI FOR A TOTAL OF 100-150 SQUARE METERS OF SURFACE IDEAL FOR
EXCHANGE…..ESSENTIALLY, THE RESPIRATORY SYSTEM IS THE LARGEST INTERFACE OF OUR
BODY WITH THE OUTSIDE WORLD, BIGGER THAN THE SKIN!!

THE RESPIRATORY SYSTEM

Without being aware of it, each minute we breathe about 15 times (a newborn breathes up to 70 times per minute!); on average, we breathe in and out about 13500 liters of air per day. The goal is to expel carbon dioxide, a toxic bi product of cellular metabolism, and to substitute it with oxygen, the substance necessary to conduct all the chemical reaction that allows us to extract chemical energy found inside food.

The respiratory system plays the main role in this important exchanging process, and closely collaborates with the circulatory system ►176, in collecting carbon dioxide from the body, bringing it to the lungs and discharging it outside the body, along with distributing oxygen collected inside the lungs, to the entire body. Also it plays an important role in the process of speech, thanks to a series of specialized structures inside the pathways.

THE STRUCTURES OF THE RESPIRATORY SYSTEM

The respiratory system is made up of many hollow organs (mouth and nose, larynx and pharynx, lungs) and of air canals (trachea, bronchi, bronchioles), which allow the body to constantly perform gaseous exchanges with the outside environment. It is divided into:

-*Air pathways* or *respiratory tracts*: made up of nasal cavities and paranasal sinuses, of the mouth, the pharynx, the larynx, the trachea and of the bronchial pathways (bronchi and bronchioles), which branch out and distribute to the lungs: all these structures have a bony or cartilaginous skeletal structure, which guarantees the easy passage of air. The mucous, that covers the wall of these hollow organs, has many functions: it warms up the air, thanks to its high vascularization; it wets it, thanks to the secretion of many glands distributed inside; it filters it, thanks to the mucous, which "glues" dust carrying it outside, through a constant movement of ciliate cells;

-The *lungs*: due to their cavity-filled structure (pulmonary alveoli or respiratory cells) the lungs are very spongy looking, and they collect the air, which is inspired through the pulmonary airways. Just like the wall of the blood capillaries of the pulmonary arteries and veins, the alveolar wall is also very thin: this facilitates the passive diffusion of the respiratory gasses (based on the degree of the concentration).

The lungs are very elastic organs, and they are able to easily stretch and contract. Each lung is protected by pleura, a membrane divided by a visceral leaflet, attached to the external surface of the lung, and a parietal leaflet, stretched out on the wall of the pulmonary recess, closely connected to both the rib cage and the superior surface of the diaphragm.

This way, each lung is surrounded by a space (pleural cavity) filled with an extremely thin liquid layer (pleural liquid) at a pressure lower than the atmospheric one. Besides wetting the pleural leaflets, facilitating their sliding movement, it guarantees the adherence of the leaflets, and indirectly that between lungs, rib cage and diaphragm: in this way it allows thoracic movements to be transmitted to the lung. If for any reason, air reaches the pleural cavity (pneumothorax), the elasticity of the pulmonary tissue is completely reduced and there is an almost total loss of the respiratory capacity.

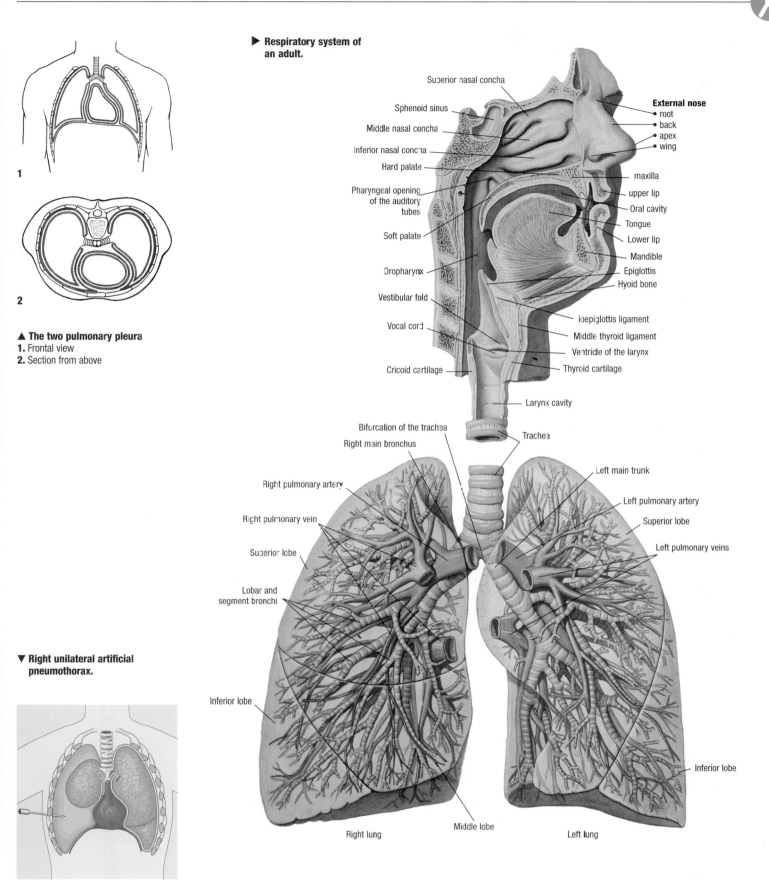

▶ Respiratory system of an adult.

Superior nasal concha

Sphenoid sinus

Middle nasal concha

Inferior nasal concha

Hard palate

Pharyngeal opening of the auditory tubes

Soft palate

Oropharynx

Vestibular fold

Vocal cord

Cricoid cartilage

External nose
• root
• back
• apex
• wing

maxilla

upper lip

Oral cavity

Tongue

Lower lip

Mandible

Epiglottis

Hyoid bone

ioepiglottis ligament

Middle thyroid ligament

Ventricle of the larynx

Thyroid cartilage

Larynx cavity

Bifurcation of the trachea

Right main bronchus

Trachea

Right pulmonary artery

Left main trunk

Right pulmonary vein

Left pulmonary artery

Superior lobe

Superior lobe

Left pulmonary veins

Lobar and segment bronchi

Inferior lobe

Inferior lobe

Right lung

Middle lobe

Left lung

▲ The two pulmonary pleura
1. Frontal view
2. Section from above

1

2

▼ Right unilateral artificial pneumothorax.

161

INSPIRATION (TO INHALE)

The lungs are protected by the rib cage ▶46, a highly articulated and extendible structure: ribs, sternum, and the vertebra can all mutually move, thanks to the activity of the thoracic muscular system ▶60-62. At the bottom, the lungs touch the diaphragm ▶63, which, stimulated by the vagus nerve ▶110, rhythmically contracts. Because the contraction of the diaphragm occurs in concomitance with the expansion of the rib cage, the lungs expand: thanks to the pleurae, their attachment to the pulmonary cavity is guaranteed, and they are pulled down by the diaphragm and toward the outside by the rib cage. This creates an internal pressure, lower than the outside pressure: this depression is re-balanced when we breathe in air from the mouth and the nose. Then the air reaches the pharynx, a cavity where both the nasal fossae and the oral cavity flow. From here it passes through the larynx, arrives at the trachea, the bronchi reaching all the way to the farthest alveoli. Inspiration is determined by the nervous impulses elaborated inside the inspiratory centers of the encephalic bulb, transmitted through the parasympathetic network of the vagus nerve, and through the orthosympathetic thoraco-lumbar one (pulmonary plexus) ▶116. While the vagus nerve carries bronchoconstrictor and vasodilator stimuli, the sympathetic nerves carry bronchodilator and vasoconstrictor stimuli: whilst breathing in, the action of the vagus nerve prevails in the vessels, and the sympathetic one prevails in the bronchi.

EXPIRATION

Unlike breathing in, breathing out is a passive movement: the muscular system of the thorax and the diaphragm relax, and the lungs, being elastic structures, spontaneously empty out, going back to a "basic" tension. At this point, the air held inside is poor in oxygen but very rich with carbon dioxide, due to the alveolar-vasal exchange that has occurred during inspiration. This air is pushed to the outside in the opposite direction to breathing in. Breathing in is established by nervous impulses elaborated inside the encephalic bulb (expiration center), transmitted by both the vagus nerve, and by the thoraco-lumbar ▶116-122 orthosympatetic network: during breathing in, the action of the vagus prevails at the level of the bronchi (bronchoconstriction), while the action of the sympathetic is key at the level of the blood vessels (vasodilatation).

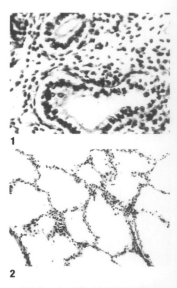

▲ **Fetal and adult pulmonary tissue**
In a fetus that has not yet begun to breath, the lungs are filled with amniotic fluid and the alveoli are not yet expanded.
1. Section of a fetal lung: the cubic simple epithelium has not stretched yet;
2. Section of an adult lung: the simple cubic epithelium is stretched out, becoming therefore a simple pavement epithelium.

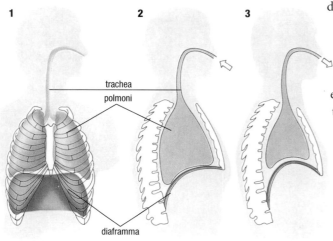

trachea
polmoni
diaframma

RESPIRATION

The respiratory frequency is primarily established by the concentration of carbon dioxide and oxygen inside the blood: in fact, an increase in carbon dioxide and a decrease in oxygen, stimulate the cerebral, glossopharyngeal, and vagus respiratory centers, stimulating the acceleration of the breathing pattern; a decrease in carbon dioxide causes a slowing down of the breathing pattern. Spontaneous breathing can also be voluntarily modified, acting mostly on the thoracic muscles: this determines the presence of three quantities of air:

-*Running air*: obtained through normal inspiration and expiration, autonomously managed by the secondary nervous system.

-*Inspiratory reserve volume*: it's obtained through forced respiration;

-*Residual air*: it's the air that remains inside the lungs after forced expiration.

Adding all these three together, we obtain the anatomical lung capacity (on average 1600 cm_ in a man, and about 1300 cm_ in a woman); adding running air and the reserve volume we obtain the vital capacity of the lung.

▼ Involuntary movements
1. sneeze; 2. hiccups

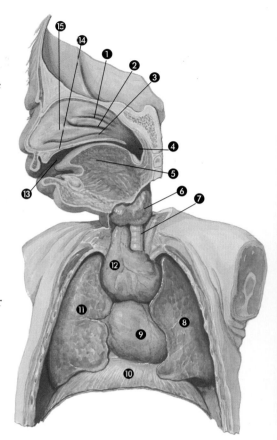

HICCUPS, LAUGHTER, CRYING, COUGHING, YAWNS AND SNEEZES

These are exceptional forms of respiration, evoked by physical and emotional stimuli. Corpuscles, intense odors, smoke, are all physical factors that irritate the nasal cavities and can cause sneezes and coughing. In the case of a sneeze, the irritation of coating of the nasal fossae stimulates a reflex movement, which closes the glottis (see figure: in red). This way, as the glottis opens, the air is compressed and forced out: often the tongue blocks the posterior part of the mouth, and the air is pushed out through the nose. The process of coughing has a similar movement. It is caused by a foreign body or by an excessive production of mucous: the irritation of the trachea or of the bronchi stimulates the reflex that closes the glottis, that contracts the lungs and forces air out (coughing). On the other hand, a yawn can be caused by many different factors: while hunger and sleepiness related yawns have a physiological nature, those produced by boredom or by seeing someone else yawning have a clear psychosomatic origin. Similarly, even laughter and crying have a psychosomatic origin. They consist of a normal inspiration followed by an explosion of short and consecutive expirations. Hiccups are the result of a spasmodic contraction of the diaphragm, stimulated by the vagus nerve. Initially relaxed (see figure, green portion), the diaphragm suddenly contracts, causing, at the same time, a forced inspiration and the closing of the glottis: this causes the characteristic sound of a hiccup

1

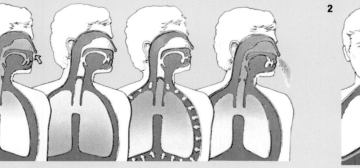

2

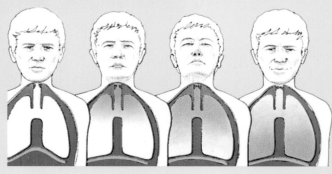

MOUTH AND NOSE

They are the airways cavities most directly in contact with the outside environment. While the mouth is used for breathing, only in particular cases, the nose is the preferred way to inhale. The nostrils are found at the entrance of the nasal cavities, which are two parallel airways separated by the bone and cartilaginous septum. They are covered with hair (**vibrissae**), whose role is to trap the biggest dust particles entering the nose.
The respiratory membrane is characterized by an epithelium filled

with ciliate cells, alternated by partially serous and partially mucous glands, that not only trap the smallest particles, but also play an anti-infection role, due to its high content of lysozyme and of immunoglobuline [178-199]. On the inside, the nasal cavities enlarge into a vault, a pavement, a medial and lateral wall, and they finally open into the most anterior

part of the pharynx called with the choanae. Para nasal cavities also open into each nasal cavity, acting as a sound resonance box. Rather than being involved in the respiration, they help keeping the skull lighter. In fact they lighten many bones of the visceral cranium: the sphenoid, the frontal, the ethmoid, the lachrymal, the upper maxillary and the palatine. Often, many secondary minor cavities branch out from these main ones.

▼ **Nasal cavity**
Left lateral section.

▶ **Oral vestibule**
Left lateral section.

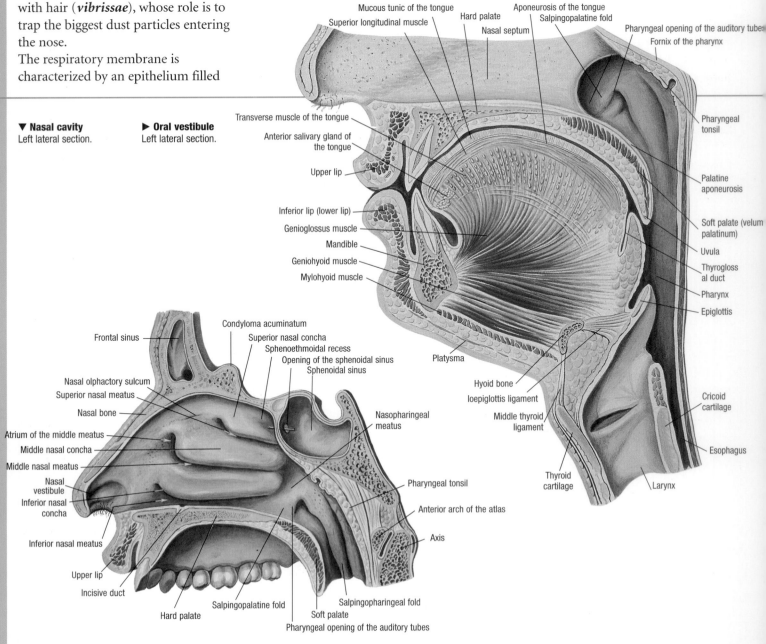

PHARYNX AND LARYNX

They are both convergent median odd ducts of the neck. The pharynx is frontally connected to the nasal cavities (defined by the choanae), with the mouth (defined by the isthmus of the fauces), with the larynx (through the pharyngeal orifice) and with the auditory tubes (which connect the tympanic box with the outside environment, through the pharyngeal orifices: this is why when we swallow or when the air pressure changes, we are able to compensate the sensation of pressure that our ear feels when we are inside a plane or while underwater diving).

Although the front of the pharynx is incomplete, due to a high number of openings, it is divided into a nasal part (rhino pharynx), a buccal part (oropharyngeal) and a laryngeal part (laryngopharynx), into a vault (the upper extremity), an anterior and a posterior wall, two lateral walls which delimit the mandibolopharyngeal space where veins, arteries and the main nervous fibers of the neck run, a superior extremity or fornix, connected at the base of the skull, and an inferior extremity, which dorsally meets the 6th cervical vertebra.

The pharynx plays a very important role: it is able to decide when a piece of food must go through the esophagus (where it stretches into ►148) or when a breath of air must go through the larynx. Muscles surround the pharynx and it is innervated by the branches of the pharyngeal plexus of the sympathetic and by the fibers of the vagus nerve (parasympathetic), which manage in a synchronized and progressive fashion all of the movements of the deglutition. Inside the pharyngeal mucosa there are many lymphatic vessels, many of which are in contact with the tonsils. The tonsils protect the first airways and the first digestive pathways from infections. The palatine tonsils, the tubal and the pharyngeal tonsils, do also assist in this important role of the pharynx. The larynx

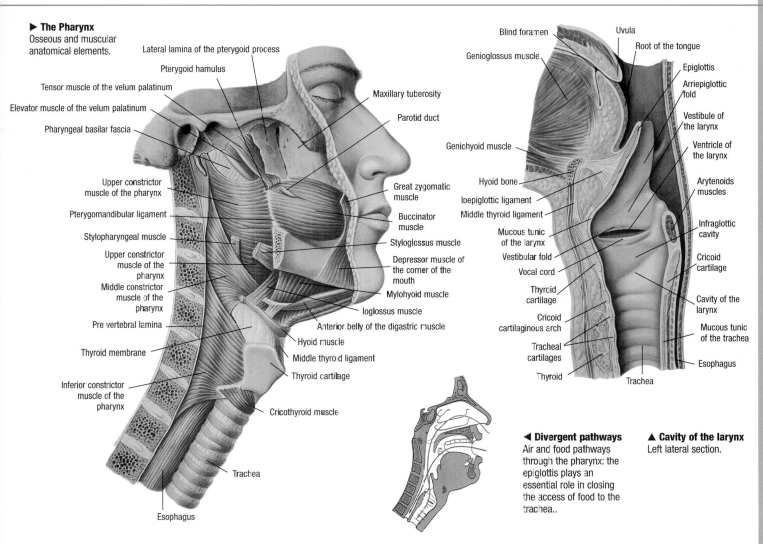

▶ **The Pharynx**
Osseous and muscular anatomical elements.

Lateral lamina of the pterygoid process
Pterygoid hamulus
Tensor muscle of the velum palatinum
Elevator muscle of the velum palatinum
Pharyngeal basilar fascia
Upper constrictor muscle of the pharynx
Pterygomandibular ligament
Stylopharyngeal muscle
Upper constrictor muscle of the pharynx
Middle constrictor muscle of the pharynx
Pre vertebral lamina
Thyroid membrane
Inferior constrictor muscle of the pharynx

Maxillary tuberosity
Parotid duct
Great zygomatic muscle
Buccinator muscle
Styloglossus muscle
Depressor muscle of the corner of the mouth
Mylohyoid muscle
Ioglossus muscle
Anterior belly of the digastric muscle
Hyoid muscle
Middle thyroid ligament
Thyroid cartilage
Cricothyroid muscle
Trachea
Esophagus

Blind foramen
Uvula
Root of the tongue
Genioglossus muscle
Epiglottis
Arriepiglottic fold
Vestibule of the larynx
Ventricle of the larynx
Geniohyoid muscle
Hyoid bone
Ioepiglottic ligament
Middle thyroid ligament
Arytenoids muscles
Mucous tunic of the larynx
Vestibular fold
Vocal cord
Thyroid cartilage
Cricoid cartilaginous arch
Tracheal cartilages
Thyroid
Infraglottic cavity
Cricoid cartilage
Cavity of the larynx
Mucous tunic of the trachea
Esophagus
Trachea

◀ **Divergent pathways**
Air and food pathways through the pharynx: the epiglottis plays an essential role in closing the access of food to the trachea..

▲ **Cavity of the larynx**
Left lateral section.

originates in the anterior part of the pharynx, immediately behind the tongue, continuing into the trachea. A specific cartilage modulates the opening of the pharyngeal orifice of this organ: when closing, the epiglottis prevents food from being inhaled during swallowing (deglutition), and depending on the space left open, it contributes to the modulation of sounds.

In fact, besides being an access organ to the deeper airways, the larynx is also the organ responsible for the emission of sounds. Made up of many cartilaginous ossicles, kept together by strong muscles and ligaments, the larynx, coated by a thick mucous tunic, is able to actively and passively lift and lower itself during deglutition, respiration and phonation,

according to the nervous signals which arrive from both the central nervous system ▶80) (voluntary stimuli) and the peripheral nervous system ▶110 (involuntary stimuli of the vagus nerve and of the sympathetic nerve). The cartilages that make up the larynx are the following: the thyroid cartilage (the biggest one), the cricoid cartilage (it supports the others and it offers an attachment to the most important muscles), the arytenoids cartilages (connected to the vocal process) the epiglottis (shaped as an oval leaf, whose convex part is turned toward the pharynx), and the corniculate accessory cartilages

(or Santorini's) and the cuneiform ones (Wrisberg's or Morgagni's).

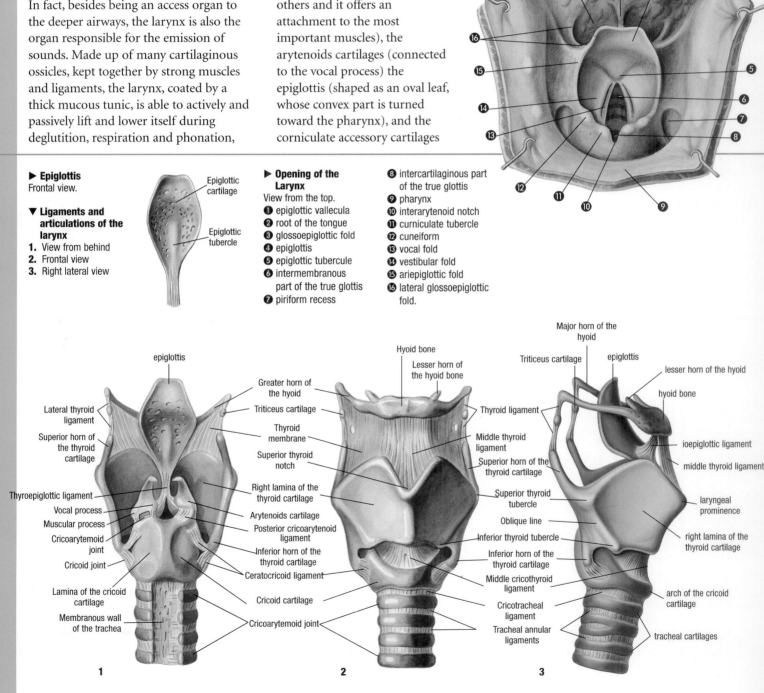

▶ **Epiglottis**
Frontal view.

Epiglottic cartilage

Epiglottic tubercle

▼ **Ligaments and articulations of the larynx**
1. View from behind
2. Frontal view
3. Right lateral view

▶ **Opening of the Larynx**
View from the top.
❶ epiglottic vallecula
❷ root of the tongue
❸ glossoepiglottic fold
❹ epiglottis
❺ epiglottic tubercule
❻ intermembranous part of the true glottis
❼ piriform recess
❽ intercartilaginous part of the true glottis
❾ pharynx
❿ interarytenoid notch
⓫ curniculate tubercle
⓬ cuneiform
⓭ vocal fold
⓮ vestibular fold
⓯ ariepiglottic fold
⓰ lateral glossoepiglottic fold.

1

epiglottis

Lateral thyroid ligament
Superior horn of the thyroid cartilage
Thyroepiglottic ligament
Vocal process
Muscular process
Cricoarytemoid joint
Cricoid joint
Lamina of the cricoid cartilage
Membranous wall of the trachea

Greater horn of the hyoid
Triticeus cartilage
Thyroid membrane
Superior thyroid notch
Right lamina of the thyroid cartilage
Arytenoids cartilage
Posterior cricoarytenoid ligament
Inferior horn of the thyroid cartilage
Ceratocricoid ligament
Cricoid cartilage
Cricoarytemoid joint

2

Hyoid bone
Lesser horn of the hyoid bone
Thyroid ligament
Middle thyroid ligament
Superior horn of the thyroid cartilage
Superior thyroid tubercle
Oblique line
Inferior thyroid tubercle
Inferior horn of the thyroid cartilage
Middle cricothyroid ligament
Cricotracheal ligament
Tracheal annular ligaments

3

Major horn of the hyoid
Triticeus cartilage
epiglottis
lesser horn of the hyoid
hyoid bone
ioepiglottic ligament
middle thyroid ligament
laryngeal prominence
right lamina of the thyroid cartilage
arch of the cricoid cartilage
tracheal cartilages

SPEECH

If language is a communication system, then many animals can speak. However, no animal not even a chimpanzee, which is able to use abstract symbols to communicate with scientists who study him, will ever be able to speak like we do.

Anatomy is the reason for that. In fact, the larynx plays a central role in both the production of sounds and in the modulation of the resonance box that modifies them (pharynx).

The internal cavity of the larynx, delimited by cartilage, ligaments and muscles, is a lot smaller than the outer circumference. Two anterior-posterior relieves called **folds** (**ventricular or superior, and vocal or inferior**) or **vocal cords**, divide it into three segments:

-The **superior segment** or **vestibule** is delimited by the posterior surface of the epiglottis and it communicates with the pharynx.

-The **middle segment** (the narrowest part), which includes the folds: the false glottis is found inside the ventricular fold, while the true glottis is found inside the vocal folds. The width and the shape of the true glottis change depending on the sex of the individual, and on the respiration and phonation phases.

-The **inferior segment**, which continues toward the bottom with a cylindrical shape. The quality and the pitch of a voice depend on the length, the size and the tension of the vocal cord (and therefore on the amplitude of the true glottis). The intensity is determined by the pressure of the airflow and the timbre (tone) is determined almost exclusively by the supralaryngeal airways: the **tongue**, the **soft palate** and the **lips** are essentially important in the articulation of speech, while the role of the **pharynx** is to be a resonance box. By changing the position of the neck (raising and bending down), the larynx changes the amplitude of the box, drastically modifying the emission of sounds.

The position of the larynx affects also the way we breathe and swallow: in humans and in animals such as a monkey, it is located very high on the neck, and it blocks the

nasopharynx, allowing us to drink and to breathe at the same time. However, such a high larynx reduces the pharyngeal "resonance box" in such a way that it is almost impossible to speak; in order to articulate different sounds a monkey uses mostly its lips and mouth. The same happens with a newborn; but as the baby grows, the larynx develops and progressively moves toward the bottom: within two years, the way a baby swallows and breaths drastically changes, and vocalization becomes possible. This is still a mysterious process that involves not only the larynx and the pharynx, but also other vital structures: speaking is so essential for human beings that to be able to speak we even alter our breathing pattern ▶163; carbon dioxide is exhaled at such a different rhythm, that if we were to breathe in such a way when we do not talk, we would soon end up hyperventilating. However, we will never notice a change in the rhythm of our conversation: nobody ever gets "tired" of speaking!

But this is not all. Primarily, the way we articulate our language has a mental component.

Observe a very young baby when you talk to them: you will notice that he reacts to your words by moving his arms, by staring at you, and by prattling. His body is shaken by ever so light coordinated muscular movements, which are a clear sign of a cerebral activity stressed by verbal language.

The cerebra, motor and verbal activities remain closely connected even in an adult: we have all experienced doing something challenging while keeping our tongue between our teeth, or speaking and describing what you are doing with one's hands, even if you are alone.

▶163

◀ Phonation
During respiration ① the vocal cords are separated; to begin speaking ②, the laryngeal cartilages moved by voluntary muscles, get closer and create a fissure or change the inclination of the cords: high pitch notes are produced by highly tense cords ③, while the low notes are produced by ④ looser cords.

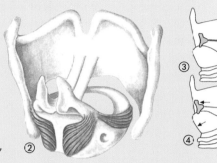

▼ Elastic cone
There are only few laryngeal elements involved in spoken language. They are:
❶ thyroid cartilage;
❷ true glottis;
❸ vocal process;
❹ arytenoids cartilage;
❺ posterior cricoarytenoid ligament.;
❻ cricoid cartilage; ❼ elastic cone; ❽ vocal ligament

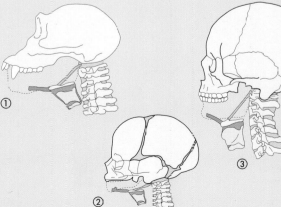

◀ Position of the larynx and ability to speak
The position of the larynx and the ability to vocalize differ from a monkey ①, to a baby ②, to an adult ③.

TRACHEA AND BRONCHI

The trachea is 10-12 cm long, with a diameter of 16-18 mm. It is an almost vertical, elastic and extensible canal that follows the larynx and bifurcates into the bronchi. It is kept locked at the point of bifurcation (sealed to the phrenic center by the diaphragm), and it is mobile in its upper extremity, which follows the swallowing movements and those of the phonation of the larynx. It has a succession of cartilaginous rings called tracheal rings, altered with annular ligaments, which connect on the back to the membranous wall of the trachea. The trachea is divided into cervical, which includes the first 5-6 tracheal rings and contains the pre tracheal lymph nodes ▶200, and the thoracic, which contains the tracheal lymph nodes ▶200.

At the level of the 4th and 5th vertebra, the trachea divides into two branches, one whose axis is inclined of 20º, and the other with an axis inclined of 45-50º: they are respectively the right and the left bronchus. They are characterized by a cartilaginous ring structure, similar to that of the trachea. The first one has a diameter of about 15 mm, and the other one of about 11 mm. This is due to the fact that the right lung occupies a bigger volume than the left, and it has more respiratory capacity. Even their length is different: before they bifurcate, the right bronchus is 2cm long, and the left one is 5 cm long. After the first bifurcation, the bronchi branch out into an arborization, which is contained inside the lungs: the "secondary" branches are therefore called intrapulmonary bronchi, and they are named after the corresponding position they are in.

The bronchi are covered by a mucosa rich with muciparous glands and with ciliate cells, which constantly create an outward flow of mucous. The involuntary muscles that envelop them are controlled by the pulmonary plexus ▶116 and by the vagus.

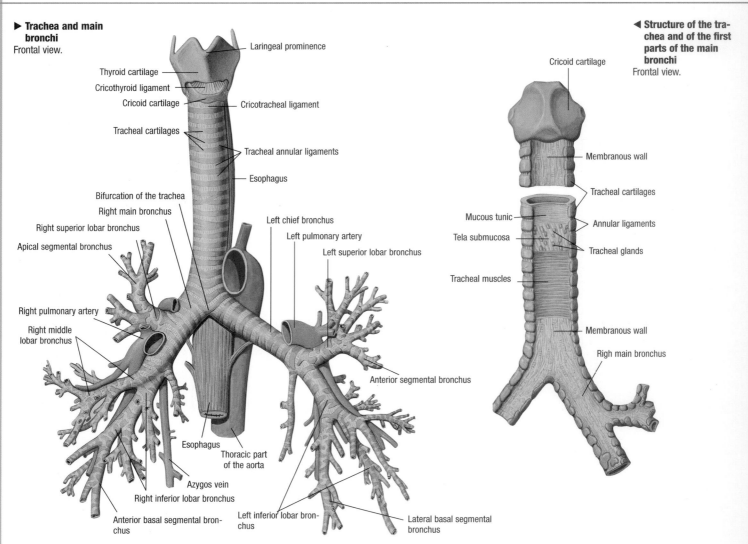

▶ **Trachea and main bronchi**
Frontal view.

Laringeal prominence
Thyroid cartilage
Cricothyroid ligament
Cricoid cartilage
Cricotracheal ligament
Tracheal cartilages
Tracheal annular ligaments
Esophagus
Bifurcation of the trachea
Right main bronchus
Left chief bronchus
Right superior lobar bronchus
Left pulmonary artery
Apical segmental bronchus
Left superior lobar bronchus
Right pulmonary artery
Right middle lobar bronchus
Anterior segmental bronchus
Esophagus
Thoracic part of the aorta
Azygos vein
Right inferior lobar bronchus
Left inferior lobar bronchus
Anterior basal segmental bronchus
Lateral basal segmental bronchus

◀ **Structure of the trachea and of the first parts of the main bronchi**
Frontal view.

Cricoid cartilage
Membranous wall
Tracheal cartilages
Mucous tunic
Annular ligaments
Tela submucosa
Tracheal glands
Tracheal muscles
Membranous wall
Righ main bronchus

THE LUNGS

Here the gaseous exchanges between air and blood take place: the unique structure of these organs, filled with highly vascularized cavities, is in fact associated with this function.

The lungs are divided into a right lung, bigger in size and subdivided into 3 lobes, and a left lung, subdivided into 2 lobes. Both of them are located inside the pulmonary recess of the rib cage, and are separated by a space included between the sternum and the spinal column, called *mediastinum*, where we can find the heart, the thymus, the trachea, the bronchi, the esophagus and the biggest blood vessels (such as the aorta). Each lung is coated by the pleura, a serous membrane made up of two leaflets (visceral, that adheres to the surface of the lung, and parietal, which adheres to the surface of the pulmonary recesses), which delimitate the pleural cavity, inside which there is a negative pressure which causes the lungs to expand during respiration [163].

Each lung is divided into:

-a *base* or *diaphragmatic surface*, which is inclined toward the bottom and on the back, and has a crescent and concave shape, and is moulded to the convexity of the diaphragm.

-a *lateral* or *cost vertebral surface*, convex in shape, and marked with numerous costal imprints.

- a *medial surface* or *mediastinal*, concave and vertical, is between the anterior and posterior margin and shows a hollow area called hilum, where the bronchi and the nerves penetrate the lung, and the blood vessels exit. Inside the hilum are also lymph nodes (called *hilari*) [199]. In front and under the hilum there is the hollow surface of the cardiac fossa, more pronounced in the left lung. Finally, near the posterior margin there are the imprints of the great vessels: the azygos vein, the aorta, the superior vena cava and the left anonymous vein; -An *apex*, made up of all those rounded parts of the lung located above the superior margin of the rib II; on the right side, the apex is fron-

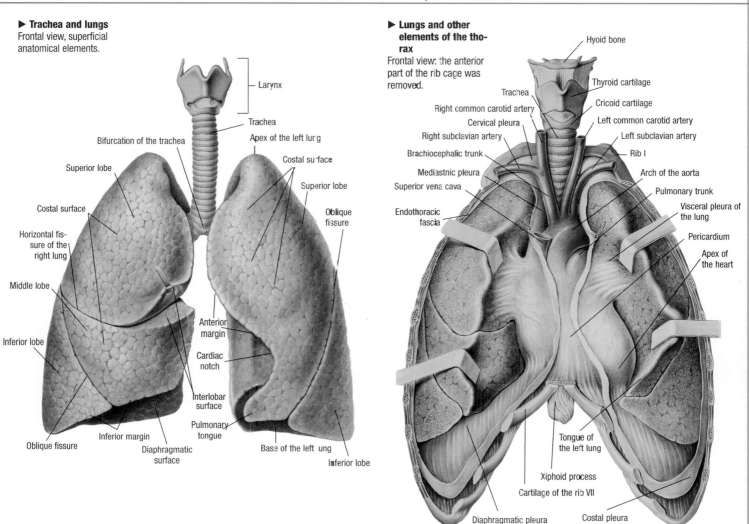

▶ **Trachea and lungs**
Frontal view, superficial anatomical elements.

Larynx
Trachea
Apex of the left lung
Costal surface
Superior lobe
Oblique fissure
Anterior margin
Cardiac notch
Interlobar surface
Pulmonary tongue
Base of the left lung
Inferior lobe

Bifurcation of the trachea
Superior lobe
Costal surface
Horizontal fissure of the right lung
Middle lobe
Inferior lobe
Oblique fissure
Inferior margin
Diaphragmatic surface

▶ **Lungs and other elements of the thorax**
Frontal view: the anterior part of the rib cage was removed.

Hyoid bone
Thyroid cartilage
Cricoid cartilage
Left common carotid artery
Left subclavian artery
Rib I
Arch of the aorta
Pulmonary trunk
Visceral pleura of the lung
Pericardium
Apex of the heart
Tongue of the left lung

Trachea
Right common carotid artery
Cervical pleura
Right subclavian artery
Brachiocephalic trunk
Mediastinic pleura
Superior vena cava
Endothoracic fascia
Diaphragmatic pleura
Xiphoid process
Cartilage of the rib VII
Costal pleura

rib II; on the right side, the apex is frontally and medially curved, while on the left side, it is less distinct from the rest of the organ. Marked by the imprint of the subclavian artery, the apex connects to the supreme intercostals artery and the internal mammary, while from the back, it touches the inferior cervical ganglion of the sympathetic ►[111].

Fissures that reach the hilum and outlining the lobes, cross the surface of each lung. There are 2 fissures inside the right lung: the main one, which reaches the base and cuts the organ obliquely; and the *secondary* one, which detaches from the main one at the level of the VI Rib and horizontally crosses the lateral surface, reaching the hilum in a upward oblique direction. In the left lung, the only existing fissure is similar to the main one of the

right lung. A branch of the main bronchus called primary bronchus crosses each lobe. Each lobe is further subdivided into territories called *zones* or *pulmonary segments*: they are pyramidal in shape, with the base turned toward the outside and the apex toward the hilum, and are structured around a *segmentary bronchus* (or *second degree* or *zonal*) and a *segmentary artery*, besides being drained by a *perisegmentary vein*.

Each segment includes hundreds of pulmonary lobules recognizable also from the outside, and delimited by thin polygonal lines of pigmented connective tissue, supplied with lobular bronchioles (1mm in diameter). Inside the lobule, each bronchiole branches out into intratubular bronchioles (0, 3 mm), which

in turn, branches out further, into 10-15 *terminal* or *minimal bronchioles*. At the end of each one of them there is a *pulmonary acinus*.

In each acinus, the terminal bronchiole originates 2 *respiratory* or *alveolar bronchioles*. Along their course they show some hemisphere extroflexions called pulmonary alveoli. As we reach the distal extremity of the bronchiole, they become more numerous, until the bronchiole splits into 2-10 *alveolar ducts* Their wall is made up of a succession of alveoli and it ends with an alveolus. This terminal part of the airways is called infundibulum or alveolar sac. The bronchiole's wall is covered with muscular fascicles in the area where each alveolus attaches, creating a sort of cuff, which regulates the inflow of

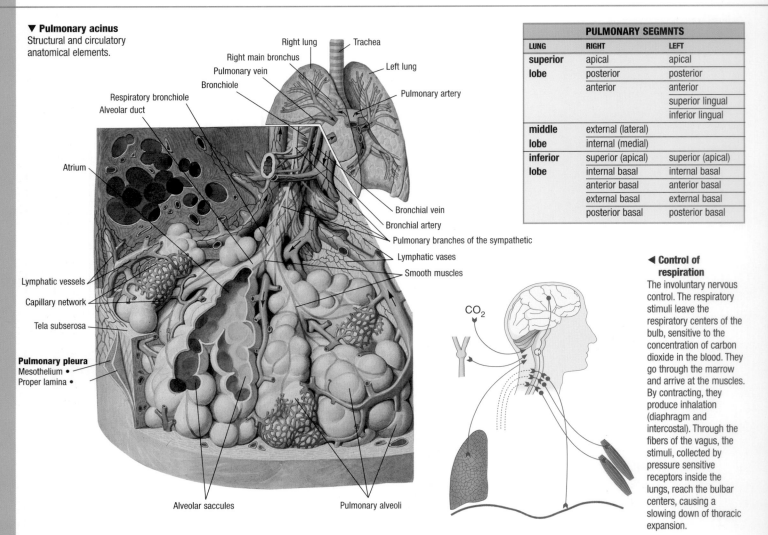

▼ Pulmonary acinus
Structural and circulatory anatomical elements.

Right lung
Trachea
Right main bronchus
Pulmonary vein
Left lung
Bronchiole
Pulmonary artery
Respiratory bronchiole
Alveolar duct

Atrium

Bronchial vein
Bronchial artery
Pulmonary branches of the sympathetic
Lymphatic vases
Smooth muscles

Lymphatic vessels
Capillary network
Tela subserosa

Pulmonary pleura
Mesothelium •
Proper lamina •

Alveolar saccules
Pulmonary alveoli

PULMONARY SEGMNTS		
LUNG	**RIGHT**	**LEFT**
superior	apical	apical
lobe	posterior	posterior
	anterior	anterior
		superior lingual
		inferior lingual
middle	external (lateral)	
lobe	internal (medial)	
inferior	superior (apical)	superior (apical)
lobe	internal basal	internal basal
	anterior basal	anterior basal
	external basal	external basal
	posterior basal	posterior basal

CO_2

◄ Control of respiration
The involuntary nervous control. The respiratory stimuli leave the respiratory centers of the bulb, sensitive to the concentration of carbon dioxide in the blood. They go through the marrow and arrive at the muscles. By contracting, they produce inhalation (diaphragm and intercostal). Through the fibers of the vagus, the stimuli, collected by pressure sensitive receptors inside the lungs, reach the bulbar centers, causing a slowing down of thoracic expansion.

GASSEOUS EXCHANGES

The alveolar surface used for gaseous exchanges is 40 times bigger than the external surface of our body: here the passage of gas goes on continuously (we are interested mainly in the respiratory one: the oxygen O2 and the carbon dioxide CO2. However, all gasses, even pollutants, follow the same rule) from the physiological liquids such as blood, the inside of the cells and the mucous, into the air, and vice versa. In general, the passage of a gas through a membrane depends on both the permeability of such membrane and on the partial pressure of the gas on each side of the membrane. The permeability of both membranes (the alveolar epithelium and the capillary endothelium) that separate the blood from the air that reaches the alveolus is big enough not to limit the diffusion of gasses. Therefore, all of the gaseous exchanges are regulated by different partial pressures of each gas, that occur, each time, on each sides of the alveolar membrane.

▶ **Gaseous exchanges at the level of the pulmonary alveoli**

OXYGEN

In the atmosphere and under normal conditions (pressure and room temperature) the partial pressure of oxygen (pO2) is 21,2 kPa. However, as it enters the lungs, this air mixes in with" already" breathed in air (less oxygerated): the pO2 decreases to 13,5 kPa.

The blood that flows inside the pulmonary capillaries is venous, and the average pO2 is 5,3 kPa. Therefore, whilst this difference in pressure persists, there is oxygen flow, which enters the blood from the alveolar air. Generally, the length of one inhales is sufficient to increase the blood pO2, almost to the level registered inside the alveolus: 13,3 kPa (oxygenated arterial blood).

CARBON DIOXIDE

The partial pressure of carbon dioxide (pCO2) in the atmosphere, under normal conditions (1 atm and 37° C) is 0.04 kPa. However, as it enters the lung, this air mixes with "already breathed" air, full of carbon dioxide: the pCo2 increases to about 5, 2 kPa.

The blood that runs through the pulmonary capillaries is venous, and pCO2 is on average 6, 1 kPa. As a result, whilst this difference in pressure persists, the flow of carbon dioxide moves to the air of the alveolus. Usually the length of one inhalation allows for enough exchanges to lower the pCO2 in the blood almost to the level registered in the alveolus: 5, 3 kPa (oxygenated arterial blood).

▼ **Anatomical relationship between the lungs and the heart. In blue we can see the venous blood; in red, the arterial blood.**

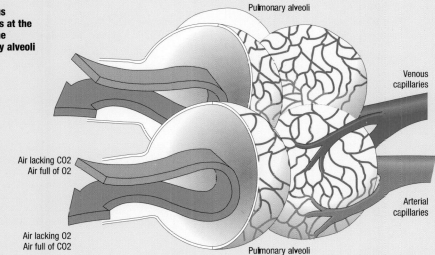

Pulmonary alveoli

Air lacking CO2
Air full of O2

Air lacking O2
Air full of CO2

Venous capillaries

Arterial capillaries

Pulmonary alveoli

PARTIAL PRESSURE OF RESPIRTORY GASES					
GAS	ATMOSPHERIC AIR		ALVEOLAR		VENOUS BLOOD	ARTERIAL BLOOD
	%	kPa	%	kPa	kPa	kPa
O_2	20,94	21,2	14,2	13,5	5,3	13,3
CO_2	0,04	0,04	5,5	5,2	6,1	5,3
N_2	79,02	80,0	80,3	76,4	76,4	76,4
TOTAL	100,00	101,24	100,0	95,1	87,8	95,0

air that is moved by the involuntary impulses originating from the anterior and *posterior pulmonary plexi* of the vagus and of the thoracolumbar ▶116 sympathetic nerves: just like the bronchi, even the nerves, the blood and the lymphatic vessels branch out into the alveoli.

THE PULMONARY CIRCULATION AND THE BRONCHIAL CIRCULATION

The two branches of the pulmonary artery, for example, enter the lungs at the level of the hilum, and following the bronchial branching they keep dividing until they become alveolar arterioles. The thick network of capillaries they create crosses the alveolar walls and flows into the *venules*, which run inside the *interlobular septi*, meeting inside the *venous branches*. These run alongside the course of the bronchi (usually on the opposite side of the arteries) and make up the two *pulmonary veins*. They exit the hilum and meet inside the left atrium of the heart ▶185.

But the blood vessels that enter the lungs constitute 2 different systems: a *functional one* (small circulation), just described, which allows gas to be exchanged between blood and air, and a *nutritional one* (great circulation), which brings nutrients to the cellular structures of the lungs and of the bronchi. It is made up of the bronchial arteries, which originate from the aorta, and that branch out

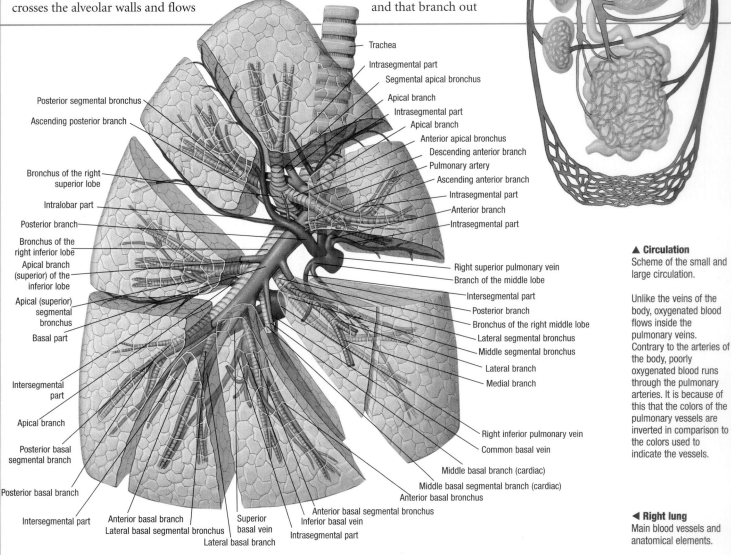

Trachea
Intrasegmental part
Segmental apical bronchus
Apical branch
Intrasegmental part
Apical branch
Anterior apical bronchus
Descending anterior branch
Pulmonary artery
Ascending anterior branch
Intrasegmental part
Anterior branch
Intrasegmental part

Posterior segmental bronchus
Ascending posterior branch
Bronchus of the right superior lobe
Intralobar part
Posterior branch
Bronchus of the right inferior lobe
Apical branch (superior) of the inferior lobe
Apical (superior) segmental bronchus
Basal part
Intersegmental part
Apical branch
Posterior basal segmental branch
Posterior basal branch
Intersegmental part
Anterior basal branch
Lateral basal segmental bronchus
Lateral basal branch
Superior basal vein
Inferior basal vein
Intrasegmental part
Anterior basal segmental bronchus
Anterior basal bronchus
Middle basal segmental branch (cardiac)
Middle basal branch (cardiac)
Common basal vein
Right inferior pulmonary vein
Medial branch
Lateral branch
Middle segmental bronchus
Lateral segmental bronchus
Bronchus of the right middle lobe
Posterior branch
Intersegmental part
Branch of the middle lobe
Right superior pulmonary vein

▲ Circulation
Scheme of the small and large circulation.

Unlike the veins of the body, oxygenated blood flows inside the pulmonary veins. Contrary to the arteries of the body, poorly oxygenated blood runs through the pulmonary arteries. It is because of this that the colors of the pulmonary vessels are inverted in comparison to the colors used to indicate the vessels.

◀ Right lung
Main blood vessels and anatomical elements.

in the bronchi's wall with 2 networks of capillaries (a deep one, for muscles and glands, and a superficial one for the mucosa). In turn, the arteries meet to make up the bronchial veins. As they exit the hilum, they flow into the azygos and the hemiazygos veins. However, the two pulmonary circulatory systems are not completely independent: some bronchial veins flow into the pulmonary veins, and some branches of the pulmonary arteries are connected to those of the bronchial arteries, through small transversal branches, connected at times even to the bronchial veins. Finally the blood that flows inside the alveolar capillaries can flow in both bronchial and pulmonary venules.

THE ALVEOLI

The alveolar wall creates a barrier between air and blood, and it is made up of different cellular layers:
-The **alveolar epithelium** is made of pneumocytes of the 1st and 2nd type, and of macrophagocytes. Particularly, the pneumocytes of the 2nd type have microvilli directed toward the lumen of the alveolus, and they produce surfactant lipoproteins secreted and stratified on the internal surface of the alveolus. The alveolar macrophagocytes ▶178-179, have amoeba like movements, and are located inside the intra alveolar septi, free or inside the epithelium, and in the alveolar lumen: their role is to protect and clean (in fact, they contain carbon phagocytized granules).

-*The basal lamina of the alveolar epithelium*;
-*The basal lamina of the capillary endothelium*;
-*The endothelium of the blood capillary*.
The capillaries are extremely thin: their lumen is 5-6 _m, and only one red blood globule can go through it. The delimiting endothelium is continuous, without any pores or unevenness, while the pericapillary stroma is minimal, made up of both collagen elastic and connective fibers.
In some spots, the basal laminae are fused, while in others, connective fibers and cells separate them: the total thickness of the alveolar wall can vary from 0,2 to 0,7 _m.

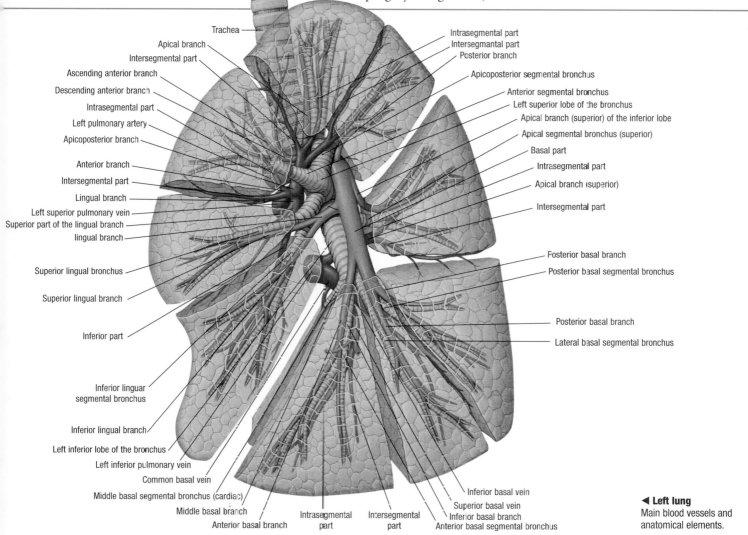

◀ **Left lung**
Main blood vessels and anatomical elements.

173

THE DENTITION

With the term dentition we refer to the teething process (deciduous or milk teeth) in children. The development of teeth begins around the 7th week of pregnancy: when the baby is born, the germs of the deciduous and of the permanent teeth are already present inside the dental sockets, and dentition generally begins around 6-7 months, even if the first tooth may appear later.

As teeth begin to appear, salivation increases and the baby tends to drool a lot; along with drooling, the baby seems to have a strong need to chew things (he puts everything in his mouth and tries to bite it), a higher irritability and agitation, and to cry a lot. Often his sleeping pattern is altered: all these symptoms become apparent few days before each tooth appears. At around 18 months, all the milk teeth have all generally grown out. However, dentition is a normal process: diseases and problems that develop during this time must be attributed to other causes and must not be neglected. Later on, at the age of 5-6, a new dentition takes place. Milk teeth are replaced by permanent teeth. Inside the maxillary and mandibular arches of a child's mouth, besides the already visible 20 milk teeth, we can begin to see the outline of the 20 permanent teeth (which will substitute them), and that of the 8 new teeth that coincide with the milk teeth: they are the premolars (I and II) and the molars (I and II).

This process in which larger teeth grow inside the mouth is strictly connected to the development of the skull and to the consequent and progressive increase of available space inside the mouth. Permanent teeth replace the milk teeth: the root of a milk tooth is in fact slowly reabsorbed because of the pressure exercised by the outline of permanent teeth located on the back, while their crown becomes thinner and falls off. Generally this occurs when underneath there is already a new tooth ready to come out.

▼ **Fetal teeth**
At birth, only a very low percentage of newborns show already formed teeth (usually the two inferior incisor teeth). In this image we can see the development of the dental outlines, which leads to a primary dentition (or "milk" dentition).

▼ **Teeth of a 6 years old baby**
The external side of the dental arches was partially removed in order to show the location of the permanent teeth, not yet surfaced.

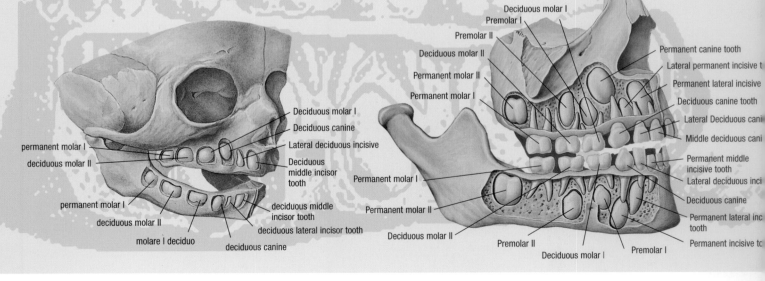

174

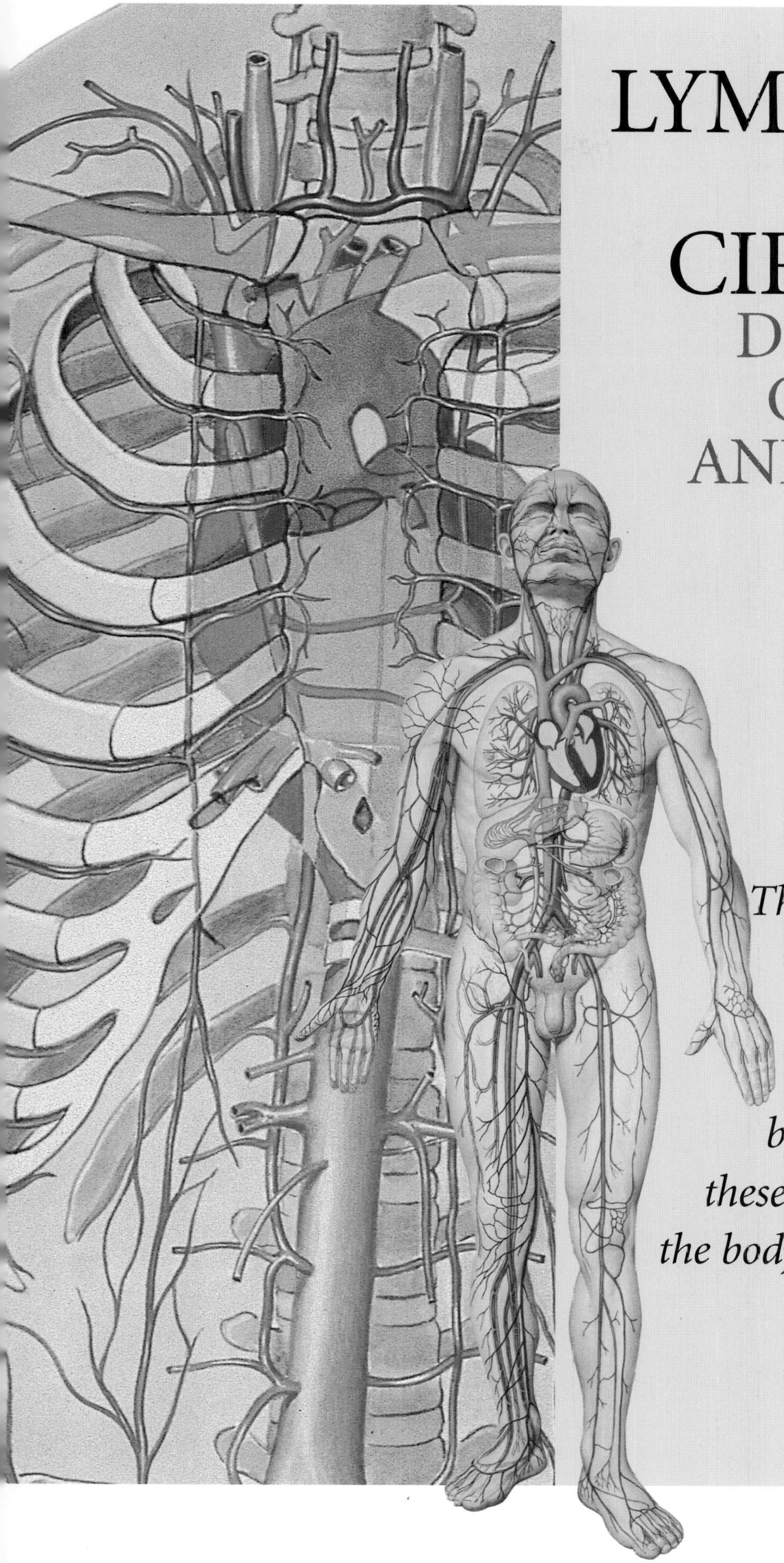

LYMPHATIC AND BLOOD CIRCULATION
DISTRIBUTION, COLLECTION AND PROTECTION

*Thousands of cells circulating
inside two networks
of blood vessels,
and transported
by a fluid rich in nutrients:
these are the systems which protect
the body, distribute vital substances and
collect waste products.*

HEART, ARTERIES, VEINS, CAPILLARIES......AND MORE CAPILLARIES, VESSELS AND LYMPH NODES....THOUSANDS OF CELLS TRANSPORTED BY A LIQUID CONNECTIVE TISSUE (THE PLASMA OR LYMPH) MAKE UP THE CIRCULATORY AND THE LYMPHATIC SYSTEMS, THAT HAVE MANY DIFFERENT FUNCTIONS.

THE CIRCULATORY SYSTEM AND THE LYMPHATIC SYSTEM

In order to survive, our body must carry on a series of activities: it must renew those cells that gradually get destroyed, it must reintegrate the energy used to keep certain organs functioning, it must keep the level of the substances inside the body constant, it must take in new material for its growth and development, and eliminate all of the non recyclable "waste products", that if accumulating, could damage the normal metabolic functions of out body.

Because the processes of obtaining material and energy, and of excreting waste substances are carried on primarily by specialized organs and systems (digestive, respiratory and excretory...), it becomes necessary to have a system that can collect and distribute vital substances to the entire body, and that can direct all the waste products to the excretory organs. The circulatory and the lymphatic systems are responsible for collection and distribution: they are two complex networks of different size canals (the vessels), inside which, fluids filled with cells flow. These are respectively blood and lymph. While the blood system is a closed circuit where blood circulation takes place thanks

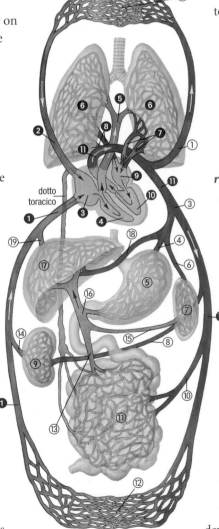

dotto
toracico

to continuous pressure produced by the contractions of a hollow organ (then heart), the lymphatic system is an open circuit, which "passively" drains the interstitial fluid from the tissues: the lymph, pushed by the muscular movements of the body, runs through the lymphatic vessels, from the periphery toward the main ducts that pour into the large veins, at the base of the neck.

The blood system is made up of the heart, and of different size and different functional vessels (*arteries and veins, capillaries*), the lymphatic system has *lymphatic nodules* or *lymph nodes* running along its entire course. They can be very small organs (some of them even microscopic; located at the periphery and called *interrupting lymph nodes*,)), or they can be very large ones, often grouped into lymph centers, where the lymph of big areas of the body meets.

Other important organs connected to the circulatory and lymphatic systems are the *bones, the thymus and the spleen*: while many different types of blood and lymphatic cells (*white blood cells, lymphocytes, red blood cells or erythrocytes, and macro phagocytes*) originate from the bone marrow, the first stage of differentiation of the lymphocytes takes place inside the thymus. Here they acquire specific functional and morphological characteristics (*lymphocytes T*). Finally, the spleen helps regulate the volume of the circulating blood mass. (Here the 'old" red blood cells begin their destructive process), and also developing different defense mechanisms for the body,

▼ Circulatory system
Frontal view of the main vessels
In **blue**: the veins; in **red**: the arteries

▼ Lymphatic system
Frontal view of the main vessels
and nodules.

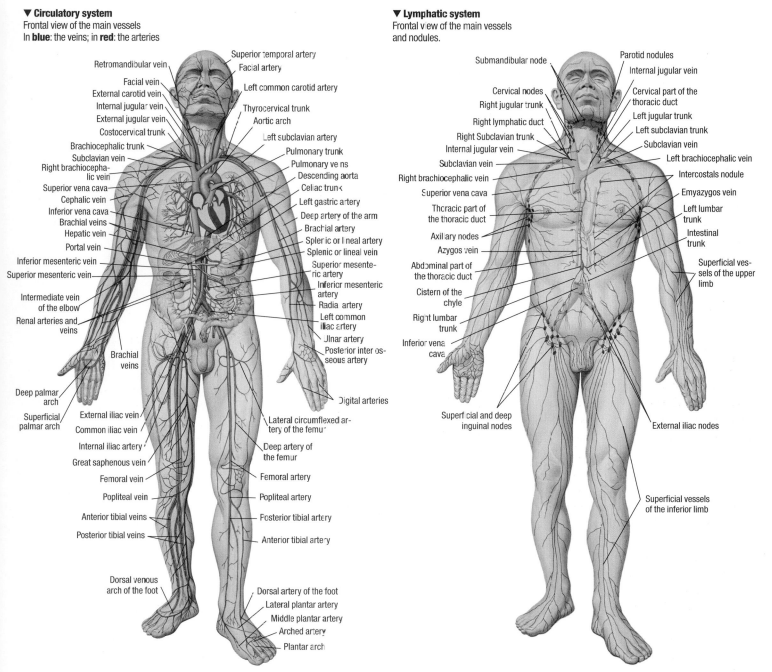

Circulatory system labels:
Retromandibular vein
Facial vein
External carotid vein
Internal jugular vein
External jugular vein
Costocervical trunk
Brachiocephalic trunk
Subclavian vein
Right brachiocepha-lic vein
Superior vena cava
Cephalic vein
Inferior vena cava
Brachial veins
Hepatic vein
Portal vein
Inferior mesenteric vein
Superior mesenteric vein
Intermediate vein of the elbow
Renal arteries and veins
Brachial veins
Deep palmar arch
Superficial palmar arch
External iliac vein
Common iliac vein
Internal iliac artery
Great saphenous vein
Femoral vein
Popliteal vein
Anterior tibial veins
Posterior tibial veins
Dorsal venous arch of the foot

Superior temporal artery
Facial artery
Left common carotid artery
Thyrocervical trunk
Aortic arch
Left subclavian artery
Pulmonary trunk
Pulmonary veins
Descending aorta
Celiac trunk
Left gastric artery
Deep artery of the arm
Brachial artery
Splenic or lineal artery
Splenic or lineal vein
Superior mesenteric artery
Inferior mesenteric artery
Radial artery
Left common iliac artery
Ulnar artery
Posterior interosseous artery
Digital arteries
Lateral circumflexed artery of the femur
Deep artery of the femur
Femoral artery
Popliteal artery
Posterior tibial artery
Anterior tibial artery
Dorsal artery of the foot
Lateral plantar artery
Middle plantar artery
Arched artery
Plantar arch

Lymphatic system labels:
Submandibular node
Cervical nodes
Right jugular trunk
Right lymphatic duct
Right Subclavian trunk
Internal jugular vein
Subclavian vein
Right brachiocephalic vein
Superior vena cava
Thoracic part of the thoracic duct
Axillary nodes
Azygos vein
Abdominal part of the thoracic duct
Cistern of the chyle
Right lumbar trunk
Inferior vena cava
Superficial and deep inguinal nodes

Parotid nodules
Internal jugular vein
Cervical part of the thoracic duct
Left jugular trunk
Left subclavian trunk
Subclavian vein
Left brachiocephalic vein
Intercostals nodule
Emyazygos vein
Left lumbar trunk
Intestinal trunk
Superficial vessels of the upper limb
External iliac nodes
Superficial vessels of the inferior limb

◀ Great circulation and pulmonary circulation
The first one originates inside the left side of the heart, the second one on the right hand side. **Blue**: Non-oxygenated blood; red: oxygenated blood; lilac: the portal system. **Yellow**: the lymphatic ducts.
Pulmonary circulation, or small circulation: ❶ inferior vena cava; ❷ superior vena cava; ❸ right atrium; ❹ right ventricle; ❺ pulmonary trunk; ❻ pulmonary ramifications; ❼ left pulmonary veins; ❽ right pulmonary veins; ❾ left atrium; ❿ left ventricle; ⓫ aorta

by promoting the proliferation and the differentiation of the lymphocytes B.

In fact, the circulatory and the lymphatic systems are not only similar because they both collect and distribute vital substances, but also because they protect the body: in fact, there are many cellular components (macro phagocytes, lymphocytes T and B and platelets), and many substances dispersed inside the blood and the lymph (antibodies, exc. complements) which are necessary to cicatrize wounds, to protect from protozoans, bacterial and viral infections.

Great circulation (or body circulation):
① Common carotid;
② Cerebral ramification;
③ Celiac trunk; ④ Gastric artery;
⑤ Gastric ramifications;
⑥ Splenic or lineal artery;
⑦ Splenic ramifications; ⑧ Renal artery;
⑨ Renal ramifications; ⑩ Mesenteric artery; ⑪ Intestinal ramifications;
⑫ Peripheral ramifications;
⑬ Mesenteric veins; ⑭ Renal vein;
⑮ Splenic or lineal vein; ⑯ Portal vein;
⑰ Common hepatic artery; ⑱ Hepatic ramifications; ⑲ Hepatic veins.

BLOOD AND LYMPH

These are the two fluids that circulate inside our body through blood and lymphatic vessels. They have similar characteristics: they are made of a cellular and a liquid part, which play diversified roles.

THE LIQUID ELEMENTS

If referring to blood it is called plasma, and it is an aqueous solution of proteins, salts and other substances (sugars, fats, urea, amino acids, vitamins and hormones)

The percentage of proteins inside the plasma is about 7%, while the percentage of fat is less than 1%. Inside the lymph, the ratio is the opposite: less protein and more

fats. The concentration of the substances varies depending on the metabolic activity, on the part of the body, on the cellular functionality and on the general conditions…… Through a morphological and chemical analysis we are able to obtain a lot of information on our overall health.

THE CELLULAR ELEMENTS: THE BLOOD

Real cells and parts of them make up 45% of the volume of the circulating blood. Cells are divided into various groups, with different characteristics, functions and origins:

-The leukocytes or white blood cells are

also called granulocytes, since their nucleus is often lobed and colored with dark granules. They all derive from the myeloblasts of the bone marrow and are further distinguished, depending on their histological characteristics, into neutrophil leukocytes (about 65% per mm_), eosinophil leukocytes (about 3% per mm_) and basophile leukocytes (about 1% per mm_). They are able to autonomously move (amoeboid movements) and to phagocytize digested cells and particles, actively participating in defending the organism.

-The monocytes or macro phagocytes or hystiocites (a total of 6% per mm_) have a large spherical nucleus shaped like a horse shoe and they are probably produced inside the endothelial system, which is found

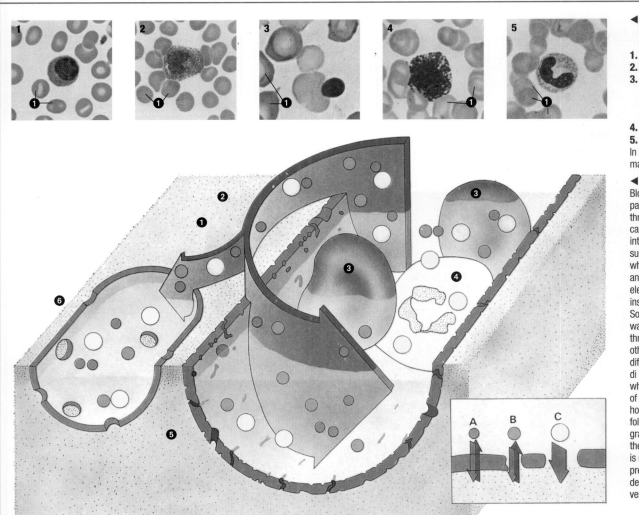

◀ **Cellular elements of the blood and the lymph**
1. Lymphocyte
2. Eosinophile
3. Formation of a erythrocyte: the nucleus (purple) is expelled
4. Basophile
5. Neutrophile
In each image we can see mature erythrocytes ❶.

◀ **Lymph formation**
Blood pressure pushes part of the plasma through the walls of the capillaries (❶) inside the interstitial spaces of the surrounding tissue (❷), while the erythrocytes (❸) and other cellular elements (❹)) remain inside the blood vessels. Some substances such as water (A), pass freely through the endothelium; others, go through different size pores (❺) di varia grandezza (B, C) which interrupt the walls of the blood vessels: however, each substance follows the pressure gradient. The majority of the excess interstitial fluid is reabsorbed where the pressure of the vessels decreases: the lymphatic vessels (❻) drain the rest.

throughout the organism. They can move from one tissue to the next, moving through intercellular spaces, and they can also phagocytize and digest foreign or faulty particles and cells: they play an essential role in "cleaning" and defending the body.
-The lymphocytes (about 25% per mm_) have a very large spherical nucleus and very little cytoplasm. They originate from the lymphoblast found inside the lymphoid tissue and inside the spleen. They carry the immune memory and regulate the massive production of antibodies, thanks to which, the organism is able to eliminate foreign substances (antigens) and aggressive organisms.
The cells which circulate inside the blood, are:
-The erythrocytes or red blood cells, pro-

duced by the bone marrow. After filling up with hemoglobin, they expel the nucleus: they are responsible for the transportation of oxygen and carbon dioxide.
-The platelets, are irregular cytoplasmatic particles, about 2 Ïm long, they detach from the cells of the bone marrow and are involved in the blood coagulating process.

THE CELLULAR ELEMENTS: THE LYMPH

Besides the macro phagocytes, many gangliar lymphocytes (produced by the

lymphatic nodules) are located inside the lymph.
During their stay inside the lymphoid organs (thymus and spleen ▶202) they go through complex changes, and become the main protagonists of the active defense of the organism (lymphocytes T and B).

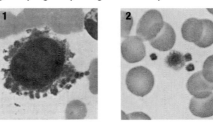

◀ **1. A megakaryocytic produces numerous platelets**
◀ **2. Two sizes platelets**

ORIGIN AND ENDING OF BLOOD AND LYMPHATIC CELLS

The majority of blood cells originate from stem cells located inside the bone marrow: therefore the most "productive" areas correspond to the areas where red bone marrow is abundant (32, 38-39): in an adult, this is found mostly inside cranial bones, vertebrae, ribs, inside the pelvis, the sternum and in the femurs. During infancy, all types of bones are involved, while during the fetal period, even the spleen and the liver are involved as well. Here, through long chains of cellular division and of successive differentiation, basophile white blood cells, neutrophyles and eosinophiles, most of the lymphocytes and numerous monocytes and macrophagocytes, are constantly forming.

Here, even red blood cells and platelets (cellular particles that transport gas and are responsible for coagulation) reach full "maturation". Lymphocytes (abundant inside the lymph) are also produced in the liver, the spleen and in the lymph nodes.

All the blood and lymph node cells have a predetermined lifespan: for example, after going from the bone marrow into the blood circulation, a red blood cell has an average life of about 120 days. During this time, it covers about 15000

km through the entire circulatory system and it usually ends up inside the liver or the spleen, which allows for the recycling of iron needed to make up hemoglobin (180).

If we consider that an average of 4 and _ million red blood cells circulate inside the body of an adult, and that every 120 days new ones are circulating, we are able to understand the efficiency and the speed of the cellular reproductive processes active inside the bone marrow, and the potential damage that this tissue can be exposed to, due to mutagen substances. Even the other cellular components have a set life span: after a relatively brief life (a shorter one if infections or inflammations occur) they are phagocytized by the located inside spleen, the liver and the lymph nodes, digested by their enzymes and "recycled" as raw materials.

▲ **Formation of the lymphatic-blood cells and of the circulating cellular particles**
The totipotential staminal cells ②, develop from the undifferentiated staminal cells found inside the bone marrow ① These produce the two lymphoid and myeloid lines. The first one generates lymphocytes ③ and plasma cells; The second one produces red blood cells ④, megacriocytes that produce platelet ⑤, basophile white blood cells ⑥, neutrophiles ⑦, eosinophiles ⑧ and monocytes ⑨. macrophagocytes originate from the last ones.

◀ **Forming areas of the cellular components of blood and lymph.**
Black dots: the liver
In purple: the spleen
In Yellow: the main lymph nodes

THE TRANSPORTATION OF GASSES AND THE DEFENSES OF THE BODY

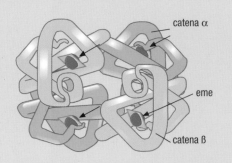

◄ A hemoglobin molecule
The two chains (· and ,) are in two different colors .The arrow indicates the point where they bind with the oxygen.

catena α

eme

catena ß

Both these essential activities of the lymph-circulatory system are based on the particular characteristics of some "special" proteins: the hemoglobin and the antibodies.

The hemoglobin is formed by two couples of different protein chains, and by 4 prosthetic groups called "heme". Each of these chains contains a bivalent iron atom. Hemoglobin is the pigment that gives red blood cells their typical red color. Iron has a very high affinity with oxygen (O_2), and a lesser affinity with carbon dioxide (CO_2): when the partial pressure of the oxygen (ppO_2) is higher than the pressure of carbon dioxide ($ppCO_2$), which happens inside the pulmonary alveoli, the iron of the hemoglobin binds to the O_2, and releases the CO_2 to which it is connected and vice versa, if the $ppCO_2$ is higher than the oxygen's ppO_2, a condition that occurs inside the peripheral capillaries. The iron binds preferably with the Co_2, and it releases O_2, which spreads to the surrounding cells.

Even the antibodies (Á globulin) are made up of 4 protein chains: 2 longer ones and 2 shorter. They are arranged around each other, in a spiral direction, shaped like a Y. Unlike the 4 hemoglobin chains, which differ in pairs only by few amino acids, the anti bodies have a very high variability zone in both the short and the long chains. Such variability is found inside all the chains, in the terminal part of the 'arms" of the Y: this is the zone in which each antibody chemically "recognizes" the specific substance to which it binds, in other words, its antigen.

Thanks to the activity of the lymphocytes T and B, the body produces thousands and thousands of different antibodies, any of which is able to bind, though the arms of the Y, to a specific substance, to a specific protein, to a toxin or to an element of the cellular surface.

The antibodies that appear on the surface of their lymphocytes are also secreted inside the blood circulation. At any moment, about 2 trillion lymphocytes are circulating inside the blood and the lymphatic system, and are ready to produce specific antibodies for the most unusual substances: from the pollen of some plants to the virus that causes influenza, or from bacteria or cat hair….

While the circulating antibodies bind to the antigens, allowing the other defenses of the body to intervene (macro phagocytes phagocytize and digest foreign substance; the complements of the blood destroy cells of hostile organisms), the antibodies found on the surface of the lymphocytes act as an "immune memory", stimulating these cells to produce high volumes of globulin, if and when needed. The processes involved in the specific protection of the body, mediated by the antibodies, are very complex: How the "immune memory" is created, how the body can distinguish its own cells from foreign ones, how the immune system can trigger different responses upon different situations (almost every substance can produce its own antibodies), are only a few of the questions scientists are still trying to answer.

► An antibody
This computer-generated image of an amino acid sequence shows the typical Y shape. In red: the antigen tied to the antibody.

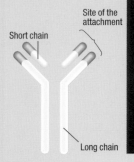

Site of the attachment

Short chain

Long chain

▲ Structural scheme of an antibody

► Specificity for the antigens:
An antibody can specifically interact with::
a. An element of the surface of a specific cell.
di una particolare cellula
b. An area of a specific protein
c. A small part of protein, even if outside its chemical contest.

a   b   c

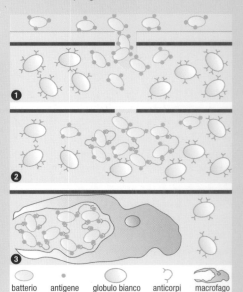

batterio   antigene   globulo bianco   anticorpi   macrofago

▲ Scheme of a defense mediated by antibodies
❶ Bacterial invasion.
❷ Recognition of the antibodies found on the membrane of the lymphocytes B, the triggering of a massive production of antibodies and the agglutination of bacteria done by the free plasmatic antibodies
❸ Intervention of the macrophagocytes and bacterial destruction.

✠ BLOOD TYPES AND TRANSFUSIONS

The protective reaction of the body is also triggered when it comes in close contact with an extraneous tissue, even if human: in fact, people do not only differ in appearance or psychology, but also in their molecular structure. In particular, on the membrane of the red blood cells, we can find glycoproteins and phospholipids commonly found inside the cellular membranes, and molecules that the body can recognize as antigens, and that trigger the production of specific antibodies. The "variations" of these proteins number more than 20, and they are genetically predetermined. However, the most important and the better known ones (relative to blood transfusions), are the variants A and B of the AB0 and the Rh systems. During a transfusion, unless our blood type is similar to that of the donor, our body will produce antibodies for the red blood cells received: by adhering to the "fo-

reign" red blood cells, our antibodies cause the agglutination and the phagocytosis by macrophagocytes, which results in severe problems that can even lead to death. The AB0 system was described, for the first time, in 1900 by the Austrian Karl Landsteiner, and it distinguishes 4 combinations for the presence of antigens on the erythrocytes: A, B, AB, 0. Because each person can only produce antibodies for specific antigens different from theirs, whomever possesses the antigen A can only react to the blood that contains antigens B, and so on; as described in the table. Therefore, whoever belongs to the group 0, is a universal donor, because he possesses erythrocytes without antigens; who belongs to the AB group is a universal receiver, because he possesses all of the antigens and does not produce antibodies.

In the Rh system (from Rhesus, a type of monkey in which this antigen was first discovered), the erythrocytes have and do not have the Rh factor. Similarly to the AB0 system, the presence of the Rh antigen inside the transfused erythrocytes, determines the production of antibodies by those individuals that do not have them (Rh-).

DONOR		RECEIVER	
BLOOD GROUP	ANTIGENS PRESENT	BLOOD GROUP	ANTIBODIES PRESENT
A	A	A	none
B	B	A	anti-B
AB	A and B	A	anti-B
0	none	A	none
A	A	B	anti-A
B	B	B	none
AB	A and B	B	anti-A
0	none	B	none
A	A	AB	none
B	B	AB	none
AB	A and B	AB	none
0	none	AB	none
A	A	0	anti-A
B	B	0	anti-B
AB	A and B	0	anti-A and anti-B
0	none	0	none

✠ VACCINATIONS AND ALLERGIES

The body will always react against foreign bodies, but the efficiency with which it will react is not guaranteed. Therefore, in order to guarantee a safe and efficient reaction to the pathogenic agents of diseases that might not even have a cure yet (viral diseases) or that might have a long and risky course, vaccination is necessary. This technique "trains" our body to recognize dangerous viruses or bacteria, putting it in contact with those same organisms (inactive or dead) or with part of them. Today, thanks to genetic engineering, we are even able to produce artificial organisms with a series of antigens that are characteristic of different pathogens: once injected, these "inexistent bacteria" produce a diversified immune response, which establishes a level of immunity against a wide variety of diseases, without jeopardizing the health of a person (the opposite happens when inactivated pathogens are utilized). However, in some cases, it is necessary to stop the immune reaction: the body overreacts in the presence of even the smallest concentration of a foreign substance, even if it is not particularly dangerous to the body. In this case we say that the body is allergic to that particular substance (allergen). The allergic reaction, that can at times even be fatal (anaphylactic shock), develops in 4 fast primary phases:

a. The allergen penetrates the body through contact, inhalation, ingestion or injection;

b. It binds to the specific antibody found on the surface of some lymphocytes B;

c. This stimulates the immediate production by the lymphocytes of histamine;

d. The histamine triggers the allergic symptoms: depending on the way it has entered the body, the symptoms will range from itching, urticaria, asthma, lachrymal and nasal secretions, fever, headaches…

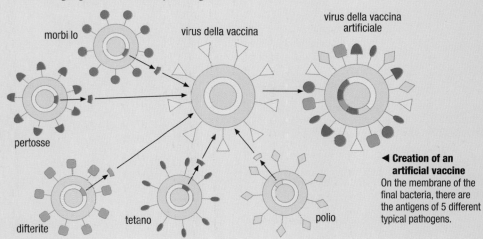

morbi lo
virus della vaccina
virus della vaccina artificiale
pertosse
difterite
tetano
polio

◄ **Creation of an artificial vaccine**
On the membrane of the final bacteria, there are the antigens of 5 different typical pathogens.

ARTERIES, VEINS AND LYMPHATIC VESSELS

They are the "pipes" of the circulatory and the lymphatic systems, inside which flows respectively blood and lymph, bodily fluids made up and containing cells with similar functions.

THE ARTERIES

They are the vessels that transport the blood from the heart in a centrifugal movement: therefore, except for the pulmonary arteries (which transport poorly oxygenated blood from the right ventricle of the heart to the lungs) all of the arteries transport oxygena-

ted blood, easily recognized for its deep red color. The blood inside the arteries is at a very high pressure, sustained and transmitted toward the periphery of the walls of these vessels, which are slightly elastic, and made up primarily of smooth mucous muscle tissue. The wave of pressure that follows each contraction of the heart spreads easily: The arterial walls sli-

ghtly collapse during the passage of the blood, contributing to its movement. The aorta is the biggest artery of all: it has a diameter of about 2,5 cm and it departs from the left ventricle of the heart, and arches toward the back. The large arteries that carry blood to the head (carotids) and to the arms (subclavian arteries subclavicular and brachial arteries) branch out immediately from the aorta. Then the aorta descends in front of the spinal column until it reaches the abdomen, dividing itself in many branches, which nourish the kidneys, the liver, the intestines and the lower limbs. In turn, all of the arteries are further divided into ramifications, whose diameter is progressively smaller, creating therefo-

Right common carotid artery
Brachiocephalic trunk
Transversal artery of the neck
Superficial branch
Dorsal scapular artery
Superior intercostals artery
Superior thoracic artery

Ascending cervical artery
Inferior thyroid artery
Deep cervical artery
Vertebral artery
Tireocervical trunk
Left subclavian artery
Suprascapular artery

Right axillary arteryramo
clavicular branch
Deltoid branch
Acromial branch

Circumflex artery of the scapula
Pectoral branch

Posterior circumflex artery of the humerus
Pericardiacophrenic artery

Posterior circumflex artery of the humerus
Pericardiacophrenic artery

Interior thoracic artery
Inferior vena cava
Posterior intercostals arteries
Right ventricle
Superior phrenic artery
Inferior phrenic artery
Common hepatic artery
Inferior adrenal artery
Right renal artery
Testicular artery
Abdominal part of the aorta
Right common iliac artery
Superior gluteal artery
Lateral sacral artery
Obturator artery
Inferior gluteal artery
Superior iliac circumflex artery
Internal pudendal artery
Ascending branch
Acetabular branch of the middle circumflex artery of the femur
External pudendal arteries
Deep artery of the femur

Aortic arch
Lateral thoracic artery
Descending part of the aorta
Ascending part of the aorta
Posterior circumflex artery of the humerus
Anterior circumflex artery of the humerus
Subscapular artery
Thoracodorsal artery
Brachial artery
Left coronary artery
Deep artery of the arm
Left ventricle
Muscolophrenic artery
Thoracic part of the aorta
Superior epigastric artery
Left gastric artery
Celiac trunk
Lineal artery
Superior mesenteric artery
Lumbar arteries
Inferior mesenteric artery
Middle sacral artery
Ileolumbar artery
• Lumbar branch
• Iliac branch
Inferior epigastric artery
Interior iliac artery
External iliac artery
Superficial epigastric artery
Deep iliac circumflex artery
Pubic branch
Spinal branches
Femoral artery
Inguinal branches
Lateral circumflex artery of the arm
• Ascending branch
• Descending branch
Middle circumflex artery of the femur

▶ **Main arteries of a male body**
Frontal view.

re a thick network of arterioles, which carry oxygenated blood to every part of the body. In turn the arterioles end into a network of arterial capillaries, minuscule vessels, about 1mm long, with a diameter of about 100th of a mm: enough to let a red blood cell pass through very tightly. The exchanges of gasses and substances from the blood to the cells of the body, and vice versa, take place in the wall of these capillaries. The muscular fibers that make up the wall of the arterioles contract and relax independently from the cardiac impulse, and they contribute in regulating the blood flow into the various areas of the body, depending on the need: their activity is regulated not only by the peripheral nervous

system, but also by certain hormones such as the adrenaline produced by the adrenal glands ▶133.

THE VEINS

They are the vessels that transport the blood toward the heart: therefore, except for the pulmonary veins (which transport oxygenated blood from the lungs to the right auricle of the heart), all of the other veins transport non oxygenated blood, whose color is typically dark red. Veins

hold approximately 50-60% of the total blood volume: here the pressure is much lower than the pressure inside the arteries, and the blood flow toward the heart is primarily stimulated by the movements of the muscles and of the arteries, adjacent to the venous vessels. In fact, veins are sectioned off by semi lunar valves, which allow the blood to flow in only one direction (toward the heart), and they have a muscular wall which is much thinner and flexible than that of the arteries. As blood runs through them, even the veins expand and contract. The venous circulation takes place in a sequence of vessels, whose diameter progressively increases: the venous capillaries, which anastomose with

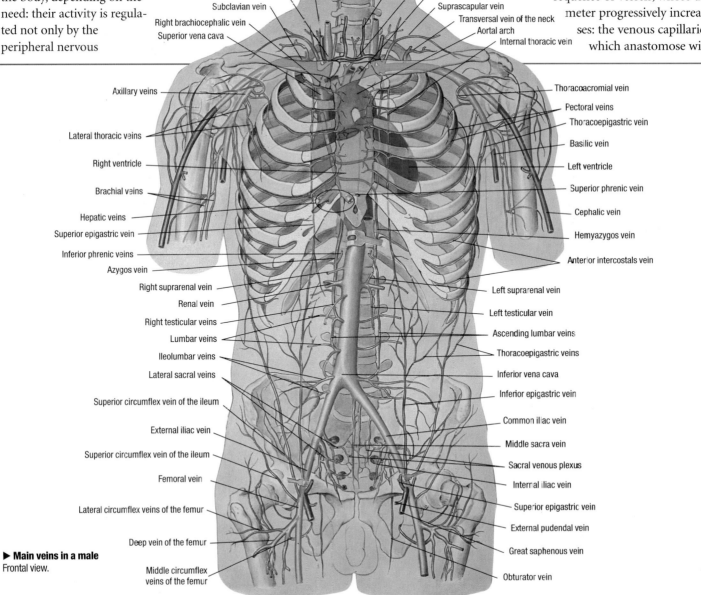

▶ **Main veins in a male**
Frontal view.

183

the arterial capillaries, collect the poorly oxygenated blood which is full of waste products. the arterial capillaries flow inside the venules, which in turn make up increasingly bigger veins, all the way to the main veins, which bring blood to the heart. For example, the vena cava, collects the venous blood arriving from the trunk, the jugular vein collects the blood arriving from the head, and the subclavian vein collects the blood arriving from the arms.

THE LYMPHATIC VESSELS

They transport the lymph from the periphery of the body to the point where the jugular vein and the subclavicular vein meet. Unlike the blood vessels, the lymphatic vessels are not much different in diameter. Also they are not distributed as evenly throughout the body as the

blood vessels are: for example, they are completely missing from the liver, around the renal tubules, inside the septi of the pulmonary alveoli, and so on; however, they are particularly numerous inside the intestine and around the arteries. The lymphatic capillaries, whose diameter is equal to that of the blood vessels, have a much thinner wall, and at times the wall is missing. They start from the blind fundus, draining fluids from the tissues, and then they come together inside the precollectors, short and fragile connecting trunks located between the absorbing peripheral network and the system of out flowing lymphatic pathways. The precollectors are divided into pre and post lymph nodals, depending if they arrive or depart to and from a lymph node. The lymph nodes are

lymphoid organs distributed everywhere inside the lymphatic network: small, medium size, large, grouped, connected to each other, they all play an important role in the maturation process of the lymphocytes ▶178-179. The collectors that depart from the most important lymph nodes, or groups, meet inside the main lymphatic trunks, which pour the lymph inside the blood system, through the thoracic duct. More so than the veins, the lymphatic vessels have a collapsing wall (the muscular tunic is extremely reduced): the lymph is pushed in the direction of the blood flow, only by the movements of those muscles through which the lymphatic vessels run. The semi lunar valves, similar to the venous ones, but more frequently distributed along the lymphatic vessels,

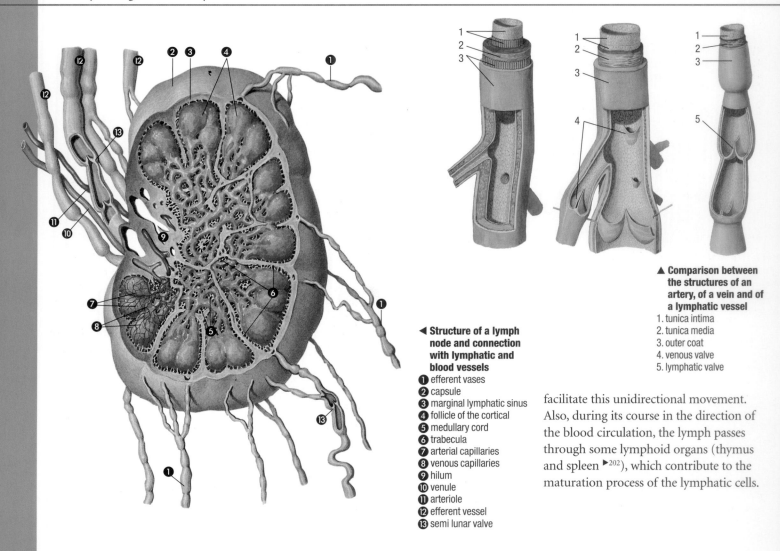

▲ **Comparison between the structures of an artery, of a vein and of a lymphatic vessel**
1. tunica intima
2. tunica media
3. outer coat
4. venous valve
5. lymphatic valve

◀ **Structure of a lymph node and connection with lymphatic and blood vessels**
❶ efferent vases
❷ capsule
❸ marginal lymphatic sinus
❹ follicle of the cortical
❺ medullary cord
❻ trabecula
❼ arterial capillaries
❽ venous capillaries
❾ hilum
❿ venule
⓫ arteriole
⓬ efferent vessel
⓭ semi lunar valve

facilitate this unidirectional movement. Also, during its course in the direction of the blood circulation, the lymph passes through some lymphoid organs (thymus and spleen ▶202), which contribute to the maturation process of the lymphatic cells.

THE HEART

A unique hollow muscle, whose primary role is to continuously pump blood through the body. Within 24 hours, about 4000 liters of blood can go through the heart; during 70 years of one's life, our heart contracts and relaxes on average 2 and _ billion times, and it pumps the 5 and _ liters of blood inside our body, through almost 96000 Km of blood vessels.

Its structure is perfect for this function: the cardiac tissue ▶28-29 has some characteristics, which are typical of both smooth and striated muscular tissues. This is a peculiarity that guarantees such constant, long lasting and difficult activity.

The heart is placed in the middle of a cir-cuit shaped like an 8, and it is structurally and functionally compartmentalized in such a way that it creates a double pump: the left half pushes the oxygenated blood arriving from the pulmonary veins, inside the aorta and to the entire body; the right half pushes the poorly oxygenated blood that arrives from the vena cava, inside the pulmonary arteries and toward the lungs. Each half of the heart is subdivided into 2 chambers: the upper one is called atrium or auricle, and it acts as a collecting point of inflowing blood; the lower one is called ventricle, and it strongly contracts, giving the blood the necessary push to follow, the general or the pulmonary circulation. Because the general circulation is much more developed than the pulmonary one, the left ventricle has a higher muscular mass: its concentration is responsible for the contraction that conveys the apical beat to the entire heart. From each ventricle departs an artery, whose access is regulated by a semilunar valve, which prevents the backflow of blood. The two cardiac halves (right and left) are separated by a septum that on the top is called interatrial septum, and

▼ **The heart**
Frontal view.
❶ Left common carotid.
❷ brachiocephalic trunk
❸ aortic arch
❹ superior vena cava.
❺ right pulmonary artery.
❻ right superior and inferior pulmonary veins.
❼ right atrium.
❽ inferior vena cava.
❾ coronary sulcus.
❿ right ventricle.
⓫ inferior or diaphragm surface.
⓬ left ventricle.
⓭ coronary sinus
⓮ left atrium.
⓯ left superior and inferior pulmonary veins
⓰ left pulmonary artery.
⓱ arterial ligament
⓲ aorta
⓳ left subclavian aorta.

base

apice

▶ **Section of the heart**
right frontal view

superior vena cava
terminal crest
sinus of the venae cava
intervenous tubercle
oval fossa
opening of the inferior vena cava
inferior vena cava
valve of the inferior vena cava
Valve of the inferior vena cava
opening of the coronary sinus
tendineous cords
posterior papillar muscle
trabeculae
opening of the superior vena cava
aorta
right auricle
limb of the oval fossa
foramens of the minimal veins
pectinate muscles
right atrium
cardiac vessels
right atrio-ventricular valve.
• posterior cuspid.
• septal cuspid
• anterior cuspid.
myocardium
anterior papillary muscle
papillary septal muscles
right ventricle
serous visceral lamina of the pericardium
Apex

▶ **Section of the heart**
Left frontal view

Oval fossa
Opening of the inferior vena cava
Right atrium
Pectinate muscles
Valve of the coronary sinus
Coronary sinus
Right atrioventrivular valve.
Septal cuspid •
Posterior cuspid.
Tendineous cords
Papillary muscles
Right ventricle
Myocardium
Trabeculae
Right superior pulmonary vein
Opening of the right inferior pulmonary vein
Left atrium
Left pulmonary vein
Opening of the left pulmonary vein
Interatrial septum
Cardiac vessels
Left atrioventricular valve
Interventricular septum
• Membranous part
• Muscular part
Tendineous cords
Papillary muscles
Left ventricle
Myocardium
Trrabeculae
Endocardium
Epicardium

on the bottom is called interventricular septum. The atrium and ventricle, are then separated by an atrioventricular valve: a membranous septum anchored to non elastic tendinous cords, which keep it in position and that branch out of the papillary muscles, made up of an extension of the ventricular walls. Ventricles and papillary muscles contract simultaneously. This causes the closure of the valves, and it prevents the backflow of blood from one chamber to the next.

THE CARDIAC CYCLE

The cardiac cycle is the repeated sequence of events, which are characteristic of the activity of the heart. Each beat, corresponds to a precise sequence of contraction phases (systole) or relaxation phases (diastole) in the four parts of the cardiac muscle:

1. Once relaxed, the atria fill up with blood.
2. The atrial pressure increases, as they get fuller; this triggers the opening of the atrioventricular valves;
3. The atria contract, triggering the sudden filling of the ventricles;
4. Even the ventricles begin to contract: this triggers the closing of the atrioventricular valves;
5. The pressure inside the ventricles drastically increases above the existing blood pressure inside the arteries;
6. This causes the semi lunar valves to open: the blood flows into the arteries from the ventricles that empty out and relax.
7. The others, already relaxed, are re-filling with blood: and the cycle continues.

The contraction of the cardiac muscular cells is spontaneous: even if all the nerves that reach the heart were interrupted, it would continue to contract! The senoatrial nodule, a group of cardiac cells placed in the superior area of the right atrium, which "initiate" the cardiac cycle, regulates the coordinated and progressive contraction of all the cells. They produce an "excitation wave" that spreads inside both atria, triggering their simultaneous contraction; once the atrioventricular nodule, a group of cells of the right atrial pavement, is reached, this wave fades enough to allow the atria to contract completely, before the ventricles begin to contract.

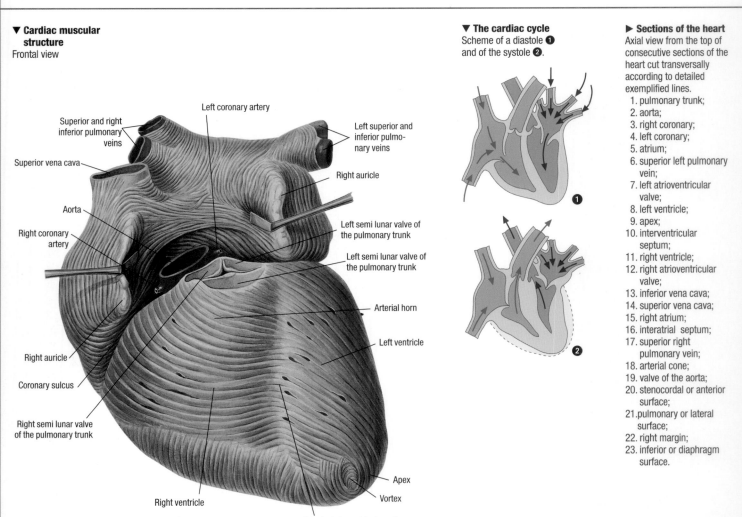

▼ **Cardiac muscular structure**
Frontal view

Superior and right inferior pulmonary veins
Left coronary artery
Superior vena cava
Left superior and inferior pulmonary veins
Aorta
Right auricle
Right coronary artery
Left semi lunar valve of the pulmonary trunk
Left semi lunar valve of the pulmonary trunk
Arterial horn
Left ventricle
Right auricle
Coronary sulcus
Right semi lunar valve of the pulmonary trunk
Right ventricle
Apex
Vortex
Anterior interventricular sulcus

▼ **The cardiac cycle**
Scheme of a diastole ❶ and of the systole ❷.

❶
❷

▶ **Sections of the heart**
Axial view from the top of consecutive sections of the heart cut transversally according to detailed exemplified lines.
1. pulmonary trunk;
2. aorta;
3. right coronary;
4. left coronary;
5. atrium;
6. superior left pulmonary vein;
7. left atrioventricular valve;
8. left ventricle;
9. apex;
10. interventricular septum;
11. right ventricle;
12. right atrioventricular valve;
13. inferior vena cava;
14. superior vena cava;
15. right atrium;
16. interatrial septum;
17. superior right pulmonary vein;
18. arterial cone;
19. valve of the aorta;
20. stenocordal or anterior surface;
21. pulmonary or lateral surface;
22. right margin;
23. inferior or diaphragm surface.

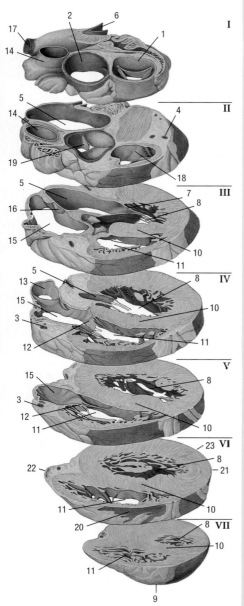

BLOOD PRESSURE

The blood pressure inside the vessels in which it is pumped, is not the same throughout the entire circulatory system, and can be extremely high at the beginning of the arteries, (right outside the ventricles), and very low inside the capillaries. Also, this pressure varies depending on the cardiac cycle: it is at its maximum during the systole (systolic pressure or "maximum") and at the minimum during the diastole (diastolic pressure or "minima"): the difference between these two values is called differential pressure or pulse. Therefore, the value of the blood pressure is proportional to the force conveyed by the cardiac systole, and it depends also on other factors such as the elasticity of the arteries, their size, the total quantity of blood present, and its composition. The condition of the arteries (their elasticity and their size) is the most important factor that influences blood pressure: it is easy to sense that if the arteries were completely rigid (like a metal pipe), the pressure would rapidly increase to a maximum level, during the systole, and then would suddenly drop during the diastole, just like when we open and close a faucet. It is also obvious that when the lumen of a vessel shrinks (due to a pathological process, or to the normal ramification of an artery) the "upstream" pressure is higher than that "downstream" pressure. As a result, the pressure measured inside the venous system is much lower than that present inside the arterial system.

In addition, the quantity of circulating blood is very important (volemia): in fact, on the same extension of the circulatory network, the higher the volume of the liquids, the higher the pressure; the lower the volume, the lower the pressure. Therefore, if the constriction of the blood vessels is always kept under control by the involuntary nervous system, the volemia is constantly kept under control by the endocrine system, which regulates the secretion and the excretion of water ▶212.

In particular, in the case of a hemorrhage (reduction of the volemia), the renal cells react by producing an enzyme that elaborates the "maturation" of angiotensine, a hormone that acts as both a vasoconstrictor and as a stimulus to secrete aldosterone ▶128, that causes the renal reabsorption of salt and water.

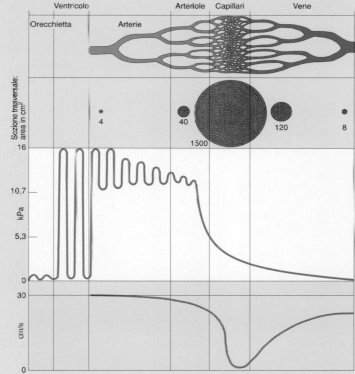

◄ **Blood pressure and blood velocity**
Size of the blood vessels; blood pressure and blood velocity variations at different levels of the general circulation.

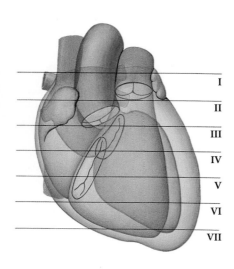

CIRCULATION INSIDE THE SKULL

The arteries and veins of the skull and of the entire body run parallel to each other. Often, the active movements of the arteries contribute to the reflux of blood into the adjacent veins. Arteries and veins often take the name of the areas into which they branch. Let's take a look at the main vessels of the circulation inside the skull.

THE ARTERIES

Brachiocephalic artery or anonymous: it is the biggest branch of the aortic arch. It moves upward in an oblique fashion, on the back and toward the right, passing first in front of the trachea, and then on its right side. It is divided into the right common carotid arteries (external and internal) and into the right subclavian artery. From the carotids, which distributes to the head and the neck, the following arteries depart: the superior thyroid artery, which branches into the neck, and the ascending pharyngeal artery, a long, thin artery which branches into the high pharynx and reaches the meninges, the lingual and the facial artery, which, among other things, makes up the occipital artery, the posterior auricular arteries, the temporal superficial arteries, the maxillary artery (which represents the most voluminous branch of the two terminals of the external carotid), the ophthalmic, the lachrymal, the etmoidal, the corioid, the cerebral and many other arteries. The left subclavian artery originates directly from the aorta, and it branches at the level of the trunk and of the arms, forming the vertebral artery, the internal thoracic (or mammary) artery, the transverse artery of the neck, the thyrocervical artery, and the costocervical trunks (short and voluminous arteries). The axillary artery extends from Rib I to the greater pectoral muscle, where it becomes the brachial artery, which runs through the arm and where it originates many collateral branches ▶193-194. In addition, even the thoracic and the subscapular arteries originate from here.

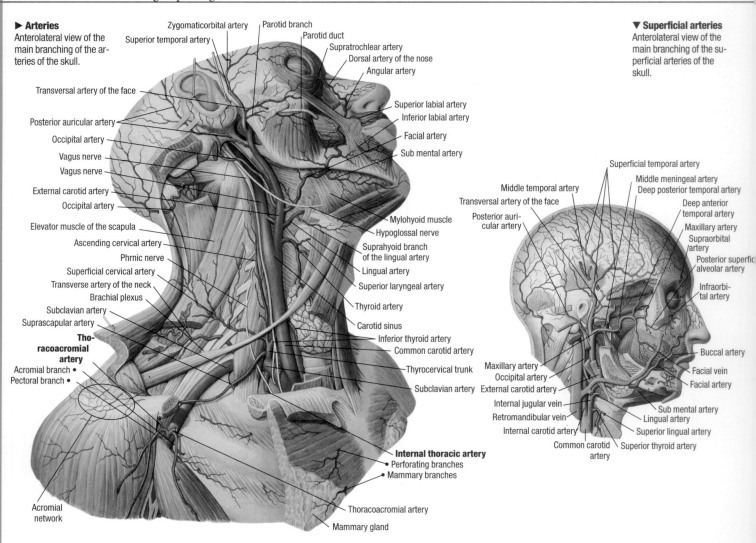

▶ Arteries
Anterolateral view of the main branching of the arteries of the skull.

Zygomaticorbital artery
Superior temporal artery
Parotid branch
Parotid duct
Supratrochlear artery
Dorsal artery of the nose
Angular artery
Transversal artery of the face
Posterior auricular artery
Occipital artery
Vagus nerve
Vagus nerve
External carotid artery
Occipital artery
Elevator muscle of the scapula
Ascending cervical artery
Phrnic nerve
Superficial cervical artery
Transverse artery of the neck
Brachial plexus
Subclavian artery
Suprascapular artery
Thoracoacromial artery
Acromial branch •
Pectoral branch •

Superior labial artery
Inferior labial artery
Facial artery
Sub mental artery
Mylohyoid muscle
Hypoglossal nerve
Suprahyoid branch of the lingual artery
Lingual artery
Superior laryngeal artery
Thyroid artery
Carotid sinus
Inferior thyroid artery
Common carotid artery
Thyrocervical trunk
Subclavian artery

Internal thoracic artery
• Perforating branches
• Mammary branches

Acromial network
Thoracoacromial artery
Mammary gland

▼ Superficial arteries
Anterolateral view of the main branching of the superficial arteries of the skull.

Middle temporal artery
Transversal artery of the face
Posterior auricular artery
Superficial temporal artery
Middle meningeal artery
Deep posterior temporal artery
Deep anterior temporal artery
Maxillary artery
Supraorbital artery
Posterior superficial alveolar artery
Infraorbital artery
Maxillary artery
Occipital artery
External carotid artery
Internal jugular vein
Retromandibular vein
Internal carotid artery
Common carotid artery
Buccal artery
Facial vein
Facial artery
Sub mental artery
Lingual artery
Superior lingual artery
Superior thyroid artery

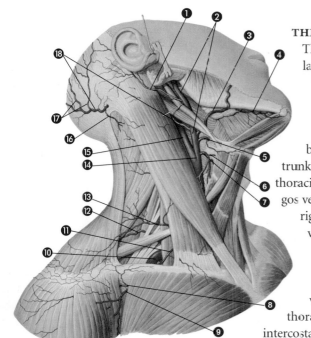

THE VEINS

The superior vena cava is a large vein that carries the blood that flows from the head, from the neck and from the upper limbs, collected by the right and left (anonymous) brachiocephalic venous trunks, and also the blood from the thoracic walls, collected by the azygos vein ▶190-191, all the way into the right atrium of the heart. Each venous trunk is the result of a confluence of an internal jugular vein and of a subclavian vein. Each trunk receives many other veins (thyroid, thoracic, phrenic, vertebral and intercostals). The venous system is also made up of numerous plexi and sinuses. The plexi are formations of anastomosed vessels and cisterns, which collect the venous blood that arrives from different parts of the body and flows toward larger veins (jugular, azygos, hemiazygos, lumbar). The sinuses are canals with a triangular, circular or semicircular sections, and with irregular lacunae (blood lakes and venous lacunae), which run into the thickness of the encephalic dura mater ▶106.

The main plexi are the posterior (external and internal vertebral plexi and the pharyngeal plexus): the cerebral and cerebellar veins, the veins of the bridge and of the medulla oblongata veins, flow into the sinuses, which are named after the cranial bones underneath which they are located.

▲ **Arteries of the head**
Right lateral view.
❶ posterior auricular artery.
❷ external carotid artery.
❸ facial artery
❹ submental artery
❺ lingual artery
❻ superior laryngeal artery.
❼ superior thyroid artery.
❽ acromial branch of the thoracoacromial artery
❾ deltoid branch of the thoracoacromial artery
❿ subclavian artery
⓫ suprascapular artery
⓬ transverse artery of the neck
⓭ superficial cervical artery
⓮ bifurcation of the carotid
⓯ internal carotid artery
⓰ descending branch of the occipital artery
⓱ occipital branches of the occipital artery
⓲ occipital artery

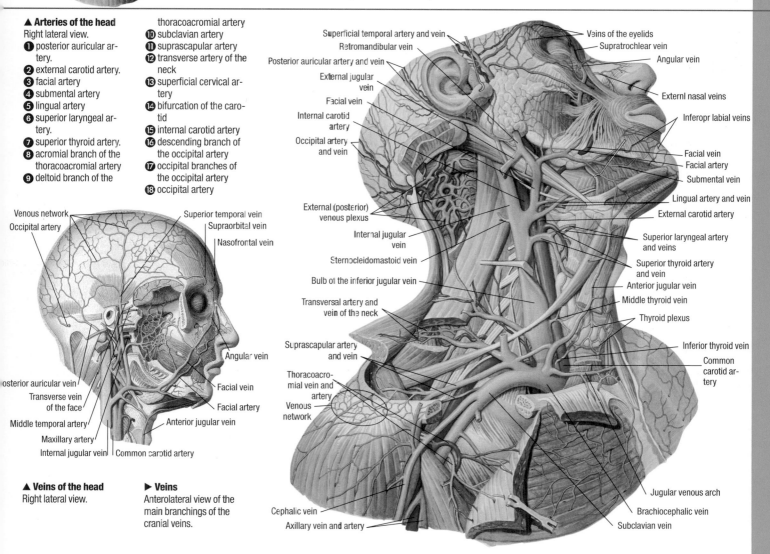

▲ **Veins of the head**
Right lateral view.

▶ **Veins**
Anterolateral view of the main branchings of the cranial veins.

CIRCULATION INSIDE THE THORAX AND THE ABDOMEN

The largest and most important vessels of the circulatory system are found in the thoraco-abdominal area: they are the vessels in which pulmonary circulation takes place (responsible for the oxygenation of the blood), those that connect the main internal organs (liver, kidneys, intestine, pancreas, spleen) to the heart and those from which all of the other vessels branch out into the cranium ▶188, the upper limbs ▶193 and the lower limbs ▶195.

Let's take a look at the main ones:

PULMONARY CIRCULATION

It is made up of arteries that transport venous blood (non oxygenated) and veins, which transport arterial blood (oxygenated). The pulmonary trunk brings venous blood from the right ventricle to the lungs: it is 5 cm long, and it divides into the right and left pulmonary arteries, which branch out to the respective lung, running parallel to the pulmo-

nary veins and to the main bronchus. Once it becomes oxygenated inside the pulmonary capillaries, the blood goes through the pulmonary veins (two per lung, an inferior and a superior), that carry it to the left atrium of the heart.

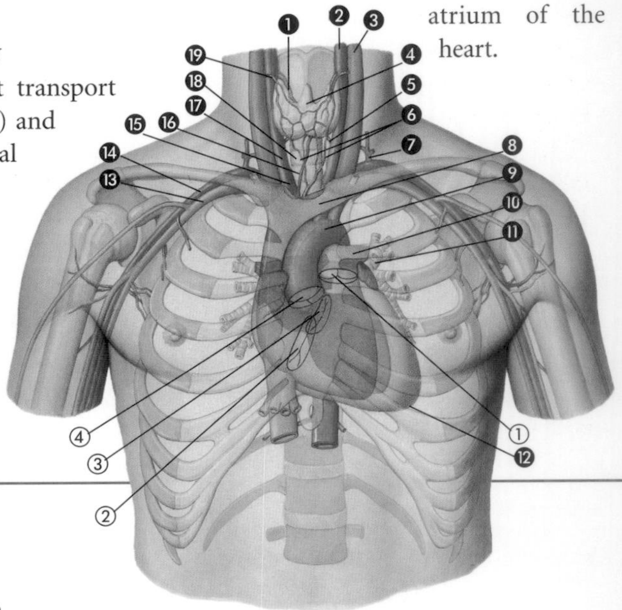

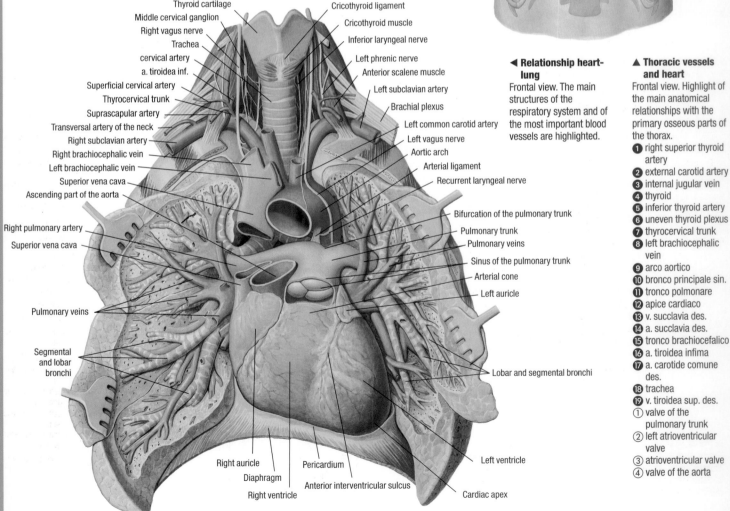

◄ Relationship heart-lung
Frontal view. The main structures of the respiratory system and of the most important blood vessels are highlighted.

▲ Thoracic vessels and heart
Frontal view. Highlight of the main anatomical relationships with the primary osseous parts of the thorax.
❶ right superior thyroid artery
❷ external carotid artery
❸ internal jugular vein
❹ thyroid
❺ inferior thyroid artery
❻ uneven thyroid plexus
❼ thyrocervical trunk
❽ left brachiocephalic vein
❾ arco aortico
❿ bronco principale sin.
⓫ tronco polmonare
⓬ apice cardiaco
⓭ v. succlavia des.
⓮ a. succlavia des.
⓯ tronco brachiocefalico
⓰ a. tiroidea infima
⓱ a. carotide comune des.
⓲ trachea
⓳ v. tiroidea sup. des.
① valve of the pulmonary trunk
② left atrioventricular valve
③ atrioventricular valve
④ valve of the aorta

Labels on lower figure:
Thyroid cartilage
Middle cervical ganglion
Right vagus nerve
Trachea
cervical artery
a. tiroidea inf.
Superficial cervical artery
Thyrocervical trunk
Suprascapular artery
Transversal artery of the neck
Right subclavian artery
Right brachiocephalic vein
Left brachiocephalic vein
Superior vena cava
Ascending part of the aorta
Right pulmonary artery
Superior vena cava
Pulmonary veins
Segmental and lobar bronchi
Right auricle
Diaphragm
Right ventricle

Cricothyroid ligament
Cricothyroid muscle
Inferior laryngeal nerve
Left phrenic nerve
Anterior scalene muscle
Left subclavian artery
Brachial plexus
Left common carotid artery
Left vagus nerve
Aortic arch
Arterial ligament
Recurrent laryngeal nerve
Bifurcation of the pulmonary trunk
Pulmonary trunk
Pulmonary veins
Sinus of the pulmonary trunk
Arterial cone
Left auricle
Lobar and segmental bronchi
Pericardium
Anterior interventricular sulcus
Left ventricle
Cardiac apex

THE SYSTEM OF THE PORTAL VEIN

The portal vein is a large venous trunk that carries all the blood arriving from the sub diaphragmatic portion of the digestive tube, from the spleen, from the pancreas and from the gallbladder, to the liver. This occurs thanks to the supply of the right and left gastric veins, of the cystic veins and of the mesenteric and lineal veins.

All of the ramifications of the venous rete mirabile originate from the portal vein. The rete mirabile crosses the liver ▶152-153 and then flows into the vena cava.

OTHER IMPORTANT VEINS

Besides the superior vena cava, from which the brachiocephalic veins depart (toward the upper limbs), the jugular (toward the neck and the head) and the cardiac (toward the heart), we must remember other important veins such as:

- The azygos vein, which collects the back-flow of blood from the thoracic walls, through the confluence of the thoracic veins, the hemiazygos, the bronchial, the intercostals, the esophageal and superior phrenic veins;

- The subclavian vein, which starts as a direct continuation of the axillary, originating the brachiocephalic venous trunk, by connecting to the internal jugular.

- The inferior vena cava, the biggest vein in the body. All of the venous vessels of the subdiaphragmatic areas meet here: the common iliac veins of the inferior limbs, the lumbar veins and the inferior phrenic veins (called parietal), the renal, surrenal, iliolumbar, obturator, gluteal, pudendal, spermatic and hepatic (called visceral) veins. The number of plexi is very high.

THE ARTERIES

The aorta is without a doubt the largest: it runs through the entire thorax, descending into the abdomen at the height of the 4th lumbar. Here it divides, and forms the right and left iliac artery which extend inside the inferior limbs. All of the other main arteries of the body originate also from the aorta. They are the arteries that enter the neck and the head (carotid and subclavian), the upper limbs (brachiocephalic) and the abdominal organs (coronary, mesenteric, renal, hepatic, gastric, lineal, genital, costal, lumbar and sacral).

THE VISCERAL VESSELS

These are the vessels that wet and drain the

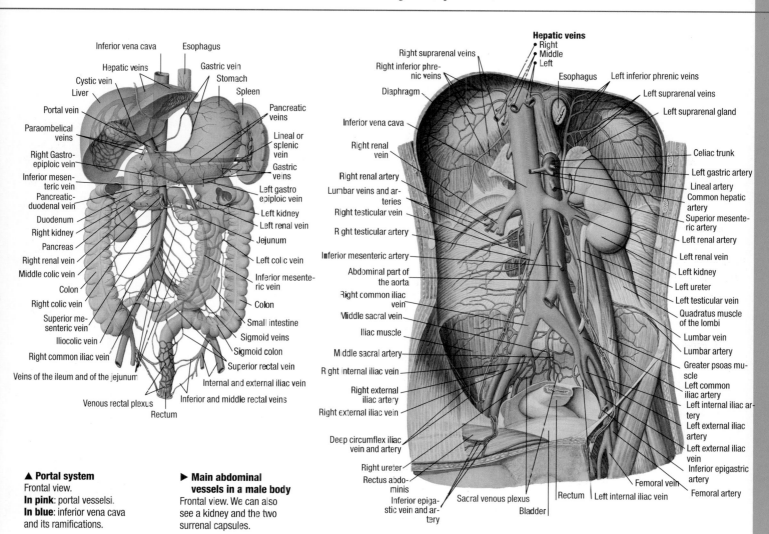

▲ Portal system
Frontal view.
In pink: portal vesselsi.
In blue: inferior vena cava and its ramifications.

▶ Main abdominal vessels in a male body
Frontal view. We can also see a kidney and the two surrenal capsules.

191

abdominal organs. They are aortal ramifications and veins that meet inside the inferior vena cava. They take on the following names:

-Celiac trunk: this is a big branch of the aorta, which divides itself into the left gastric artery, the hepatic artery and the lineal artery. It serves the lower extremity of the esophagus, the stomach, the duodenum, the pancreas, the liver and the spleen;

-The superior mesenteric artery: this vascularizes the small intestine, the right half of the large intestine, the head of the pancreas, the duodenum, the jejunum and the ileum;

-The inferior mesenteric artery: it vascularizes the left portion of the transverse colon, the descending colon, the iliopelvic colon and the rectum;

-The branches of the abdominal aorta: they vascularize the adrenal glands, the kidneys and the genitals and they originate the lumbar, sacral and iliac arteries;

-The renal vein: the middle and inferior surrenal veins, the ureteric and the internal spermatic veins (in the male) flow here.

-The hepatic veins: these are 15-20 veins that drain the backflow of blood from the liver. They are divided into major or trunk ►2-3, and minor ►10-15 and they do not have any valve.

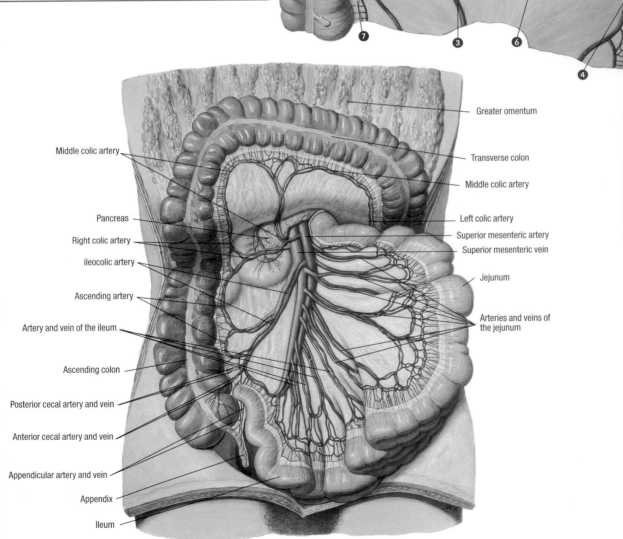

▲ Intestinal circulation

The blood vessels reach each ansa of the intestine, crossing the peritoneal membranes.
❶ right ansa of the colon
❷ transverse colon
❸ middle colic artery
❹ left colic artery.
❺ left ansa of the colon
❻ transverse mesocolon
❼ visceral peritoneum

Greater omentum

Transverse colon

Middle colic artery

Middle colic artery

Left colic artery

Superior mesenteric artery

Superior mesenteric vein

Pancreas

Right colic artery

ileocolic artery

Ascending artery

Artery and vein of the ileum

Jejunum

Arteries and veins of the jejunum

Ascending colon

Posterior cecal artery and vein

Anterior cecal artery and vein

Appendicular artery and vein

Appendix

Ileum

◄ Visceral vessels
Frontal view

CIRCULATION INSIDE THE UPPER LIMBS

All the vessels, in which the circulation of the upper limbs takes place, derive from four main vessels: the right and left subclavian artery and the right and left subclavian vein.

THE ARTERIES

The right and left subclavian arteries bring nutrients and oxygen to the bones, the muscles, the nerves, and to the joints of the upper limbs, and are divided into many important ramifications, whose number and origin can vary. Among the most constant ones there are the thyrocervical trunks, from which, among other things, the transverse arteries of the scapula (or superior scapular) originate: after reaching and going over the superior margin of the scapula, they distribute into the cutis of the acromion

Each subclavian artery continues on into the axillary artery, where it originates the following artery of the arm:

-The thoracoacromial artery: this artery further divides into the pectoral, the acromial, the deltoid and the clavicular branches, which supply those areas of the body from which they take their name;

-The lateral thoracic artery: this artery spreads into the pectoral muscles, into the anterior serratus muscle, the axillary lymph nodes and into the mammary glands;

-The subscapular artery: this is the most voluminous branch of the axillary artery. It originates the thoracodorsal and scapular circumflex arteries, wetting the triceps muscles of the arm, the small round muscle, the large round, and the upper axillary margin of the scapula;

- The posterior circumflex artery of the humerus: this supplies the head of the humerus and the scapulo-humeral joint.

In turn, each axillary artery continues on into the humeral or brachial artery, whose main branches are:

-The deep artery of the arm: it originates the radial collateral and the middle collate-

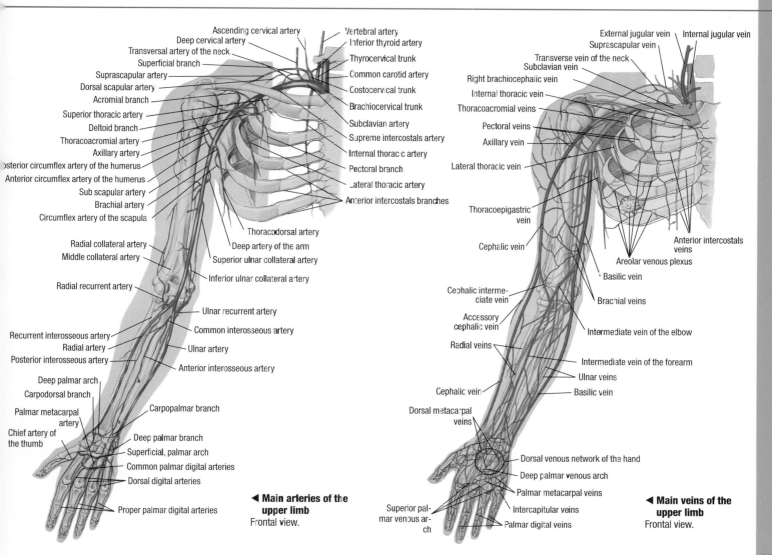

◄ Main arteries of the upper limb
Frontal view.

◄ Main veins of the upper limb
Frontal view.

ral arteries. Among other things it also wets the articulation of the elbow; -The superior and inferior ulnar collateral arteries, which are the main suppliers to the articulation of the elbow;

-The ulnar and radial arteries, which branch into the hand: when we place a finger inside the wrist we feel the movement of the radial artery.

THE VEINS

Generally they run parallel to the arteries: the blood drained from the numerous vessels of the hand, flows into the cephalic vein, the accessory cephalic and the basilic veins (deep and superficial). It also flows inside the brachial veins, some of which flow inside the cephalic vein, above the elbow. They make up the main venous vessels, which drain the forearm and the arm.

Above them, other brachial veins connect to the basilic vein creating the axillary vein. By Rib I, the cephalic and axillary veins of the arm, and the thoracodorsal vein of the thoraco-axillary area, converge and form

the subclavian vein. In turn, the subclavian vein, along with the internal jugular arriving from the neck and the over scapular, flows into brachiocephalic vein, one of the big branches of the superior vena cava.

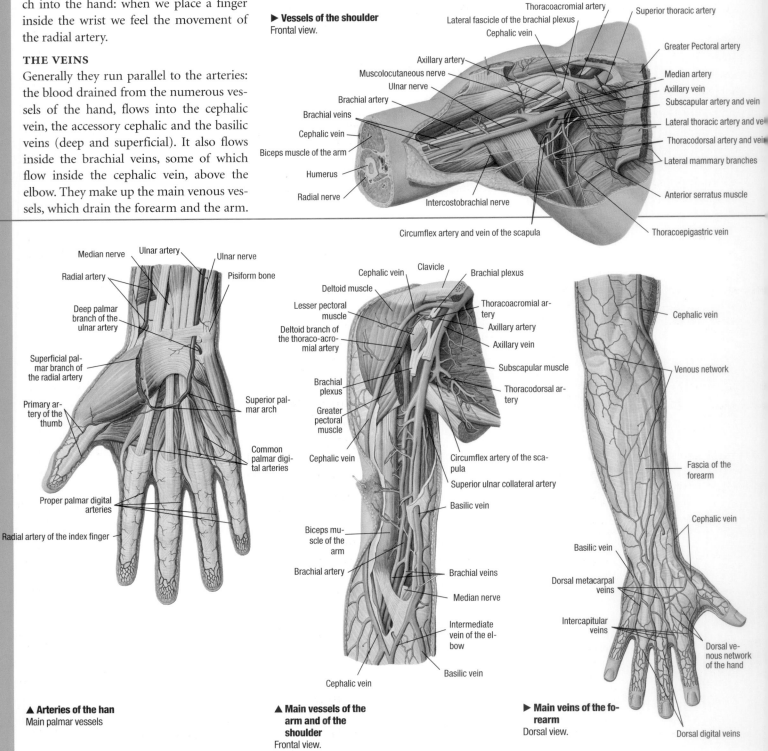

▶ **Vessels of the shoulder**
Frontal view.

▲ **Arteries of the han**
Main palmar vessels

▲ **Main vessels of the arm and of the shoulder**
Frontal view.

▶ **Main veins of the forearm**
Dorsal view.

CIRCULATION IN THE LOWER LIMBS

All the blood vessels of the lower limbs derive from the main abdominal vessels: the arteries of the abdominal aorta, whose branches are the right and left common iliac arteries (external and internal), and the veins of the inferior vena cava, which branches out inside the right and left common iliac veins (external and internal).

THE ARTERIES

The left and right common iliac arteries (external and internal) and their numerous branches feed the pelvis, the pelvic viscera (the reproductive organs, the bladder and the other elements of the urinary apparatus, the rectus intestine), the bones, the joints, the nerves and the muscles of the lower abdomen, the bottom of the pelvis and the lower limbs. The main arterial branches inside then pelvic area are the superior bladder arteries, the vesicle-deferential artery (in the male), or the uterine artery (in the female), the middle rectal artery, the vaginal artery (both in males and females), from which the following arteries originate: the obturator, the internal pudendal and the inferior gluteal or sciatic arteries. The main blood vessels that feed the area of the pelvis and of the thigh, are made up of muscular branches of the inferior gluteal artery, which supply the gluteus maximus, the coccygeal muscles, the quadratus femoris and the greater adductor; these are the iliolumbar arteries, the lateral sacral, the superior gluteal, the femoral, the circumflex and the perforating arteries. The main arteries of the leg are the popliteal artery, which supplies blood to all the posterior muscles and to the knee joints, and the anterior tibialis, which continues into the dorsal artery of the foot (or pedidia), which gives rise to the minor vessels of the foot.

THE VEINS

The venous network, which drains the foot,

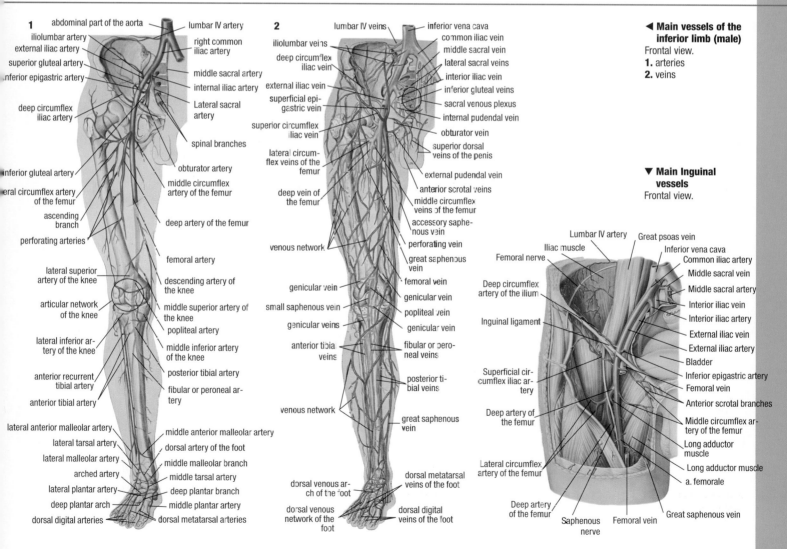

1
- abdominal part of the aorta
- iliolumbar artery
- external iliac artery
- superior gluteal artery
- inferior epigastric artery
- deep circumflex iliac artery
- inferior gluteal artery
- lateral circumflex artery of the femur
- ascending branch
- perforating arteries
- lateral superior artery of the knee
- articular network of the knee
- lateral inferior artery of the knee
- anterior recurrent tibial artery
- anterior tibial artery
- lateral anterior malleolar artery
- lateral tarsal artery
- lateral malleolar artery
- arched artery
- lateral plantar artery
- deep plantar arch
- dorsal digital arteries

- lumbar IV artery
- right common iliac artery
- middle sacral artery
- internal iliac artery
- Lateral sacral artery
- spinal branches
- obturator artery
- middle circumflex artery of the femur
- deep artery of the femur
- femoral artery
- descending artery of the knee
- middle superior artery of the knee
- popliteal artery
- middle inferior artery of the knee
- posterior tibial artery
- fibular or peroneal artery
- middle anterior malleolar artery
- dorsal artery of the foot
- middle malleolar branch
- middle tarsal artery
- deep plantar branch
- middle plantar artery
- dorsal metatarsal arteries

2
- lumbar IV veins
- iliolumbar veins
- deep circumflex iliac vein
- external iliac vein
- superficial epigastric vein
- superior circumflex iliac vein
- lateral circumflex veins of the femur
- deep vein of the femur
- venous network
- genicular vein
- small saphenous vein
- genicular veins
- anterior tibia veins
- venous network
- dorsal venous arch of the foot
- dorsal venous network of the foot

- inferior vena cava
- common iliac vein
- middle sacral vein
- lateral sacral veins
- interior iliac vein
- inferior gluteal veins
- sacral venous plexus
- internal pudendal vein
- obturator vein
- superior dorsal veins of the penis
- external pudendal vein
- anterior scrotal veins
- middle circumflex veins of the femur
- accessory saphenous vein
- perforating vein
- great saphenous vein
- femoral vein
- genicular vein
- popliteal vein
- genicular vein
- fibular or peroneal veins
- posterior tibial veins
- great saphenous vein
- dorsal metatarsal veins of the foot
- dorsal digital veins of the foot

◄ **Main vessels of the inferior limb (male)**
Frontal view.
1. arteries
2. veins

▼ **Main Inguinal vessels**
Frontal view.

- Lumbar IV artery
- Iliac muscle
- Femoral nerve
- Deep circumflex artery of the ilium
- Inguinal ligament
- Superficial circumflex iliac artery
- Deep artery of the femur
- Lateral circumflex artery of the femur
- Deep artery of the femur
- Saphenous nerve
- Great psoas vein
- Inferior vena cava
- Common iliac artery
- Middle sacral vein
- Middle sacral artery
- Interior iliac vein
- Interior iliac artery
- External iliac vein
- External iliac artery
- Bladder
- Inferior epigastric artery
- Femoral vein
- Anterior scrotal branches
- Middle circumflex artery of the femur
- Long adductor muscle
- Long adductor muscle
- a. femorale
- Femoral vein
- Great saphenous vein

converges toward the main vessels of the leg: the great and small saphenous veins, the fibular veins and the anterior and posterior tibial veins, are all relatively superficial. Inside the thigh, the tibial veins flow into the popliteal vein, which is located deeper, and is also reached by the perforating veins, which drain the blood from the surrounding muscles and bones.

Near the ileum, the lateral circumflex vein of the femur meets with the medial circumflex vein of the femur and with the great saphenous inside the femoral vein, which collects all of the blood arriving from both the venous cycle of the lower limb, and from the abdominal and genital area (subcutaneous abdominal veins).

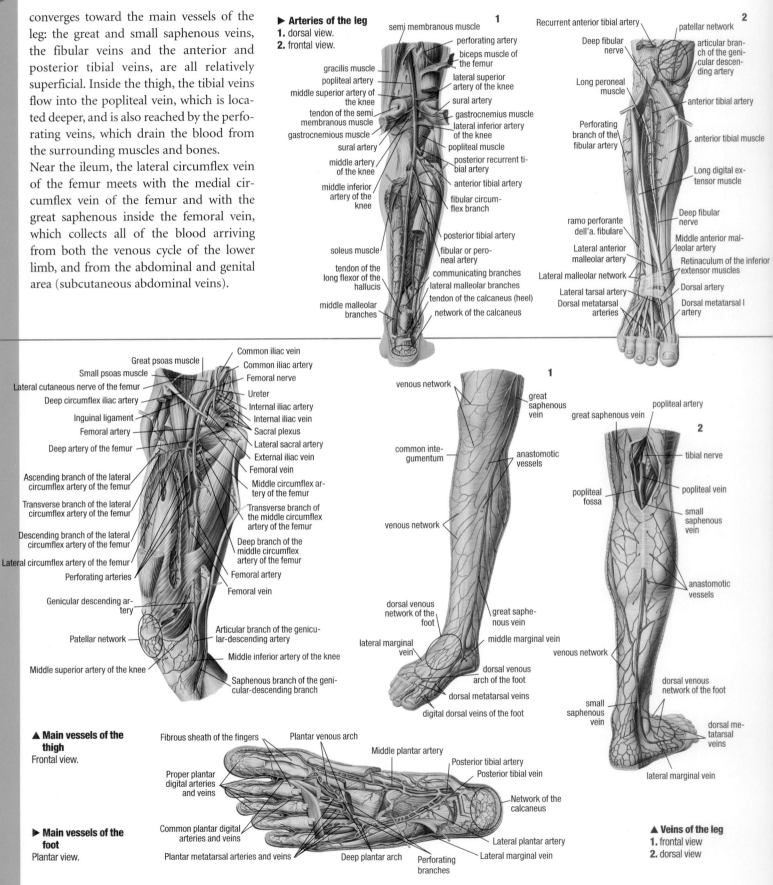

► **Arteries of the leg**
1. dorsal view.
2. frontal view.

▲ **Main vessels of the thigh**
Frontal view.

► **Main vessels of the foot**
Plantar view.

▲ **Veins of the leg**
1. frontal view
2. dorsal view

THE FETAL CIRCULATION

The circulation of a fetus is a totally different matter: it doesn't breathe through lungs, and therefore pulmonary circulation does not oxygenate the blood. Inside a fetus, the placenta acts as a lung, a kidney and as the intestine. This maternal organ is richly wetted by blood arriving from the iliac and circumflex arteries. Besides allowing for gaseous exchanges to occur, it also allows nutrients, water, salts and antibodies to reach the fetus. In the meantime, it collects metabolic waste products: in fact, the renal activity of a fetus begins only around the 4th month, and it remains quite slight until birth.

Besides having a placenta and not having a pulmonary circulation (activated only at birth, thanks to the development of the pulmonary epithelium [26,162]), the fetus has a "mixed" general circulation: in fact, until the cardiac septi remain incomplete, there is a mixing of venous and arterial blood by the heart.

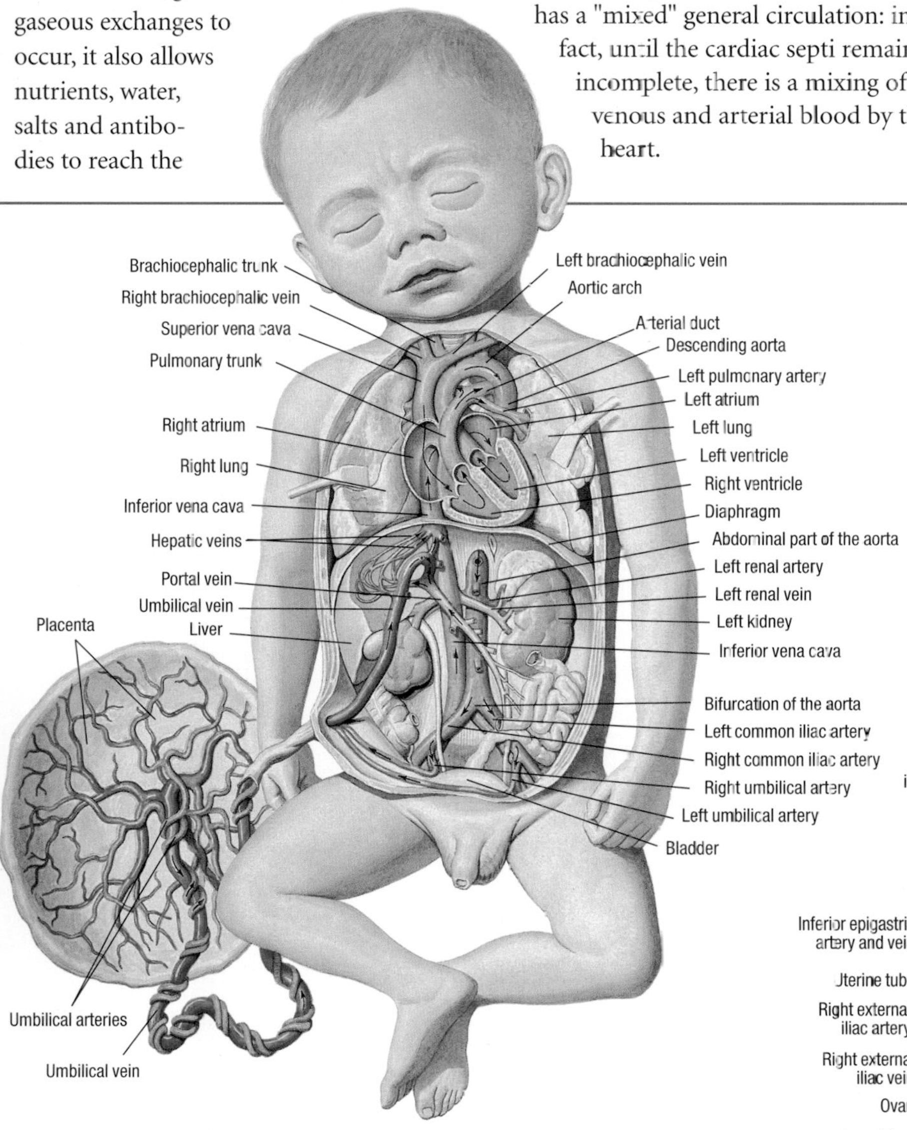

Brachiocephalic trunk
Right brachiocephalic vein
Superior vena cava
Pulmonary trunk
Right atrium
Right lung
Inferior vena cava
Hepatic veins
Portal vein
Umbilical vein
Placenta
Liver

Left brachiocephalic vein
Aortic arch
Arterial duct
Descending aorta
Left pulmonary artery
Left atrium
Left lung
Left ventricle
Right ventricle
Diaphragm
Abdominal part of the aorta
Left renal artery
Left renal vein
Left kidney
Inferior vena cava
Bifurcation of the aorta
Left common iliac artery
Right common iliac artery
Right umbilical artery
Left umbilical artery
Bladder

Umbilical arteries
Umbilical vein

▲ Fetal circulation

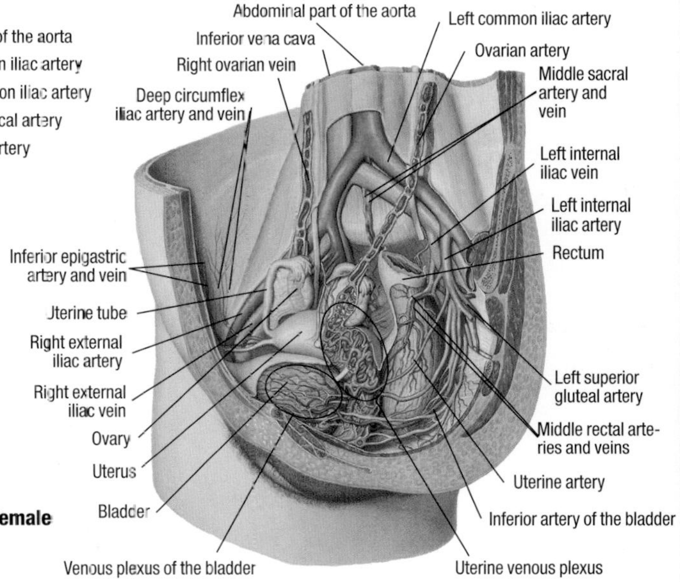

▶ Blood vessels of a female abdomen
Right lateral view

Abdominal part of the aorta
Inferior vena cava
Right ovarian vein
Deep circumflex iliac artery and vein
Left common iliac artery
Ovarian artery
Middle sacral artery and vein
Left internal iliac vein
Left internal iliac artery
Rectum

Inferior epigastric artery and vein
Uterine tube
Right external iliac artery
Right external iliac vein
Ovary
Uterus
Bladder
Venous plexus of the bladder

Left superior gluteal artery
Middle rectal arteries and veins
Uterine artery
Inferior artery of the bladder
Uterine venous plexus

THE LYMPHATIC NETWORK

The lymphatic capillaries are the "absorbing" part of the lymphatic system, and by anastomizing with each other, they make up the network of origin. They are afferent to the precollectors, often equipped with valves and with a thin smooth spiral musculature. In turn, the precollectors continue into the pre and post lymphonodal collectors, depending if they run inside the cutaneous or sub-cutaneous tissues, or if they belong to the muscle-visceral compartments. Also, they are often convergent and anastomized: the superficial collectors run independently from the blood vessels, while the deep collectors almost always accompany arteries and veins. They are afferent to many lymph nodes, and to many lymphatic organs (the thymus and the spleen ►202) and they end in the main lymphatic trunks: the most important is the thoracic duct, which starts at the level of the 2nd lumbar vertebra and discharges the lymph inside the venous system , 45 cm above the point where the internal jugular and the left subclavian meet.

The right and left jugular, are other important lymphatic trunks. They drain the lymphatic pathways of the corresponding half, between the head and the neck; the right and left broncomediastinic lymphatic trunks, which drain the viscera and the walls of the thorax, the diaphragm and even the liver.

The right lymphatic duct, or Galen's great lymphatic vein is not always present: it can develop from the confluence of the jugular lymphatic trunks, the subclavian and sometimes even the broncomediastinic trunk. The right lymphatic duct is extremely short, and it opens where the internal jugular vein and the subclavian vein

Occipital lymph nodes

Mastoid lymph nodes

Parotid

Jugulohomohyoid lymphatic node

Right internal jugular vein

Superficial lateral cervical lymph nodes

Jugular digastric lymph node

Right brachiocephalic vein

Right subclavian trunk

Subclavian vein

Axillary vein

Axillary artery

Cephalic vein

Para mammary glands

Mammary gland

Superior parotid lymph nodes

Deep parotid lymph node

Submandibular lymph nodes

Submental lympha-tic nodule

Right common caro-tid artery

Right jugular trunk

Right lymphatic duct

Superior vena cava

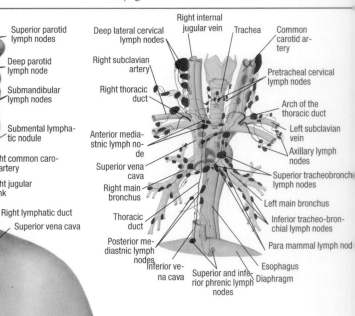

Deep lateral cervical lymph nodes

Right internal jugular vein

Trachea

Common carotid ar-tery

Right subclavian artery

Right thoracic duct

Pretracheal cervical lymph nodes

Arch of the thoracic duct

Anterior media-stnic lymph no-de

Left subclavian vein

Axillary lymph nodes

Superior vena cava

Superior tracheobronch lymph nodes

Right main bronchus

Left main bronchus

Thoracic duct

Inferior tracheo-bron-chial lymph nodes

Posterior me-diastnic lymph nodes

Para mammal lymph nod

Inferior ve-na cava

Superior and infe-rior phrenic lymph nodes

Esophagus

Diaphragm

▲ **Main lymph nodes of the deep lymphatic pathways of the neck and of the thorax**
Frontal view.

◄ **Main lymph nodes and superficial lymphatic pathways of the neck, and of the thorax of a woman.**
Right frontal lateral view.

meet.

LYMPHATIC VESSELS AND LYMPH NODES OF THE HEAD AND THE NECK

The superficial lymphatic collectors drain mostly the rich networks of origin, which correspond to the lips, the nasal wings and lobule, and the eyelids; sometimes they also drain the lymphatic networks of the tissues found underneath (muscles of the face). Often small facial and cheek lymph nodes insert as well.

The deep lymphatic collectors are absent inside the nervous organs, inside the ocular bulb and inside the inner ear, but they are numerous inside the palatine tonsil, inside the nose and the pharynx, and they are afferent to the deep cervical lymph nodes.

The lymph nodes are collected mainly into groups or chains, distributed along the main blood vessels. They are named after the neighboring areas of the skull: occipital lymph nodes, mastoid (or posterior auricular), parotid, submandibular, submental, retropharyngeal, and superior cervical, deep cervical and anterior cervical.

LYMPHATIC VESSELS AND LYMPH NODES OF THE UPPER LIMBS

The superficial lymphatic collectors drain mostly the rich networks of the fingers and the palm of the hand. Three groups join inside the forearm: the medial, lateral and anterior collectors; they all move toward the axillary region, terminating into the brachial group of the axillary lymph nodes.

The deep lymphatic collectors drain bones, joints, muscles and aponeurosis, and are satellites of the deep veins and arteries. They meet inside the brachial collector of the arm, which also ends inside the brachial group of the axillary lymph nodes.

The lymph nodes are distributed along the entire arm, into groups: the main ones are the lymph nodes of the palm of the hand, the radial, the interosseous, the cubital and the brachial. However, most of them are collected inside the axillary lymph center, made up of a variable number of lymph nodes ▶10-60 arranged into groups and chains: the brachial group (or lateral) drains the limb; the thoracic group (or pectoral) drains the thorax; the mammary

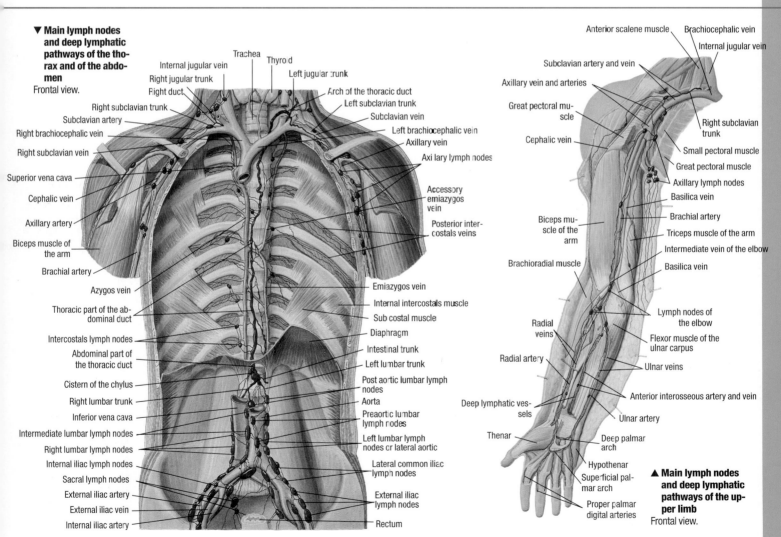

▼ **Main lymph nodes and deep lymphatic pathways of the thorax and of the abdomen** Frontal view.

▲ **Main lymph nodes and deep lymphatic pathways of the upper limb** Frontal view.

gland, the tegumentum and the muscles of the abdominal wall and of the supraombelical area; the sub scapular group drains the posterior part of the thorax; the central group and the sub scapular group (or pical), where all the other groups come together inside the subclavian lymphatic trunk. The lymph nodes are divided into parietal (they drain the bones and the muscles) and visceral (they drain the internal organs); they are all named after the area that they drain. The lymph nodes include: sternal (anterior wall of the thorax, mammary gland, epigastric region and diaphragm), medial or lateral intercostals, anterior diaphragmatic (right and left: pleura, heart, pericardium, lungs, thymus; of the transverse chain: thymus, thyroid, trachea; diaphragmatic diaphragm and liver); posterior mediastnic (walls of

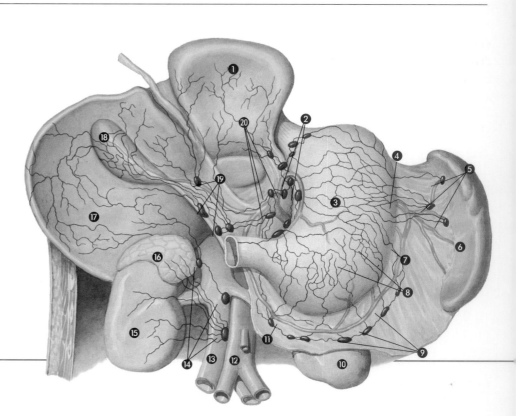

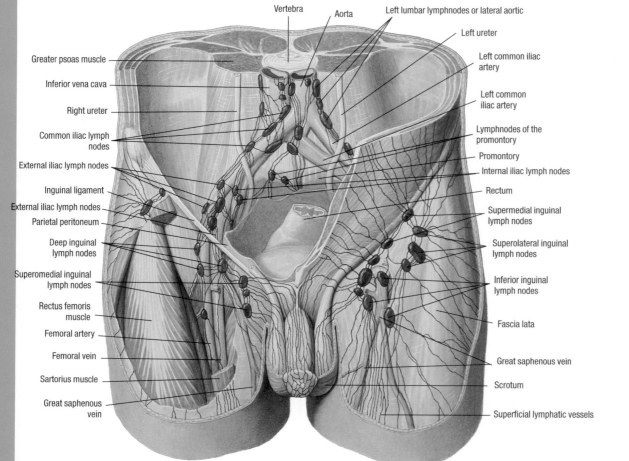

Vertebra Aorta Left lumbar lymphnodes or lateral aortic

Left ureter

Left common iliac artery

Left common iliac artery

Lymphnodes of the promontory

Promontory

Internal iliac lymph nodes

Rectum

Supermedial inguinal lymph nodes

Superolateral inguinal lymph nodes

Inferior inguinal lymph nodes

Fascia lata

Great saphenous vein

Scrotum

Superficial lymphatic vessels

Greater psoas muscle

Inferior vena cava

Right ureter

Common iliac lymph nodes

External iliac lymph nodes

Inguinal ligament

External iliac lymph nodes

Parietal peritoneum

Deep inguinal lymph nodes

Superomedial inguinal lymph nodes

Rectus femoris muscle

Femoral artery

Femoral vein

Sartorius muscle

Great saphenous vein

▲ **Main lymph nodes and lymphatic pathways of the superior abdomen**
Frontal view.
❶ left hepatic lobe.
❷ left gastric lymph nodes.
❸ stomach
❹ splenic or lineal artery.
❺ lineal or splenic lymph nodes
❻ spleen
❼ left gastroepiploic artery.
❽ gastric lymphatic plexus
❾ right gastro humeral lymph nodes.
❿ left kidney.
⓫ right adrenal gland.
⓬ aorta
⓭ inferior vena cava
⓮ lumbar lymph nodes
⓯ right kidney
⓰ right surrenal gland
⓱ right hepatic lobe
⓲ gallbladder
⓳ hepatic lymph nodes
⓴ Celiac lymph nodes

◀ **Main lymph nodes and male inguinal lymphatic pathways**
Frontal view.

the aorta, esophagus, diaphragm, pericardium),
Anterior mediastnic, bronchial (tracheobronchial, of the

bronchial bifurcation of the pulmonary hilum).

The lymph nodes of the intestinal lymphatic pathways are numerous and also very important: the lumboaortic lymph nodes, make up the lumboaortic plexi, and whose efferent converge inside the intestinal lymphatic trunk; the mesenteric, the mesobiliary, the gastric, the hepatic and the pancreatic-lineal.

lymphatic collectors and lymph nodes of the abdomen make up lymphoid chains in the shape of plexi, which follow the course of the blood vessels. They are distinguished into the internal and external iliac plexi that lead into the common iliac plexus, afferent to the lumboaortic plexus. The collectors inside the limbs are satellites of the veins named after them, and they meet inside the inguinal lymph nodes.

LYMPHATIC VESSELS AND LYMPH NODES OF THE PELVIS AND OF THE LOWER LIMBS

As for the abdomen, the

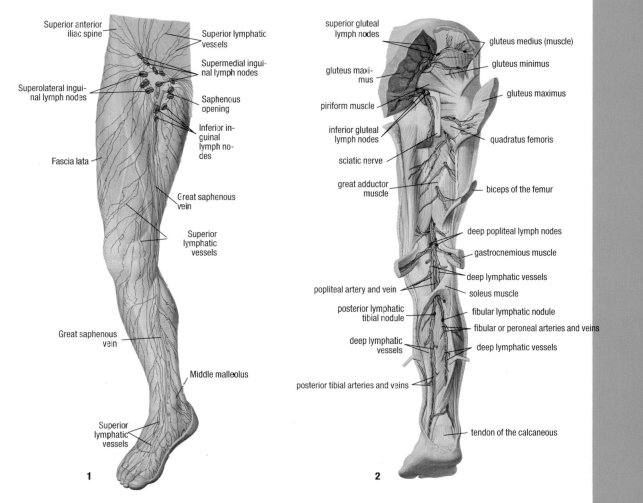

▲ **Main lymph nodes and intestinal lymphatic pathways**
❶ middle biliary lymph nodes
❷ transverse colon
❸ superior mesenteric artery.
❹ adrenal gland
❺ left biliary lymph nodes.
❻ aorta
❼ left kidney.
❽ common iliac lymph nodes
■ inferior mesenteric lymph nodes.
❾ sigmoid lymph nodes
❿ superior rectal lymph nodes.
⓫ ovary
⓬ rectum
⓭ bladder
⓮ uterus
⓯ inferior vena cava.
⓰ lumbar lymph nodes
⓱ right kidney.
⓲ small intestine
⓳ Isuperior mesenteric lymph nodes.
⓴ right biliary lymph nodes.

▶ **Main lymph nodes and lymphatic pathways of the lower limb**
We can see how the blood vessels and the lymphatic vessels run parallel to each other.
1. frontal view, superficial anatomical elements.
2. dorsal view, deep anatomical elements.

Superior anterior iliac spine
Superior lymphatic vessels
Supermedial inguinal lymph nodes
Superolateral inguinal lymph nodes
Saphenous opening
Inferior inguinal lymph nodes
Fascia lata
Great saphenous vein
Superior lymphatic vessels
Great saphenous vein
Middle malleolus
Superior lymphatic vessels

1

superior gluteal lymph nodes
gluteus medius (muscle)
gluteus maximus
gluteus minimus
gluteus maximus
piriform muscle
inferior gluteal lymph nodes
quadratus femoris
sciatic nerve
great adductor muscle
biceps of the femur
deep popliteal lymph nodes
gastrocnemius muscle
deep lymphatic vessels
popliteal artery and vein
soleus muscle
posterior lymphatic tibial nodule
fibular lymphatic nodule
deep lymphatic vessels
fibular or peroneal arteries and veins
posterior tibial arteries and veins
deep lymphatic vessels
tendon of the calcaneous

2

THE SPLEEN AND THE THYMUS

They are the internal organs most directly involved in the production and maturation process of the lymphocytes, and of those cells circulating inside the blood and the lymph, which defend the body (immune responses). Two lymphocytic populations produce a double immune response (humoral immunity, due to the presence of antibodies inside the blood plasma [180], and the cellular immunity, due to the presence of cells which are able trigger an immune response): the lymphocytes B,

divided into plasma cells that are able to secrete antibodies, and the lymphocytes T, responsible for the "immune memory". The two lymphocytic populations have different origins: the lymphocytes B derive from staminal cells, which primarily differentiate from the spleen; the lymphocytes T derive from staminal cells, which differentiate inside the thymus. Even if they occupy two different areas, these two cellular types coexist inside the

peripheral lymphatic organs (spleen and lymph nodes).

THE SPLEEN

It regulates the cellular and humoral hematic composition, and the volume of the circulating mass, thanks to a particular vascular architecture and to its characteristic lymphoid tissue set. Here, the old red blood cells are "demolished", the lymphocytes B begin to differentiate and to proliferate, and to interact with the lymphocytes T (characteristic of the immune response). Inside the spleen there are different territories (the red pulp and the white pulp), delimited by trabeculae, and richly vascularized by lineal veins and arteries. The white pulp is made up of arterioles that, as soon as they emerge from the trabeculae, are covered with a thick sheath

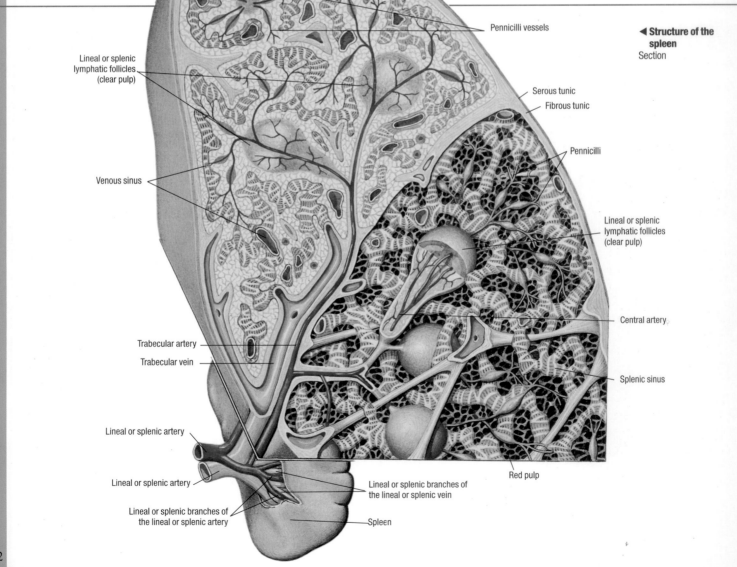

Pennicilli vessels

Lineal or splenic lymphatic follicles (clear pulp)

Venous sinus

Trabecular artery

Trabecular vein

Lineal or splenic artery

Lineal or splenic artery

Lineal or splenic branches of the lineal or splenic artery

◀ **Structure of the spleen**
Section

Serous tunic

Fibrous tunic

Pennicilli

Lineal or splenic lymphatic follicles (clear pulp)

Central artery

Splenic sinus

Red pulp

Lineal or splenic branches of the lineal or splenic vein

Spleen

of lymphoid tissue. Here we are able to find the germinative centers contained inside the Malpighi's nodules or corpuscles, made of lymphoid tissues. We are also able to find lymphocytes B, T and macrophagocytes (dendritic cells). The red pulp, mostly found inside the spleen, is made up of the cods of the pulp, the reticular stroma (where most of the macrophagocytes are located along with lymphocytes and other blood cells). This zone is crossed by a group of much thinner arterioles (50 Ìm in diameter), which are divided into clumps of pennicilar arterioles (15 Ìm in diameter) that end into the capillaries with a shell, small vessels surrounded by muscular fibers, which are able to block the flow of blood into the red pulp. This is the place where the red blood cells are demolished, while plasmacellular differentiation takes

place inside the lymphoid sheaths that cover the vessel of the red pulp. From here, the blood goes through the venous sinuses, which meet inside the veins of the red pulp, followed by the trabecular veins, the roots of the lineal vein.

THE THYMUS
It is a transitional organ: highly developed inside the fetus ►.44, its structure goes through a mayor revolution and it changes dramatically with age. This progressive atrophy of the thymus causes the organism to become more vulnerable. The thymus is the organ in which the totipotent staminal cells differentiate into the lymphocytes T: here they go through functional and morphological changes which characterize it irreversibly. The final maturation stage of

the lymphocytes T occurs in other peripheral lymphatic organs, thanks to the thymosine, a hormone produced by the epithelial part of the thymus. The thymus is in fact highly vascularized and divided into various parts: it is made up of 2 lobes with extensions called thymic horns. When we look at a section of it, it appears to be full of lobules, and we can distinguish a cortical and a medullar part. In reality each lobe is made up of a continuous cord of a more vascularized medullar substance, full of lymphatic vessels and nerve endings, made up primarily of epithelial cells and of the cortical substance, which envelops them in a capsule made of lymphoid cells.

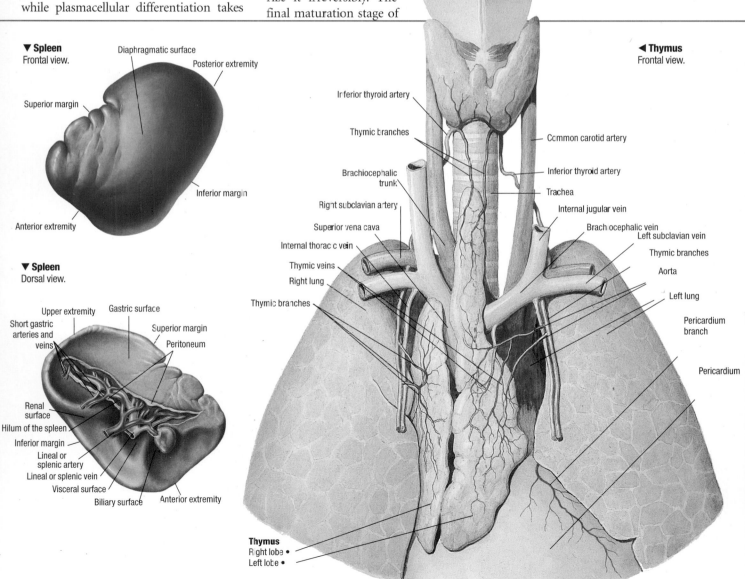

▼ Spleen Frontal view.
Diaphragmatic surface
Posterior extremity
Superior margin
Inferior margin
Anterior extremity

▼ Spleen Dorsal view.
Upper extremity — Gastric surface
Short gastric arteries and veins
Superior margin
Peritoneum
Renal surface
Hilum of the spleen
Inferior margin
Lineal or splenic artery
Lineal or splenic vein
Visceral surface
Biliary surface
Anterior extremity

◄ Thymus Frontal view.
Inferior thyroid artery
Thymic branches
Brachiocephalic trunk
Right subclavian artery
Superior vena cava
Internal thoracic vein
Thymic veins
Right lung
Thymic branches
Common carotid artery
Inferior thyroid artery
Trachea
Internal jugular vein
Brachiocephalic vein
Left subclavian vein
Thymic branches
Aorta
Left lung
Pericardium branch
Pericardium

Thymus
Right lobe •
Left lobe •

✚ THE INFARCTION

The heart is a muscle constantly at work that consumes a lot of energy. In order to function properly, it needs a very good blood supply, which is normally guaranteed by the proper functioning of the coronaries, the arteries and veins that surround it. However, these vessels often become "old": arteriosclerosis is generally the most common cause of coronary failure, a circulatory defect that reduces or even kills their functionality.

If one of the coronaries becomes blocked (for example, due to an atherosclerotic deposit or more often to a blood clot), the heart cannot receive adequate oxygen and nutrients supply, and doesn't have the adequate energy supply to be able to function properly and to maintain a uniform rhythm. This causes "a cardiac pain" that manifests itself through a radiating pain (from the chest to the back, from the arms to the neck and from the maxillary to the area above the belly button or epigastrius), which causes a suffocating sensation, cold sweats, nausea and vomiting.

In the worst cases, the part of the myocardium that doesn't receive enough nutrients dies: the muscular cardiac cells start to die and the muscle looses any vital function in that specific area: this is a myocardial infarction, better known as a "heart attack".

The effects of this sudden loss of functionality are very different, depending on the extent of the damage to the area: from a "silent" asymptomatic infarct, to a different level of permanent invalidity, all the way to possible death, which generally affects 1 out of 3 people experiencing an infarction.

Although half of the people affected by an infarction get back to a normal life in a very short time, often there are other complications that may have serious consequences: the most frequent is an abnormal cardiac rhythm, accelerated or irregular, which affects the normal activity of the heart. Many factors concur in determining this serious pathology: age, a history of the disease in the family, being male, obesity, smoke, diabetes, hypertension, and also a poor physical activity and a diet too rich in animal fats.

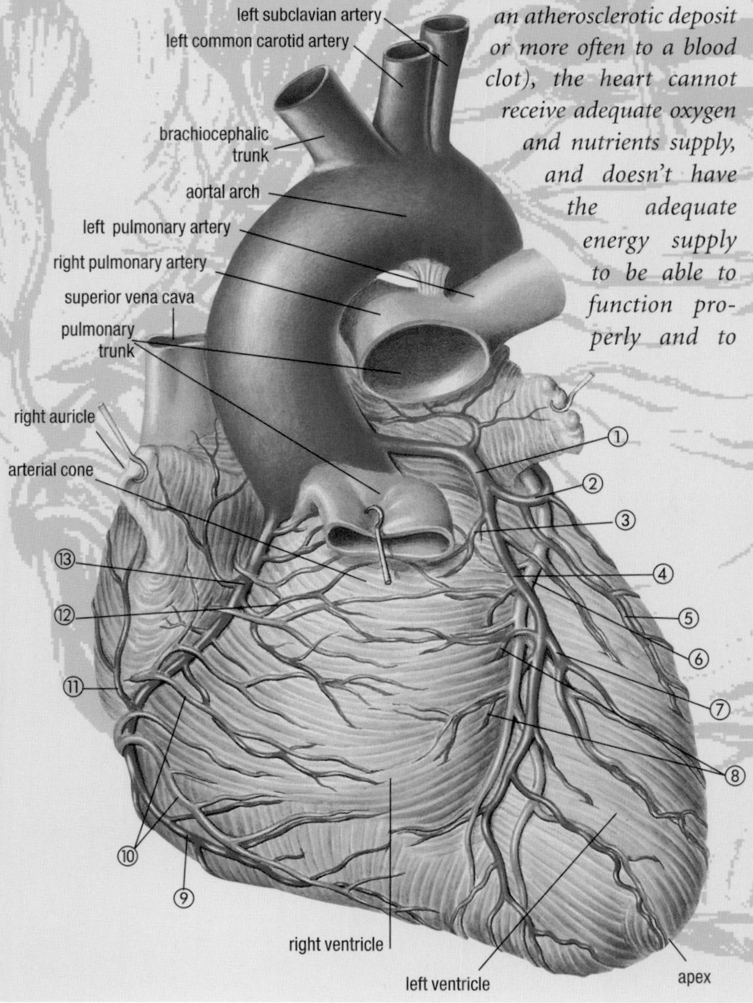

Labels (diagram)

left subclavian artery
left common carotid artery
brachiocephalic trunk
aortal arch
left pulmonary artery
right pulmonary artery
superior vena cava
pulmonary trunk
right auricle
arterial cone
right ventricle
left ventricle
apex

◀ **The coronaries:**
Main blood vessels that feed the heart.
① left coronary artery.
② circumflex branch of the left coronary artery.
③ branch of the arterial cone.
④ anterior interventricular branch of left coronary artery.
⑤ left marginal branch of the left coronary artery.
⑥ great vein of the heart.
⑦ lateral branch.
⑧ septal interventricular branches
⑨ right marginal branch of the right coronary artery
⑩ anterior veins of the heart
⑪ intermediate atrial branch
⑫ branch of the arterial cone
⑬ right coronary artery.

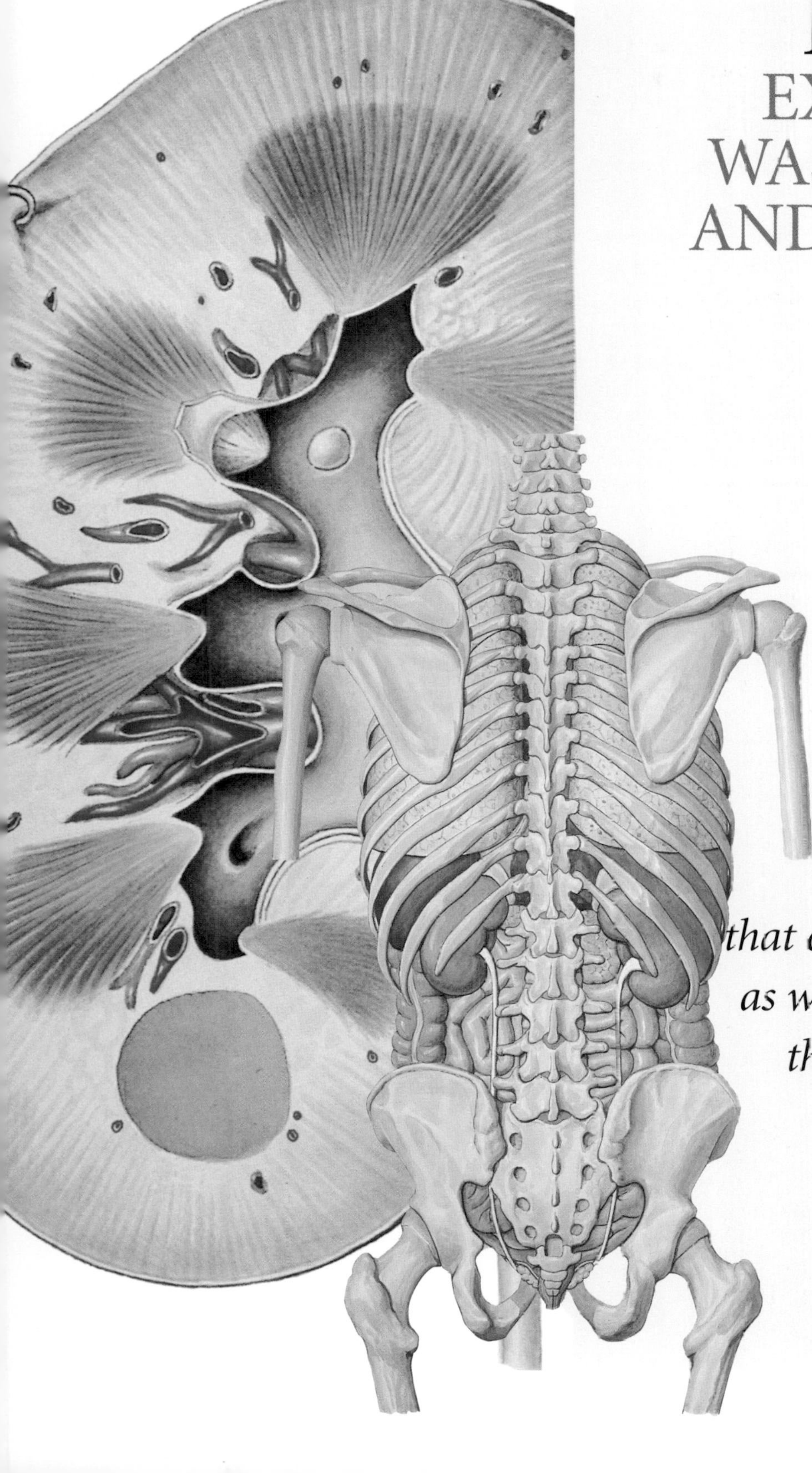

SKIN AND KIDNEYS
EXTRUSION OF WASTE PRODUCTS AND HOMEOSTASIS

*It is vitally important
to be able to eliminate
all of those substances
that could damage the organism,
as well as to maintain constant
the internal environment.*

THE EXCRETORY SYSTEM IS RESPONSIBLE FOR THE PURIFICATION OF THE BLOOD FROM WASTE PRODUCTS OF CELLULAR ACTIVITY AND FOR MANTAINING THE WATER BALANCE (HOMEOSTASIS)

THE EXCRETORY SYSTEM

The vital processes inside the cells of our body use the energy supply and the raw material provided by the digestive system [144], the respiratory system [160] and the circulatory system [176].

The result of these processes is the survival of each individual cell, its reproduction and its specific activity. At the same time, the cellular metabolic chain reaction affects the production of many "residual " substances: although the majority of subproducts are recycled to maximize the use of energy and materials, some of the them are useless or even dangerous to the cells. Therefore, to avoid their accumulation, the body uses different systems, depending on the type of waste: the process responsible for the removal of these substances from the body is called excretion, and the organs involved are called emunctories. We have already encountered some of them: the lungs, for example, are considered to be the emuntory organs of all the gaseous wastes (especially carbon dioxide), even if this type of excretion is part of the more complex respiratory process.

Even the intestine eliminates waste substances: for example the biliary pigments [152] produced by the degradation of the hemoglobin [180]. However, the true excreting organs are the skin and the kidneys: Both skin (produces sweat) and the kidneys (they produce urine), contribute in the constant elimination of waste products deriving from the breakdown of proteins, sugars and fats. At the same time, they keep the body's water and salt balance under control (homeostasis).

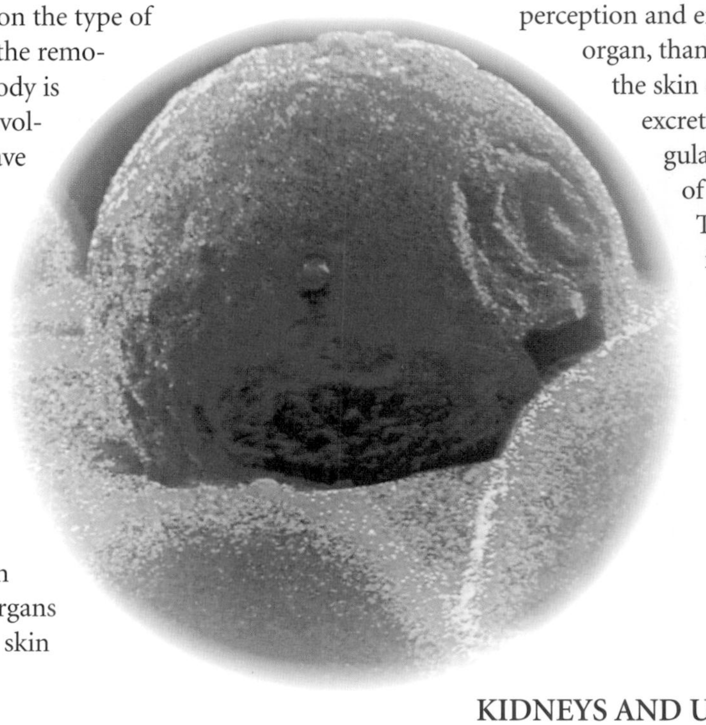

THE CUTIS OR SKIN

If, along with its annexes (hair and nails) the cutis makes up the tegument (the lining that isolates and protects the body from the outside world) and if it represents an important perception and environmental interaction organ, thanks to the nervous receptor [104], the skin can then be considered a big excreting, homeostatic and thermoregulation organ due to the presence of many sudoriparous glands.

The sudoriparous glands are irregularly scattered inside the 2 cm_ of skin that cover our body, and they produce about 20-50 cl of sweat per day.

KIDNEYS AND URINARY PATHWAYS

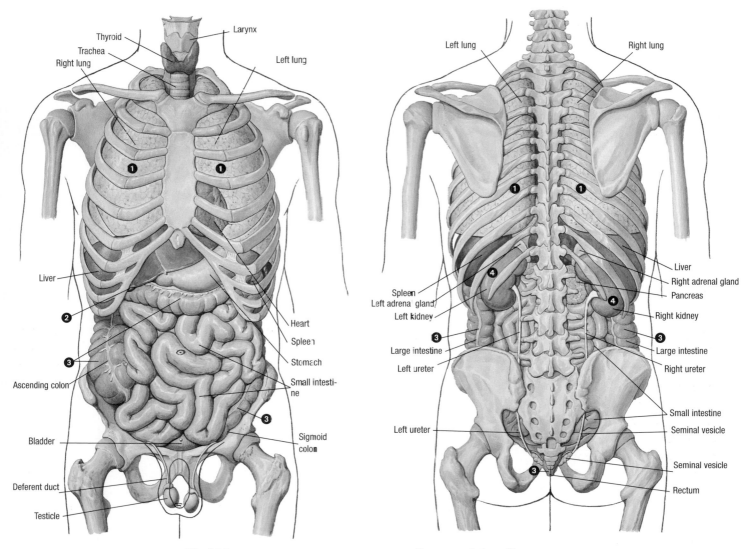

▲ **Internal organs**
Anterior and posterior view of the male abdomen. We can see the reciprocal relationship of the different organs, in particular the location of the emuntory organs: ❶ lung;
❷ biliary gallbladder; ❸ intestine;
❹ kidneys.

The kidneys are excretory organs par excellence, and they filter up to 1200 ml of blood per minute: on average, they depurate about 1700 liters of blood per day, and produce up to 1 and 1/2 liters of urine.

The excretory units of the kidneys are called nephrons, and they collect the freshly produced urine into a common renal basin. The urine passes through the ureter, a large duct made of smooth muscle and a mucous epithelium, and reaches the bladder collecting here: under normal conditions, this elastic membranous muscle contains up to 250 cm_ of fluid, but it can hold up to 450 cm_, in extreme cases. It communicates with the outside through the ureter, a duct that ends with the urinary meatus, which is lined with the ring of the sphincter muscles. By remaining contracted, the sphincter muscles block the emission of urine.

◄ **Sweat**
This is microphotography of a drop of sweat on a thumb, taken with an electron microscope. Sweating not only has an excretory function, but it is important in the thermoregulation of the body: by evaporating, sweat cools down the overheated body.

SKIN AND SWEAT

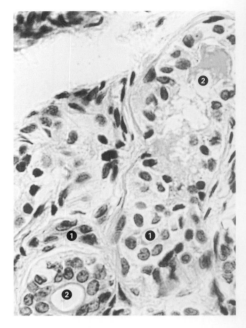

The cutis is rich in blood capillaries and it represents an important emuntory organ.

In fact, the sudoriparous glands produce a secretion that contains substances which are usually found inside the urine (urea, uric acid, phosphates, sodium salts, potassium, calcium, magnesium, etc..)
They are distinguished into:
-Apocrine glands, generally related to hair; they are glomerular tubular glands made up of cells whose apical part of the cytoplasm is expelled along with the secretion. They are located in specific parts of the body, and their function is often linked to that of the sexual hormones: they reach their maximum development during puberty and then they atrophy with old age.

-Eccrine glands: are not related to hair and are found almost everywhere in the body. They are simple glomerular tubular glands, and they can extend all the way to the hypoderm, wrapping around themselves. The cells that reach the glandular lumen have many apical microvilli, in which the selective absorption of electrolytes takes place: as a result of this, sweat becomes hypotonic compared to the plasma.

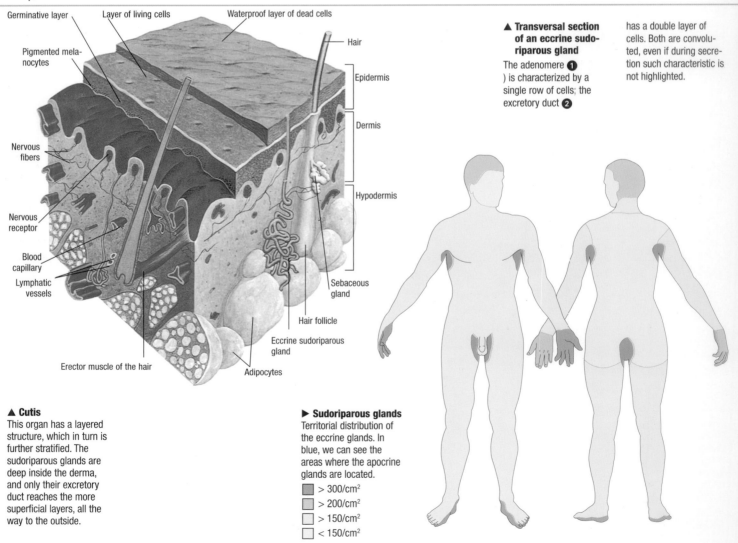

Germinative layer
Layer of living cells
Waterproof layer of dead cells
Pigmented melanocytes
Hair
Epidermis
Dermis
Nervous fibers
Hypodermis
Nervous receptor
Blood capillary
Lymphatic vessels
Sebaceous gland
Erector muscle of the hair
Hair follicle
Eccrine sudoriparous gland
Adipocytes

▲ Cutis
This organ has a layered structure, which in turn is further stratified. The sudoriparous glands are deep inside the derma, and only their excretory duct reaches the more superficial layers, all the way to the outside.

▲ Transversal section of an eccrine sudoriparous gland
The adenomere ❶) is characterized by a single row of cells; the excretory duct ❷ has a double layer of cells. Both are convoluted, even if during secretion such characteristic is not highlighted.

▶ Sudoriparous glands
Territorial distribution of the eccrine glands. In blue, we can see the areas where the apocrine glands are located.

- ▮ > 300/cm²
- ▮ > 200/cm²
- ▯ > 150/cm²
- ▯ < 150/cm²

ORGANIC FLUIDS AND THIRST

The body of an adult person consists of 65% of water: this water is found inside (intracellular liquid) and outside the cells (extra cellular liquid), and it represents, in both situations, the means by which many substances, such as mineral salts and proteins get dissolved. The intra cellular liquid represents 63% of the total body weight, and its chemical composition is fairly constant. On the other hand, the extra cellular liquid represents 37% of the body weight, and its composition varies, mainly in its protein content, depending on the types of functions. So, depending on its characteristics, it divides into the following: an interstitial fluid, which wets the tissues and occupies the entire space between the cells; a plasmatic fluid (or plasma), which circulates inside the blood vessels; a lymphatic fluid (or lymph), which circulates inside the lymphatic system; a cephalorachidian fluid found inside the nervous system; a synovial fluid, found inside the bone joint, and so on. The composition of the extra cellular fluid is very different that that of the intra cellular fluid: this difference is managed by the membranous phenomena and by the metabolic process. The total amount of organic fluids remains almost constant, thanks to the precise control mechanisms that maintain the equilibrium between the amounts of water ingested and excreted. Thanks to the nervous centers of the anterior area of the hypothalamus (94) the thirst sensation regulates the amount of water taken in. Its excretion (mostly the one through the kidneys) is regulated by the action of the pituitary gland, thanks to the action of the antidiuretic hormone (130, 128). Let's give an example. When it is very hot, the skin produces large amounts of sweat that through evaporation maintains the body at an acceptable temperature. An increasing excretion of water through the skin is compensated by a lower amount of urine produced: the adrenal gland registers a water "deficit" and it secretes the antidiuretic hormone (ADH) that by reaching the renal tubes through the blood circulation, stimulates the re absorption of water from the urine. If the amount of water we drink is not enough to adequately compensate the loss of it through sweating and breathing (let's remember that while we breathe we release water vapor), the slight increase in blood concentra-

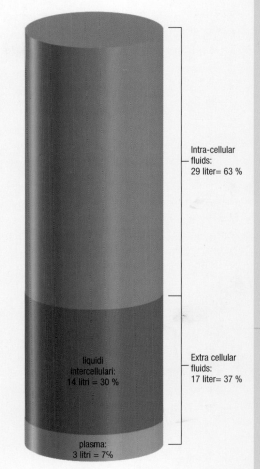

Intra-cellular fluids: 29 liter= 63 %

liquidi intercellulari: 14 litri = 30 %

Extra cellular fluids: 17 liter= 37 %

plasma: 3 litri = 7%

CHEMICAL COMPOSITION		
SUBSTANCES	concentrtation in ‰	
	PLASMA	URINE
WATER	900-930	950
FAT PROTEINS	70-90	0
GLUCOSE	1	0
UREA	0,3	20
URIC ACID	0,03	0,5
CREATININE	< 0,01	1
SODIUM	3,2	3,5
POTASSIUM	0,2	1,5
CALCIUM	0,08	0,15
MAGNESIUM	0,025	0,06
CHLORYNE	3,7	6
ION PHOSPHATE (PO_4)	0,09	2,7
ION SULPHATE (SO_4)	0,04	1,6

tion stimulates the posterior lobe of the pituitary gland to secrete more ADH, reducing a lot the loss of water inside the urine. On the other hand, if the blood is too diluted, (we have taken in too much water), the secretion of ADH is inhibited and the water inside the renal tubules is not reabsorbed: the production of urine becomes abundant, and the optimal water balance is re established.

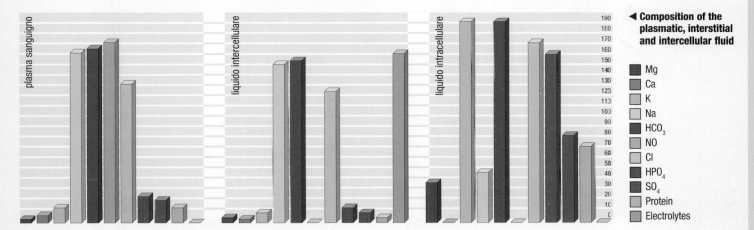

◄ **Composition of the plasmatic, interstitial and intercellular fluid**

- Mg
- Ca
- K
- Na
- HCO_3
- NO
- Cl
- HPO_4
- SO_4
- Protein
- Electrolytes

plasma sanguigno

liquido intercellulare

liquido intracellulare

THE KIDNEYS AND THE URINARY PATHWAYS

The kidneys are symmetrical to the spinal column, and they are located in the postero-superior region of the abdomen (lumbar region). They not only clean the blood: but they also eliminate useless and excess substances, dangerous to the body (mainly the azotized compounds that derive from the degradation of proteins and of drugs) they play a delicate role in the regulation of the water equilibrium inside the body, of the acid-base equilibrium and of the electrolytic composition of the blood. In particular, if the kidney is not able to keep the level of sodium (one of the most important electrolytes) constant, the result will be either water retention, or a high dehydration level, due to a loss of salt.

The kidneys are essentially two systems of vessels in close contact with each other: on one side we have blood capillaries of the renal artery loaded with substances ready to be excreted; and on the other we have the Bowman capsules and the nephrons, into which the waste products pour. The urinary pathways transport waste products on the outside, and their correspondence begins by the renal calyces and by the renal pelvis (or renal pelvis). Then they flow inside the two ureters (one per kidney: right

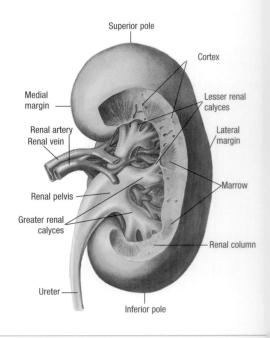

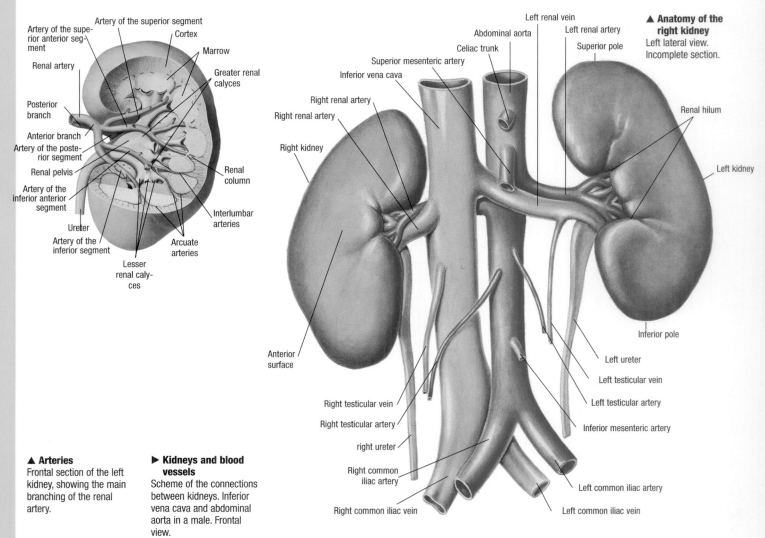

▲ Arteries
Frontal section of the left kidney, showing the main branching of the renal artery.

▶ Kidneys and blood vessels
Scheme of the connections between kidneys. Inferior vena cava and abdominal aorta in a male. Frontal view.

▲ Anatomy of the right kidney
Left lateral view. Incomplete section.

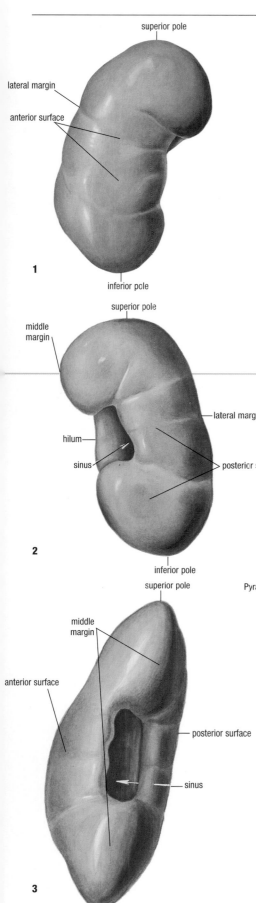

and left), which emerges inside the urinary bladder (a hollow and uneven organ) located inside the pelvic cavity that comes out with the urethra, a very narrow and short vessel in women, and larger and longer in men ►[218].

The two adult kidneys are on average 150-160 g, and they are slightly different: the left one is generally bigger than the right one. However, both of them have a smooth surface: in fact, the lobe shaped kidney of a fetus tends to disappear during the first years of one's life.

Besides having a convex anterior surface, a slightly curved flat posterior surface, a round superior pole and a pointed inferior pole, a convex lateral margin and a caved medial margin, the hilum is present in each kidney: this is a fissure 3-4 cm long, through which the main blood and lymphatic vessels, the nerves and urinary pathways (renal pelvis) run. While the venous vessels are in the front, the arterial vessels are located in the middle between the renal pelvis, located on the back. The flattened cavity where the hilum flows, is called renal sinus: Here are the first urinary pathways (the highest ones: the greater and lesser renal calyces and the renal pelvis), the branching of the renal artery, the roots of the renal artery, the lymphatic vessels and the nerves, all immersed in the adipose tissue that starting from the hilum, extends to the entire renal surface (adipose capsule).

The wall of the renal sinus are irregular due to the renal paillae, which represent the apex of the renal pyramids (or Malpighi's) interposed   by the renal columns (or

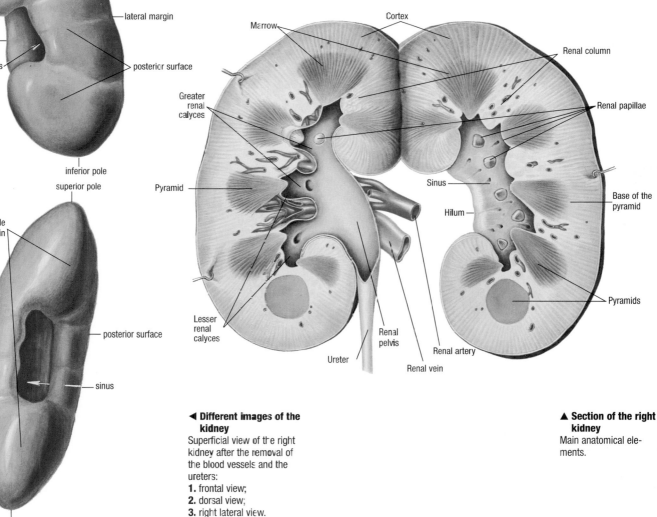

◄ **Different images of the kidney**
Superficial view of the right kidney after the removal of the blood vessels and the ureters:
1. frontal view;
2. dorsal view;
3. right lateral view.

▲ **Section of the right kidney**
Main anatomical elements.

Bertin's).

The kidneys are kept in place by the renal fascia (a thick modified connective tissue), by the vascular peduncle, which connects them to the aorta, and to the inferior vena cava, and by the abdominal pressure. Despite this, they are able to move: while inhaling, they raise, and under pathological conditions, they can permanently move downward reaching at times even the iliac fossa. Inside, each kidney is covered with a fibrous capsule, which, inside the hilum, fuses with the tunica aventitia of the calyces and of the blood vessels. Below, the kidney is covered with a muscular tunic of intertwined smooth fibers. Later there is a cortical zone that is divided into:
-a radiate part, made up of the medullary rays (or of Ferrein): these are tubules arranged into cone like fascicles that start from the base of the pyramids, and reach the cortex, becoming thinner and stopping close to the renal surface;
-a convoluted part that occupies the space in between the medullary rays, forming the renal columns and the outer cortical fascia. This part is made up of renal corpuscle (or Malpighi's) and of the contorted tubules.
A medullary zone follows the cortical zone, and it is divided into 8-18 conical formations (the pyramids): they extend from the cortical substance to the renal papillae, whose free end has 15-30 papillaries by the outlet of the papillary ducts (or Bellini's). Along with the collector ducts, they run around the pyramids in an axial direction.

The kidney is also divided into:
-lobes, which are made of a pyramid and of the corresponding layer of cortical substance;
-lobules, made up of a medullary ray and of the convoluted part that surrounds it. Each lobule is roughly outlined by the blood vessels that run the cortex in a radial direction.

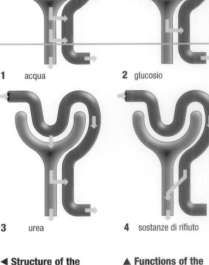

1 acqua

2 glucosio

3 urea

4 sostanze di rifiuto

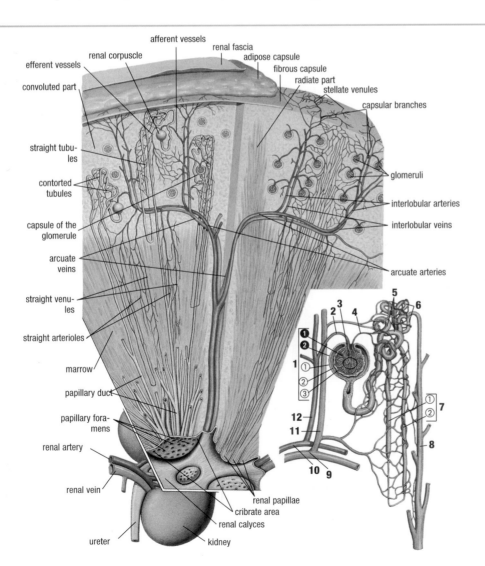

◀ **Structure of the kidney**
1. renal corpuscle:
   ❶ glomerule
   ❷ capsule of the glomerule:
      ① external part
      ② internal part
      ③ lumen
2. afferent glomerular arteriole
3. capillary network of the glomerule
4. efferent glomerular arteriole
5. proximal part of the tubule of the nephron
6. distal part of the tubule of the nephron
7. ansa
   ① descending part
   ② ascending part
8. collagen renal tube
9. arcuate vein
10. arcuate artery
11. interlobular vein
12. interlobular artery

▲ **Functions of the nephron**
The nephron plays different roles, depending on the substances.
1. Depending on the hormonal stimuli, water is passively (osmosis) or actively reabsorbed.
2. The glucose that is excreted inside the glomerule is completely reabsorbed and re enters the circulation.
3. The urea is only partially re-absorbed.
4. The waste products are not reabsorbed.

■ Blood vessel
■ Renal tubule

THE PRODUCTION OF URINE

Each kidney is made up of about 1 million nephrons: they are the functional units of the kidney. Each on of them is made up of:

-A renal corpuscle or Malpighi's shaped like a calyx, with a diameter of 150-300 Ìm, which touches the blood capillaries and forms the preurine;

-A renal tubule, 30-40 mm long, in contact as well with the blood vessels, which begins inside the renal corpuscle and flows inside the collector ducts, which go to the other renal pathways. They are distinguished into a proximal contorted tubule, an ansa of the nephron and a distal contorted tubule, further divided into an initial straight portion and a convoluted distal portion. Each part of the tubule has a different function, and transforms the preurine into urine.

Inside the renal corpuscles

The renal corpuscles are found inside the convoluted part of the renal cortex and are made up of a thin epithelial layer (glomerular capsule or of Bowman's), which encloses a capsular space in which the glomerule is located. The glomerule is a ball of 3-5 branches of an arterial capillary. The epithelial cells of the Bowman's capsule that get in contact with the capillaries (podocytes) have very unique characteristics: they are full of pedicels, which are thin and short extensions that expand all the way to the surface of the capillaries and that intertwine leaving only 250 Å wide open fissures (filtration fissures). These fissures are kept close by semi permeable filtrating membranes, 60 Å thick. Even the epithelium of the glomerular capillaries is "fenestrated": many pores with a diameter of 500-1000Å, interrupt the continuity of the blood vessel's walls, allowing what's in the blood to "ooze" into the capsular lumen.

In fact, the blood that passes through the glomerule is at a high pressure (about 9,3 kPa), since it flows into an arteriole whose diameter is smaller than the "entering" one that makes up the glomerular capillaries.

As a result of this, water, glucose, urea, mineral salts and other substances "ooze" into the glomerular space: la preurine, with the same constitution of the blood plasma ▶*209, starts to move down the tubule.*

Inside the tubule

Thanks to the intimate contact between the tubular epithelium and the capillary epithelium, the renal tubule is not a simple duct directed outward, but a highly efficient apparatus for the re-absorption of water and of other substances.

Inside the proximal tubule, not only the excretion of substances such as creatine takes place, but also over 85% of water, sodium chloride and substances such as glucose, amino acids, ascorbic acid, and proteins found inside the "glomerular filtrate", are re-absorbed. The cells of the tubule actively re-absorb the sodium ion; the water, and the chloride ion passively follows the osmotic pressure. Urine concentrates inside the ansa of the nephron, so that water can be re-absorbed. This continues along the straight portion of the distal tubule, thanks to the active re-absorption of sodium, which causes an increase in the osmotic pressure. On the other hand, inside the convoluted distal portion, the water re absorption is optional, and it occurs thanks to the antidiuretic hormone; however, the sodium absorption continues, balanced by the excretion of potassium, hydrogen and ammonia ions.

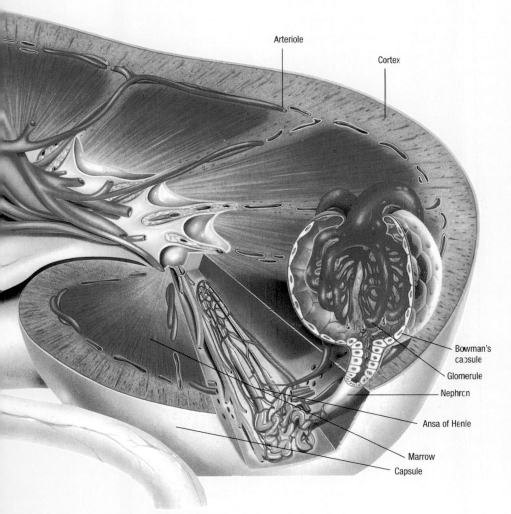

Arteriole

Cortex

Bowman's capsule

Glomerule

Nephron

Ansa of Henle

Marrow

Capsule

◀ Glomerule

It is a thick knot of blood capillaries (sometimes over 50) with a diameter of 0,1 mm. a liquid rich in water and other substances filters from them and collects inside the Bowman's capsule and then moves into the nephron (5Ìm in diameter, and up to 2,5 cm long).

Each kidney is made up of over 1 million nephrons: covered around blood capillaries, they allow for substances such as water and glucose to be re-absorbed. The nephrons converge inside the collector duct, which flow inside the renal pelvis.

✚ RENAL CALCULI

They are solid concretions, which form inside the lumen of the renal calyces or toward the exit of the renal pelvis: the calculi that form directly inside the ureter, are extremely rare. The calculi are a combination of inorganic salts (such as calcium, phosphorus, ammonia, etc.) or organic salts (such as the uric acid). We still do not know exactly how they form, even if in some cases we know the exact cause of their formation. One of these causes is gout: this metabolic disease, in which there is an incorrect catabolism of the purines, with a consequent increase in the concentration of uric acid inside the blood and in the tissues (with resulting leg pain), was very common in well-to-do families, during the 18th century, due to a diet too rich in animal proteins. In people affected by gout, the concentration of uric acid increases dramatically even in the renal secretion: as the urine concentrates in the terminal parts of the urinary pathways, this salt precipitates, forming the calculi.

A similar mechanism intervenes in the formation of precipitating calculi caused by metabolic dysfunctions: these are the calcium calculi.

However, the possible (and theoretically valid) causes for the onset of the renal calculi are numerous: a poor diet or a poor vitamin intake; an imbalance in the chemical composition of the urine due to alterations of the renal processes, to infections or to a poor drainage of one or more renal pathways; endocrine complications (mostly at the expenses of the parotid glands). It is also possible that the onset of a calculus could be due to a set of joint causes: the fact that it occurs primarily in men and not women, that the onset occurs between the ages of 40 and 60 and not among young people, and that single calculi are more common than multiple ones, strengthen this hypothesis.

Usually, the presence of these foreign bodies inside the kidney causes excruciating pains (renal colic) and determines the presence of blood or pus in the urine, compromising the renal activity. Often they move into the ureter, causing worse pains, until they are naturally expelled. However, when the calculus is too big to get expelled naturally, or when it blocks a renal pathway, or when it causes an infection or recurring acute attacks, it is often necessary to surgically intervene. While in the past, special tools were used to grab and pull out these calculi (very painful procedures), today we rely on endoscopic lithotripsy, on shock wave extracorporeal lithotripsy (ESWL) or on laser lithotripsy, all techniques which reduce a surgical intervention to a minimum. In some cases we can also rely on specific drugs: if the calculus is made up of only salts from the uric acid, a specific drug therapy can dissolve them.

Renal calculi
These are some calculi deriving from different levels of the urinary pathways. Thanks to new ultrasounds technologies, calculi are destroyed without the need of surgical intervention, and they are excreted without any difficulty. Also patients recover much faster.

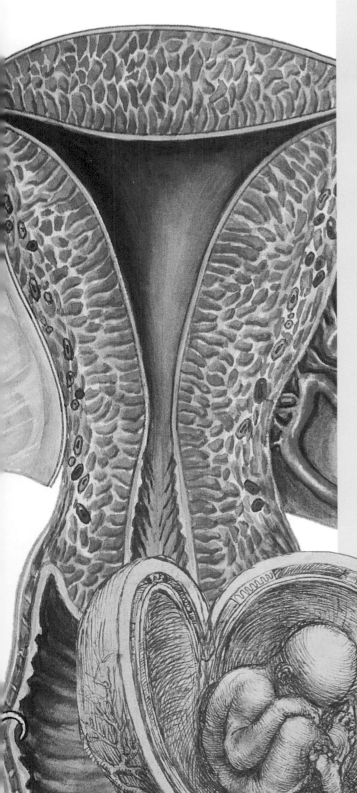

REPRODUCTION
THE KEYS
TO THE FUTURE

*The survival of the human species
is closely connected to reproductive activity,
from the fusion of
the female and male gametes
a new individual is born.*

ALTHOUGH THE GENERAL STRUCTURAL AND ORGANIZATIONAL CHARACTERISTICS ARE
THE SAME, THE REPRODUCTIVE SYSTEM OF A MALE AND A FEMALE IS DEEPLY DIFFERENT

MEN AND WOMEN

The reproductive system differentiates the two sexes, the male and female; however, both have quite noticeable structural similarities: the reproductive system is structured into homologous organs that are organized according to reproduction.

STRUCTURAL SIMILARITIES

Both systems are made up of two gametes' producing gonads. Gametes are reproductive cells with half the number of chromosomes ▶217, 224. The gonads (testicles in men and ovaries in females) produce sperm in males (can autonomously move) and ovules in the woman (unable to move). They also secrete sexual hormones ▶128, which are essential for both a correct physical development, and for the success of the reproductive processes. They are connected on the outside through a system of ducts covered by many glands, and by ciliate tissue, which facilitates the outer movement of the gametes.

FUNCTIONAL DIFFERENCES

The reproductive system of a woman is located inside the abdomen, able to receive the embryo and to let it develop: the uterus is specifically connected to these functions.
On the other hand, the male reproductive system is almost all exposed: the penis, the erectile organ responsible for laying it sperm inside the female vagina, is the terminal part in which the ducts flow, and the testicles are located inside the scrotal sac, which holds it and maintains the ideal temperature for the development of the gametes.

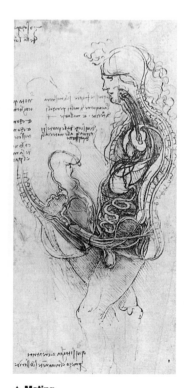

▲ Mating
Studies on mating begaviors conducted by Leonardo Da Vinci, approximately around 1493.

DETERMINING THE SEXES

The DNA, is the genetic material found inside all cells. It "guides" the harmonious development of an organism and is divided into 46 "pieces" called chromosomes. In each cell, these chromosomes are organized into pairs: in both sexes there are 22 pairs (autosomal chromosomes). The last two chromosomes can be similar or different, and their combination is different between a male and a female: they are the sexual chromosomes, and an X and a Y characterize them.

All of the female cells have 2 X chromosomes, while all of the male cells have an X and a Y chromosome. This difference produces a cascading of all other differences, which characterize the difference between male and female, and is connected to the same reproductive process: in fact, each individual is created from the fusion of a female sexual cell (the ovule 224) that contains 22 single autosomal chromosomes and only one sexual chromosome X, and a male sexual cell (the sperm 224) that contains 22 single autosomal chromosomes and a male X or Y sexual chromosome. If a sperm that has an X chromosome fecundates an ovule, a female will be born; if the sperm contains a Y chromosome, the person will be a male.

This is the genetic makeup from which all of us were created. In fact, the genes that form the sexual chromosomes "guide" the development of the sexual apparatus and of the secondary sexual characteristics: starting from an undifferentiated genital tubercle found inside a 4-7 weeks embryo, within the 12th week of pregnancy, the sexual characteristics develop. However many problems can develop during this initial developmental phase. One of the most frequent one is chryptorchism, commonly known as "testicle retention": in fact, during the embryonal development of a male, the testicles develop inside the abdominal cavity, just like the ovaries. As they develop, they move toward the scrotal sac, which they reach at birth. But sometimes this doesn't happen, and the testicles can remain high inside the abdomen. It is not always necessary to

surgically intervene: sometimes "this descent" occurs during the first year or even during puberty.

More complex and critical problems affect about 65000 children each year: many children are born as hermaphrodites. This mythological word (Hermaphrodites was the son of Hermes, the messenger of the Greek gods, and of Aphrodite, the goddess of love. He had both the female and the male

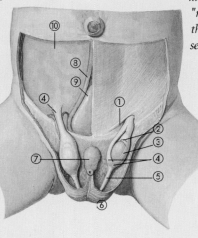

Above: Anatomy of a child affected by lateral chryptorchism:
① superficial inguinal ring, ② epididynis
③ testicle, ④ vaginal tunic, ⑤ gubernacle of the testicle, ⑥ scrotum,
⑦ penis, ⑧ line alba,
⑨ urinary bladder,
⑩ parietal peritoneum.

Below: Embryonal evolution of the external sexual organs.

organs) indicates the presence of both male and female sexual organs. Few years ago, it was common belief that these individuals suffered simply from a malformation of the reproductive apparatus, and they were surgically treated to become women. However, recent research on the biological basis of sex, show that many factors may trigger this diversity: hormones and polluting substances with para hormonal actions, a differentiated cellular sensitivity to hormones, and a "mosaic" genetic set could all be the cause of a wide range of sexual variations.

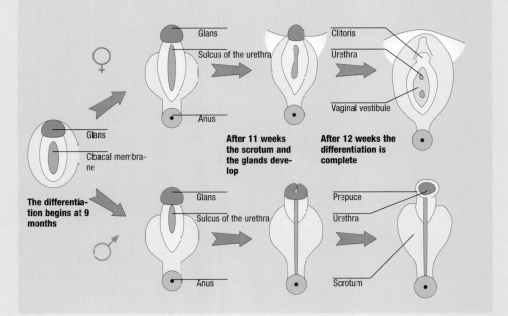

THE MALE REPRODUCTIVE APPARATUS

The testicles have an elongated shape and their size is very similar to that of the ovaries. They produce sperms and are located inside the scrotal sac, outside the abdomen. Many canals called seminiferous tubules or spermatic ducts, which converge toward a network of efferent vessels that gather to form a small body called epididymis, cross them. Here, thanks to the help of the seminal glands and of the prostate gland, located at the base of the penis, under the bladder and lateral to the urethra, the sperms mature. Right before ejaculation (the expulsion of the sperm during sexual intercourse or masturbation) the sperm moves into the ejaculatory duct where there are secretions of many different glands, among which we can find the Cowper's gland. As they become autonomous, they move all the way down to the end part of the urethra (the duct connected to the bladder, shared by both the reproductive and the excretory) where they are released.

The penis acts as a copulation organ, and to be able to enter the vagina it must have a certain level of rigidity: the two cavernous bodies (cylindrical structures of highly vascularized spongy tissue) fill up with blood and allow the erection to take place. Internal pressure helps to slow down the outflow of blood from the afferent veins, until the muscular contraction inside the arteries, at the time of ejaculation, decreases the inflowing amount of blood. At this point the penis relaxes.

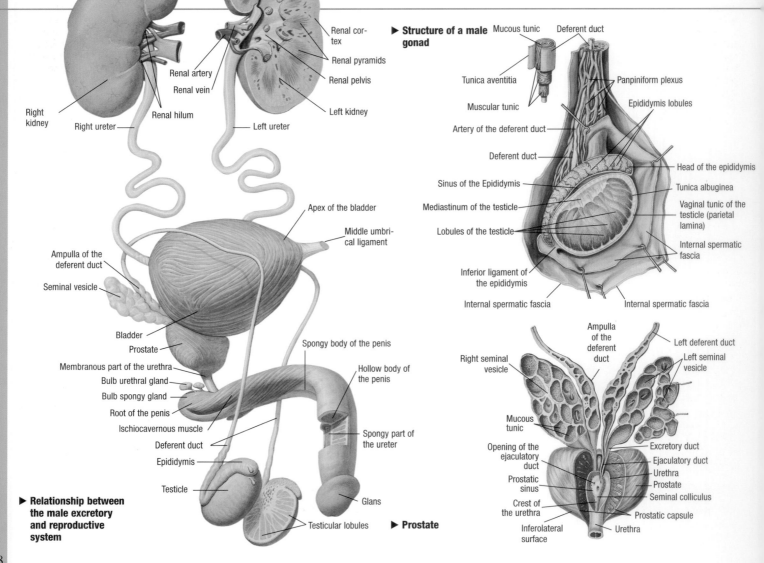

Renal cortex
Renal pyramids
Renal pelvis
Renal artery
Renal vein
Renal hilum
Right kidney
Right ureter
Left ureter
Left kidney

Apex of the bladder
Middle umbrical ligament

Ampulla of the deferent duct
Seminal vesicle
Bladder
Prostate
Membranous part of the urethra
Bulb urethral gland
Bulb spongy gland
Root of the penis
Ischiocavernous muscle
Deferent duct
Epididymis
Testicle

Spongy body of the penis
Hollow body of the penis
Spongy part of the ureter
Glans

Testicular lobules

▶ **Relationship between the male excretory and reproductive system**

▶ **Prostate**

▶ **Structure of a male gonad**

Mucous tunic
Deferent duct
Tunica aventitia
Muscular tunic
Artery of the deferent duct
Deferent duct
Sinus of the Epididymis
Mediastinum of the testicle
Lobules of the testicle
Inferior ligament of the epididymis
Internal spermatic fascia

Panpiniform plexus
Epididymis lobules
Head of the epididymis
Tunica albuginea
Vaginal tunic of the testicle (parietal lamina)
Internal spermatic fascia
Internal spermatic fascia

Ampulla of the deferent duct
Left deferent duct
Left seminal vesicle
Right seminal vesicle
Mucous tunic
Opening of the ejaculatory duct
Prostatic sinus
Crest of the urethra
Inferolateral surface
Excretory duct
Ejaculatory duct
Urethra
Prostate
Seminal colliculus
Prostatic capsule
Urethra

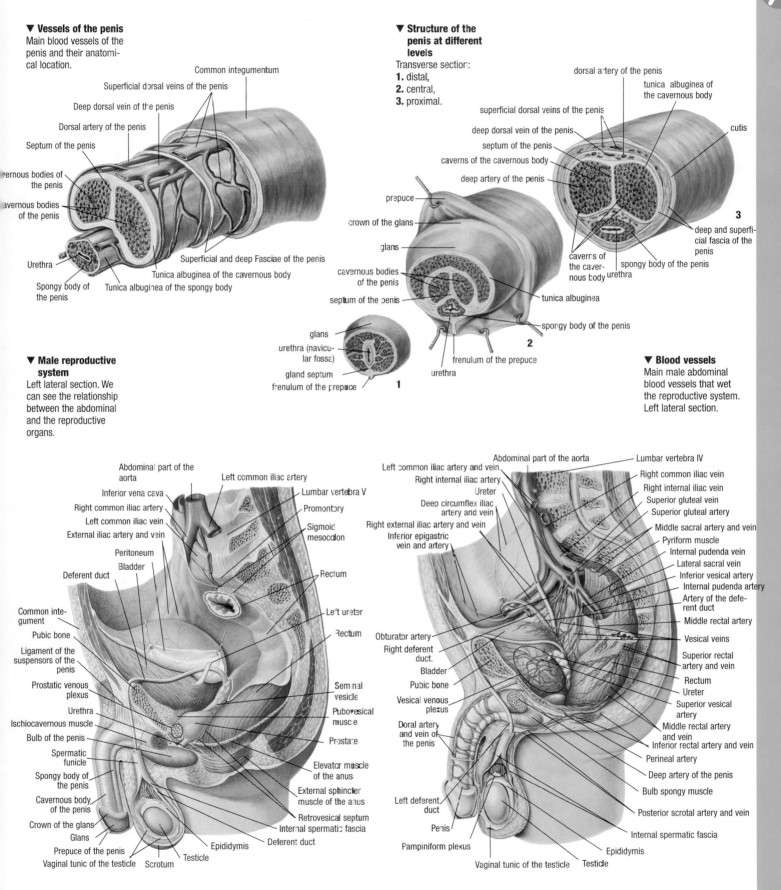

▼ Vessels of the penis
Main blood vessels of the penis and their anatomical location.

Common integumentum
Superficial dorsal veins of the penis
Deep dorsal vein of the penis
Dorsal artery of the penis
Septum of the penis
cavernous bodies of the penis
cavernous bodies of the penis
Urethra
Spongy body of the penis
Tunica albuginea of the spongy body
Tunica albuginea of the cavernous body
Superficial and deep Fasciae of the penis

▼ Structure of the penis at different levels
Transverse section:
1. distal,
2. central,
3. proximal.

dorsal artery of the penis
tunica albuginea of the cavernous body
superficial dorsal veins of the penis
deep dorsal vein of the penis
septum of the penis
caverns of the cavernous body
deep artery of the penis
cutis
prepuce
crown of the glans
glans
cavernous bodies of the penis
septum of the penis
caverns of the cavernous body
urethra
deep and superficial fascia of the penis
spongy body of the penis
tunica albuginea
spongy body of the penis
frenulum of the prepuce
urethra

glans
urethra (navicular fossa)
gland septum
frenulum of the prepuce

1
2
3

▼ Male reproductive system
Left lateral section. We can see the relationship between the abdominal and the reproductive organs.

Abdominal part of the aorta
Inferior vena cava
Right common iliac artery
Left common iliac vein
External iliac artery and vein
Peritoneum
Bladder
Deferent duct
Common integument
Pubic bone
Ligament of the suspensors of the penis
Prostatic venous plexus
Urethra
Ischiocavernous muscle
Bulb of the penis
Spermatic funicle
Spongy body of the penis
Cavernous body of the penis
Crown of the glans
Glans
Prepuce of the penis
Vaginal tunic of the testicle
Scrotum
Testicle
Epididymis
Deferent duct

Left common iliac artery
Lumbar vertebra V
Promontory
Sigmoid mesocolon
Rectum
Left ureter
Rectum
Seminal vesicle
Pubovesical muscle
Prostate
Elevator muscle of the anus
External sphincter muscle of the anus
Retrovesical septum
Internal spermatic fascia

▼ Blood vessels
Main male abdominal blood vessels that wet the reproductive system. Left lateral section.

Abdominal part of the aorta
Left common iliac artery and vein
Right internal iliac artery
Ureter
Deep circumflex iliac artery and vein
Right external iliac artery and vein
Inferior epigastric vein and artery
Obturator artery
Right deferent duct.
Bladder
Pubic bone
Vesical venous plexus
Doral artery and vein of the penis
Left deferent duct
Penis
Pampiniform plexus
Vaginal tunic of the testicle

Lumbar vertebra IV
Right common iliac vein
Right internal iliac vein
Superior gluteal vein
Superior gluteal artery
Middle sacral artery and vein
Pyriform muscle
Internal pudenda vein
Lateral sacral vein
Inferior vesical artery
Internal pudenda artery
Artery of the deferent duct
Middle rectal artery
Vesical veins
Superior rectal artery and vein
Rectum
Ureter
Superior vesical artery
Middle rectal artery and vein
Inferior rectal artery and vein
Perineal artery
Deep artery of the penis
Bulb spongy muscle
Posterior scrotal artery and vein
Internal spermatic fascia
Epididymis
Testicle

THE FEMALE REPRODUCTIVE SYTEM

The ovaries (the female gonads that produce the ovule) weigh only 30 g. and are located inside the pelvic cavity in an area between the belly button and the pubic bone. They are positioned on the side of the uterus (a muscular flipped over pear-shaped body) to which they are connected via the uterine tube or fallopian tube, or oviduct. When the mature ovule is released from the ovary it enters the peritoneal cavity and is captured by the tentacle like projections located at the end of the closest oviduct. Thanks to the cilia and to the contraction of the tube, the ovule reaches the uterine cavity.

At this point, if it is able to meet a sperm inside the tube, it gets "implanted" on the endometrial walls, a special epithelium which internally covers the uterine cavity, and begins its embryonal differentiation.

If fecundation does not take place, we have a menstrual cycle: the ovule and the degenerated endometrium are expelled through the neck of the uterus and the vagina.

Above the opening of the urethra, there is the clitoris, the main female organ of pleasure.

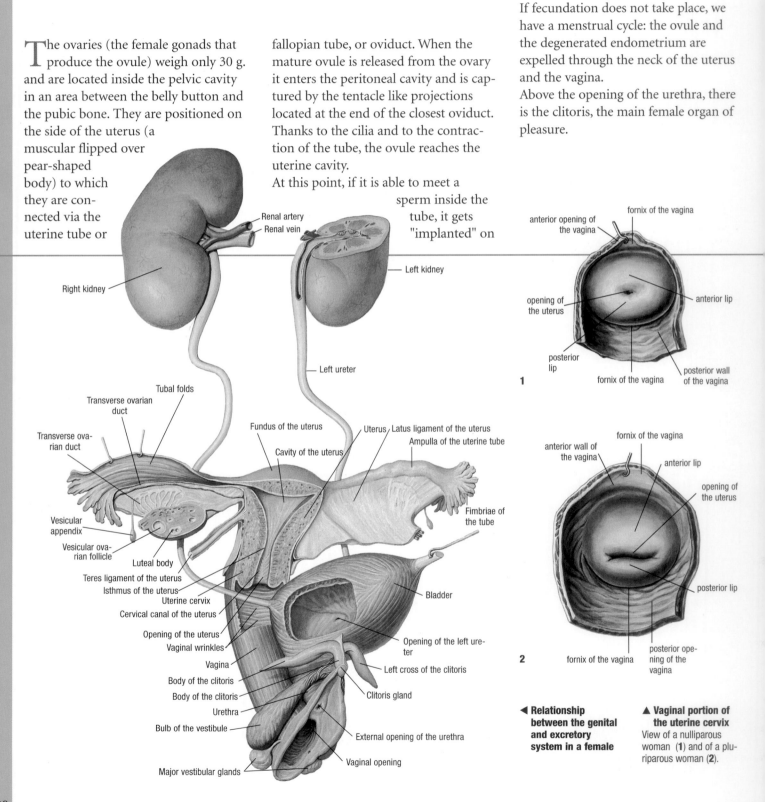

Right kidney

Renal artery
Renal vein

Left kidney

Left ureter

Tubal folds

Transverse ovarian duct

Transverse ovarian duct

Fundus of the uterus

Cavity of the uterus

Uterus / Latus ligament of the uterus

Ampulla of the uterine tube

Vesicular appendix

Vesicular ovarian follicle

Luteal body

Fimbriae of the tube

Teres ligament of the uterus

Isthmus of the uterus

Uterine cervix

Cervical canal of the uterus

Opening of the uterus

Vaginal wrinkles

Vagina

Body of the clitoris

Body of the clitoris

Urethra

Bulb of the vestibule

Major vestibular glands

Bladder

Opening of the left ureter

Left cross of the clitoris

Clitoris gland

External opening of the urethra

Vaginal opening

1

anterior opening of the vagina

fornix of the vagina

opening of the uterus

anterior lip

posterior lip

fornix of the vagina

posterior wall of the vagina

2

anterior wall of the vagina

fornix of the vagina

anterior lip

opening of the uterus

posterior lip

fornix of the vagina

posterior opening of the vagina

◄ **Relationship between the genital and excretory system in a female**

▲ **Vaginal portion of the uterine cervix**
View of a nulliparous woman (**1**) and of a pluriparous woman (**2**).

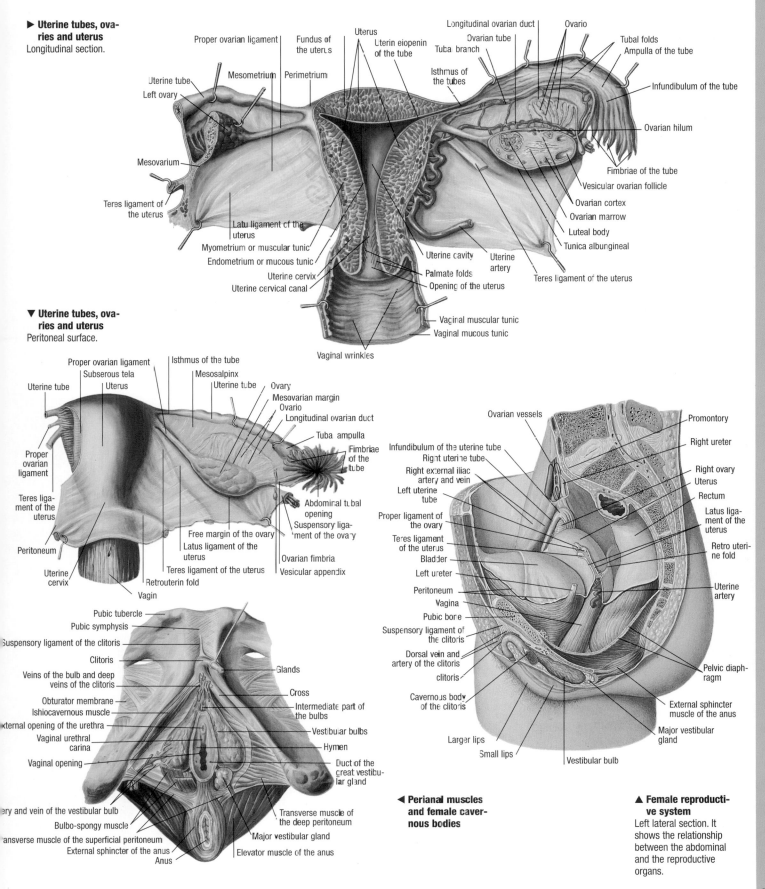

► **Uterine tubes, ovaries and uterus**
Longitudinal section.

Proper ovarian ligament
Fundus of the uterus
Uterus
Uterin eiopenin of the tube
Longitudinal ovarian duct
Ovario
Tubal folds
Ampulla of the tube
Tuba branch
Ovarian tube
Uterine tube
Mesometrium
Perimetrium
Isthmus of the tubes
Infundibulum of the tube
Left ovary
Mesovarium
Ovarian hilum
Fimbriae of the tube
Vesicular ovarian follicle
Teres ligament of the uterus
Ovarian cortex
Ovarian marrow
Luteal body
Tunica albungineal
Latu ligament of the uterus
Myometrium or muscular tunic
Endometrium or mucous tunic
Uterine cervix
Uterine cervical canal
Uterine cavity
Uterine artery
Palmate folds
Opening of the uterus
Teres ligament of the uterus
Vaginal muscular tunic
Vaginal mucous tunic
Vaginal wrinkles

▼ **Uterine tubes, ovaries and uterus**
Peritoneal surface.

Proper ovarian ligament
Subserous tela
Uterus
Isthmus of the tube
Mesosalpinx
Uterine tube
Ovary
Mesovarian margin
Ovario
Longitudinal ovarian duct
Uterine tube
Proper ovarian ligament
Tuba ampulla
Fimbriae of the tube
Teres ligament of the uterus
Abdominal tubal opening
Suspensory ligament of the ovary
Peritoneum
Free margin of the ovary
Latus ligament of the uterus
Teres ligament of the uterus
Ovarian fimbria
Vesicular appendix
Uterine cervix
Retrouterin fold
Vagin

Ovarian vessels
Promontory
Right ureter
Infundibulum of the uterine tube
Right uterine tube
Right external iliac artery and vein
Left uterine tube
Right ovary
Uterus
Rectum
Proper ligament of the ovary
Latus ligament of the uterus
Teres ligament of the uterus
Retro uterine fold
Bladder
Left ureter
Peritoneum
Uterine artery
Vagina
Pubic bone
Suspensory ligament of the clitoris
Dorsal vein and artery of the clitoris
clitoris
Pelvic diaphragm
External sphincter muscle of the anus
Major vestibular gland
Cavernous body of the clitoris
Larger lips
Vestibular bulb
Small lips

Pubic tubercle
Pubic symphysis
Suspensory ligament of the clitoris
Clitoris
Veins of the bulb and deep veins of the clitoris
Obturator membrane
Ishiocavernous muscle
External opening of the urethra
Vaginal urethral carina
Vaginal opening
Glands
Cross
Intermediate part of the bulbs
Vestibular bulbs
Hymen
Duct of the great vestibular gland
ery and vein of the vestibular bulb
Bulbo-spongy muscle
ransverse muscle of the superficial peritoneum
External sphincter of the anus
Anus
Transverse muscle of the deep peritoneum
Major vestibular gland
Elevator muscle of the anus

◄ **Perianal muscles and female cavernous bodies**

▲ **Female reproductive system**
Left lateral section. It shows the relationship between the abdominal and the reproductive organs.

221

SECONDARY SEXUAL CHARACTERISTICS

In both sexes, the maturation process that results during puberty, not only causes the development of all the structures of the reproductive apparatus, but also of the so called secondary sexual characteristics, which consist of a certain number of somatic, physiological characteristics which are profoundly different among the two sexes.

IN THE MALE

The increasing activity of the pituitary gland ▶[130], during this period, stimulates the adrenal glands and the testicles to produce androgens, and more specifically testosterone ▶[128,149]. This hormone affects the growth of the penis and of the testicles, and their darker pigmentation. At the same time, it induces the development of the somatic secondary characteristics: shoulders broaden, the muscle mass increases (more specifically the arms and the legs), hair grows thicker, especially in the pubic area, under the armpits, on the arms and on the chest. Beard, mustaches and sideburns develop on the face, and sometimes hair grows on the back as well. As it modifies, the larynx changes position on the neck ▶[167] and the voice deepens.

In addition, the cutaneous tissue changes: the sebaceous and sudoriparous glands undergo a more intense activity.

Deep psychological changes accompany also physical changes, which coincide with the development of intellectual abilities.

IN THE FEMALE

Even in women, the increasing activity of the pituitary glands, triggered by the stimuli of the hypothalamus ▶[130], induces the adrenal glands and the gonads to produce hormones: while the ovaries secrete progesterone and estrogen ▶[128,140], the hormones that regulate the ovarian cycle, the adrenal glands produce androgens, which are

HOMOSEXUALITY

Until few years ago, probably because of hundreds of years of influence of Christian ethics on the observation of nature, ethologists and western zoologists believed that, masturbation, copulation, the formation of a long lasting monogamous relationship, and courtship behaviors among same sex animals were abnormal behavior.

This belief was not based on specific observations and methodical research, but on the prejudice that human homosexuality was an aberration "against nature", "unnatural": nature was and had to be heterosexual.

It was only after the 24th International Conference of Ethology, on August 1995, that the world realized that the research on animal homosexuality was a legitimate field of investigation. The studies conducted in this field have and are still changing the attitude that science has toward this problem: animal homosexual behavior is not only popular but in some species of animals is actually the norm. The animals under investigation are all "superior animals": they are mammals such as orangutan, macaques, bison, antelopes, giraffes, chimps, gorillas, lions, seals, killer whales, dolphins, and birds such as silver seagulls, penguins and ducks.

When we speak of sexuality, the debate on what should be considered normal becomes fierce, because the results obtained from humans, directly involve ethical and moral issues. And, if in humans, culture and education are also involved with the biology and the genetics of animals as well, the sexual relationships of same sex individuals are not only a sexual factor, but they involve different behavioral levels.

What determines the sexual choice of a human being as well as an animal? This question is typical of those cultures that still view homosexuality as an anomaly: today nobody asks anymore why one prefers to use the left hand over the right one, and nobody would think to change this "defect" anymore; Nobody asks why one prefers a color to another. On the other hand, if we think of sexuality as specifically designed for reproduction, then homosexuality still must be explained. A recent convincing interpretation sees the sexuality of "superior animals" (more behaviorally and cerebrally advanced) not only as a behavior solely deigned for reproduction, but as on of the many expression of biological exuberance. Just as it occurs in all other aspects of nature, even in the sexual sphere, nature seems to be regulated on one side, by selection and limiting factors, and on the other side by an abundance and an excess of environmental

Male sea lions assume particular positions in order to stimulate each other's genitals. These positions are typical, and are never assumed when mating with females.

responsible for the somatic changes and for the characterization of the female body.

Due to the hormonal changes, the thickness of the adipose tissue increases, particularly on the hips, thighs, gluts, and the forearms and under the nipples, giving the breast its shape.

At the same time, all of the glandular tissues are stimulated to grow: particularly those of the mamilla which enlarge the breast, the sebaceous and soporiferous ones that, by increasing their activity, often cause discomfort, and the vaginal ones, that begin their lubricating activity.

Pubic and axillary hair begins to grow in women as well. The sexual organs begin to enlarge and the cutaneous tissue starts to change.

Also, just as it happens for males, psychological changes begin to appear, expanding the emotional and intellectual abilities of the person.

◄ Mamma
Right lateral section

Clavicle
Subclavian muscle
Pectoral fascia
Common integument
Lung
Intercostals muscles
Greater pectoral muscle
Smaller pectoral muscle
Rib
Mammary suspensory ligaments
Lobes of the mammary gland
Mammary papilla
Milk ducts
Mammary gland
Milk sinuses

possibilities. The homosexual and heterosexual components that are not geared solely toward reproduction could therefore be foreseen as alternatives to heterosexual behavior geared primarily toward reproduction, in other words, manifestations of the heterogeneity typical of the biological systems. At the same time, it is true that the sexual behavior of our species is heavily conditioned by "acquired" cultural elements. This is possible because the response to sexual stimuli in men is not only a conditioned reflex, but a complex nervous reaction which involves also a wide area of the cerebral cortex: memory centers, olphactory centers, auditory and tactile, associative centers….. The response evoked by the stimuli inside the hypothalamus and inside the limbic system is "mediated" by the cerebral cortex: this circumstance frees the human sexual answer from its physiological roots,

submitting it to reason, education and experience. In other words, to culture and other cultural traditions. In this way, while in some cultures homosexuality is ritualized assuming important symbolic meanings and individuals are free to express themselves, in other cultures, this behavior is condemned, and homosexuals are marginalised and their behavior is considered "unnatural".

◄ Response to a sexual stimulus
1. The stimulus causes a reflex arch (blue and red arrows)
2. The stimulus reaches the brain through sensory pathways (blue) of the spinal cord.
3. When the stimulus reaches the brain, the sexual response becomes conscious.
4. The associations between (social context, memories) and visual, tactile, olphactory signals determine the sexual response.
5. The cerebral response is transmitted by the motor pathways (red) and can reinforce or inhibit the sexual reflex.

SPERMS, OVULA AND OVARIAN CYCLE

Gametes are produced inside the female ovaries, and inside the male testicles, according to a process of reducing the genetic set which allows the fusion of an ovule and of a sperm to reconstruct a totipotent cell with a complete chromosome set.

IN THE MALE

Through a succession of cellular divisions, thanks to the production by interstitial cells of the testicles of androgen hormones, 4 sperms originate from each secondary spermatocyte. They are released inside the lumen of the seminiferous tubule that, thanks to the ciliate epithelium and to contractions, pushes them all the way to the epididymis, where the sperms are stored for about 10 days. They are continuously produced: a sexually mature person can produce up to 100.000 mm_ of ejaculation. Then the sperms move to the deferent vessel where they mature, thanks to the products of various glands.

IN THE FEMALE

The ovary is made up of different cells: inside the stroma, there are many "lethargic" ovula, surrounded by interstitial cells. They develop inside the Graaf's follicles, which are minuscule spheral cavities that, as the ovule matures, tend to enlarge and to migrate toward the surface of the ovary, opening on the outside (dehiscence of the follicle).

The ovule, surrounded by a radiated crown of follicular cells, is released inside the peritoneal cavity (ovulation).

▶ **Detail of a testicle**
The diagram, which describes the section of a seminiferous duct, shows two different enlargements of the stratified disposition of the cells from which the sperms originate. The pictures **A** and **B** show two different histological sections that allow us to see the sperm that "appears" inside the lumen of the duct.

▶ **Production of ovula and spermatozoai**
The testicular spermatogones multiply through mitosis, keeping the number of chromosomes intact. After differentiating into secondary spermatocytes, they go through meiosis, and generate spermatozoa, which are characterized by half the total number of chromosomes. On the other hand, the ovule, originates inside the oocyte with the successive extrusion of two polar types: the first one represents the nucleus of the mitosis of the oocyte; the second one is made up of a meiotic nucleus that cut the number of chromosomes in half.

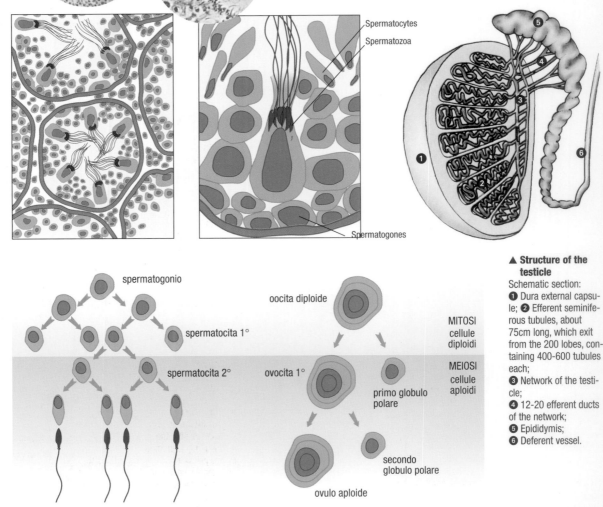

Spermatocytes
Spermatozoa
Spermatogones

spermatogonio
oocita diploide
spermatocita 1°
MITOSI cellule diploidi
spermatocita 2°
ovocita 1°
MEIOSI cellule aploidi
primo globulo polare
secondo globulo polare
ovulo aploide

▲ **Structure of the testicle**
Schematic section:
❶ Dura external capsule; ❷ Efferent seminiferous tubules, about 75cm long, which exit from the 200 lobes, containing 400-600 tubules each;
❸ Network of the testicle;
❹ 12-20 efferent ducts of the network;
❺ Epididymis;
❻ Deferent vessel.

OVARIAN CYCLE

The ovarian activity begins when the hormones of the adrenal gland ►130 stimulate the development of the follicles that, by secreting estrogen, allow the endometrium to proliferate and further activate the pituitary gland. Only one of the stimulated follicles matures each month: every 28 days, one or the other ovary releases an ovule. The peak production of estrogen triggers the production of pituitary LH (luteinizing hormone), which determines the dehiscence of the follicle. This transforms into a luteal body that by secreting progesterone, increases the development of the endometrium (menstrual cycle) and stimulates the pituitary gland that initiates a new cycle.

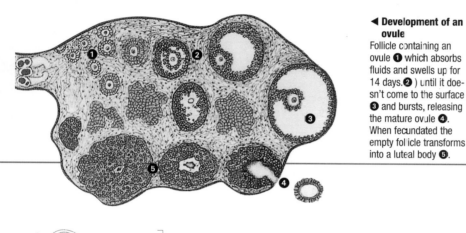

◄ Development of an ovule
Follicle containing an ovule ❶ which absorbs fluids and swells up for 14 days.❷) until it doesn't come to the surface ❸ and bursts, releasing the mature ovule ❹. When fecundated the empty follicle transforms into a luteal body ❺.

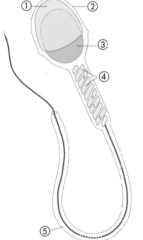

testa
(5 μm)

parte di congiunzione
(5 μm)

flagello
(50 μm)

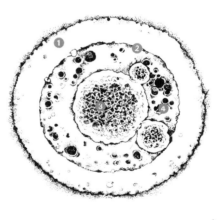

▲ Structure of a spermatozoon
Schematic section:
① Acrosome; it's a vesicle filled with enzymatic fluid, which bursts near the ovule, releasing substances that breakdown the pellucid area, allowing the spermatozoon to fecundate the ovule;
② Cellular membrane; it fuses with the membrane of the ovule allowing the
③ nucleus to enter, joining together the paternal genetic set with the maternal one.
④ Mitochondria; are numerous and necessary for the production of the energy needed by the flagellum
⑤ to move. This is a propulsion element that allows the spermatozoon to reach the ovule. The flagellum and the junction part are not part of the fecundated ovule, but they remain on the outside.

▲ Structure of an ovule during fecundation
Schematic section:
❶ Pellucid area; it's a protein coating that protects the ovule
❷ Cellular membrane
❸ Nucleus: after the fecundation, its chromosomes join with the homologous ones, contained inside the nucleus of the spermatozoon
❹ Mitochondria and other cellular structures.

CONTRACCEPTION

Those who decide not to have any children can rely on natural or artificial methods to efficiently prevent fecundation.

Natural methods
The most drastic and less practiced method is complete abstinence. The coitus interruptus (the rapid removal of the penis from the vagina right before ejaculation) requires a lot of self-control by the man, and doesn't offer a high guarantee of protection, just like the calculation of the fecund days in a woman (the so called Ogino-Krauss method): we still know very little about the survival time of the spermatozoa inside the female apparatus, and about the unstable female hormonal rhythms.

Medical-surgical methods
The male surgical sterilization or vasectomy (simpler), and the female tube resection (more complex and less certain), are usually drastic and permanent procedures. The ducts of the respiratory apparatus are cut or tied, interrupting then the regular afflux of the gametes. The IUD (Intra Uterine Device), is a plastic or metal inert device that often contains copper and with various shapes. The gynecologist inserts it inside the uterine cavity: by irritating the endometrium it prevents the ovule from implanting into it with a high success rate. The already famous birth control pill is an oral contraceptive that alters the hormonal balance of a woman. It is made up of a cocktail of hormones that prevent ovulation. There are also other drugs that interfere with the density of the cervical mucous or with the structure of the endometrium. Researchers are experimenting a new birth control pill for men, which can stop the production of spermatozoa.

Mechanical Methods
These methods are simple, safe (except for some individual hypersensitivities) and mechanically prevent fecundation: they are the condom (it is applied when the penis is erect and is the only method that protects from sexually transmitted diseases), the diaphragm and the cervical cap(they cover the top part of the cervix and the neck of the uterus). When using these devices a spermicidal cream should be used as well.

FECUNDATION, PREGNANCY AND NURSING

After ejaculation, sperm is shot out into a very hostile environment, made up of acidic vaginal secretions: following the gradient, and thanks to the vivacious movements that characterize them, they enter the uterus, but before reaching the ovule they must go through a series of physical and chemical barriers, which drastically reduces their number. We believe that only 100 out of the 350 million of sperms that pass the neck of the uterus will reach the tubes. Near the ovule, they release enzymes that destroy the protein covering that surrounds the ovule, until only one, unites with it blocking the entrance of other sperms. This process is called fecundation, and it produces the first embryonal cell: the zygote. Soon the zygote begins to rapidly split and by the time it reaches the uterus it is already made up of many cells. At one point it gets" trapped" inside one of the anfractuosities of the endometrium, it adheres to it and continues to split (implant of the zygote). Around the 7th day of fecundation, some cells penetrate inside the maternal tissue (endometrial nidation) and move through the uterine mucosa:

This is the beginning of pregnancy, the process that allows the unborn baby to develop inside the mother's womb. As the embryo grows more, it starts to differentiate: in about 9 days the vitelline sac and the amniotic sac begin to develop. They are filled with amniotic fluid, and by the end of

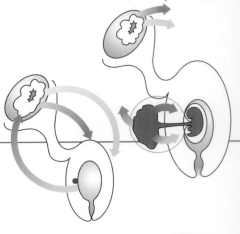

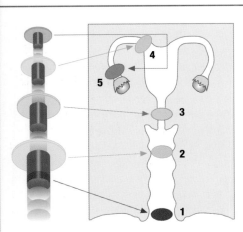

▲ Selection of the spermatozoa
Reduction of the number of sperms at different levels of the female genital apparatus:
1. up to _ of the 350 million of sperms normally contained in an ejaculation is anomalous, non competitive;
2. ;
"trap" of the cervical mucous which many sperms dissolve by freeing specific enzymes and "scarifying" themselves to let others pass through; **3.** only about 1 million go through the cervix; **4.** only a thousand sperm enter the tubes; **5.** about 100 reach the ovule.

▶ Embryonal development during the first week
❶ 2 cells
(30 h, Ø 120 µm)
❷ 12-18 cells
(4 gg, Ø 120 µm)
❸ blastula
(5 gg. Ø 120 µm)
❹ gastrula
(5 gg. e mezzo, Ø 140 µm)
❺ blastocytes
(6 gg., Ø 140 µm).

▶ Hormones
The chorionic gondotropin (orange arrow) produced by the embryo maintains the production of estrogen (green arrow) and progesterone (blue arrow) by the luteal body, replaced after the third month, by the placenta.

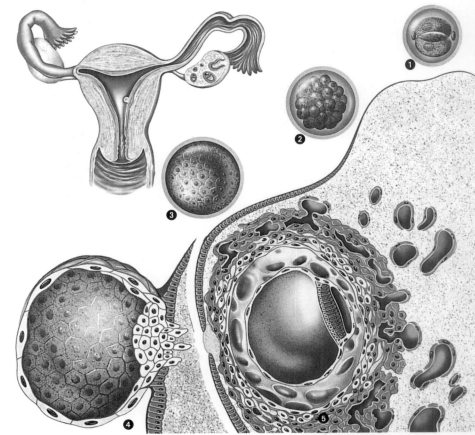

the 4th month, a mother can already feel the movements of the baby: the fetus is floating inside the amniotic fluid; it lifts and lowers its legs, opens and closes the hand, turns its head and does somersaults. The vegetative nervous system becomes active: the body, covered with a thin and transparent layer of skin, has a deep red color, typical of muscles; t is 13,5 cm long and weighs 130g. Within 40 weeks, a normal baby has already developed each of the

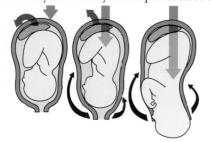

delicate and complex systems of the body: more than 200 bones, 50 dental germs, muscles and functional organs, and a brain, whose size is about _ of that of an adult; starting from a single cell, after about 9 months a new being is created, made up of about 200 million cells, whose weight is 1 billion times the weight of an ovule. Usually childbirth takes around 280 days from the beginning of the last men-

◄ Childbirth's hormones
Childbirth is induced thanks to a progressive decrease in the production of progesterone (red arrow) and by an increasing production of pituitary oxytocin (purple arrow), which stimulates muscular contraction.

strual cycle, but because conception does not occur right at the end of the menstrual cycle, it is difficult to determine the exact date of delivery. It is not clear yet what it is that regulates with so much precision the length of pregnancy: it is possible that childbirth is regulated by a convergence of maternal and fetal hormonal signals. Also, precise hormonal signals regulate the production of breast milk.

◄ Milk production
Suction ❶ induces a nervous reflex ❷ which stimulates the pituitary gland ❸ to secrete prolactin and oxytocin ❹. These hormones induce, on one side, the production of milk, and on the other its passage through the milk ducts.

STERILITY

Sometimes fecundation might not be successful: about one couple out of 6 is not able to reproduce. Generally, a couple around 20 years of age, has an 18% probability to have a baby, and a woman's fecundity decreases after the age of 30: on average, a couple needs to try for 5-6 months before getting pregnant, but even if after 2 years of attempts the result is often not successful, regardless of sterility. In fact, according to recent investigations, only 10% of people are truly sterile. For thousands of years, sterility was associated with a woman's inability to have children: it is only since 1950 that some American research has shown that male sterility is quite frequent as well (about 40% of sterility in couples is caused by male sterility, while about 10% of sterility in couples can be caused by both female and male problems)

Male sterility
There are three types of male sterility, depending on the type of anomaly, which can in turn derive from:
-An abnormal number of sperms: the number of sperms that reaches the ovule is very important, and if it is too low, fertilization does not take place. 8% of the men that

underwent a fertility test, didn't produce enough spermatozoa (azoospermia); a more diffuse problem is oligospermia: in this case the production of sperms can be inhibited or slowed down (even irreversibly) by an inflammation of the gonads, by using drugs, alcohol, barbiturates or antidepressants, and even by the exposure to some environmental agents such as radiation and industrial substances (some metal, and organic compounds) or even by using underwear that is too tight;
-Functional anomalies of the spermatozoa such as the presence of two tails ① or of a double head ②, or a reduced mobility of the flagellum, altering therefore the ability of the sperm to move and to fecundate;
-Functional anomalies of the copulatory-ejaculatory apparatus: the inability to have an erection or to contract the muscles at the base of the bladder (that push the sperm through he urethra), can be due to psychological, physical causes, or to the use of drugs and medicines.

Female sterility
If with age pregnancy becomes harder (in fact the production of ovules decreases after reaching 40), in young women, sterility can be caused by different factors: there are two main categories:

-Hormonal sterilities: they alter ovulation and characterize about 20-35% of female sterility. They are often accompanied by irregular menstrual cycles, and are connected to ovarian anomalies (absence of follicles), pituitary alterations or a deficiency of the luteal body that causes the death of the ovule before it reaches maturation (frequent in the first few years after reaching 30);
-Mechanical sterilities: are caused by obstacles that make the migration of the ovule very difficult (an obstruction of the tube can cause sterility in 25-40% of cases), by obstacles of the spermatic reception (10-15% of sterility), by endometrial alterations, which obstruct the nidation of the zygote.

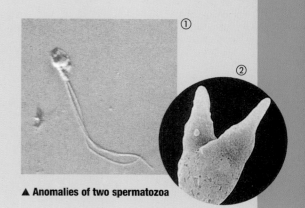

▲ **Anomalies of two spermatozoa**

✛ THE BIONIC MAN

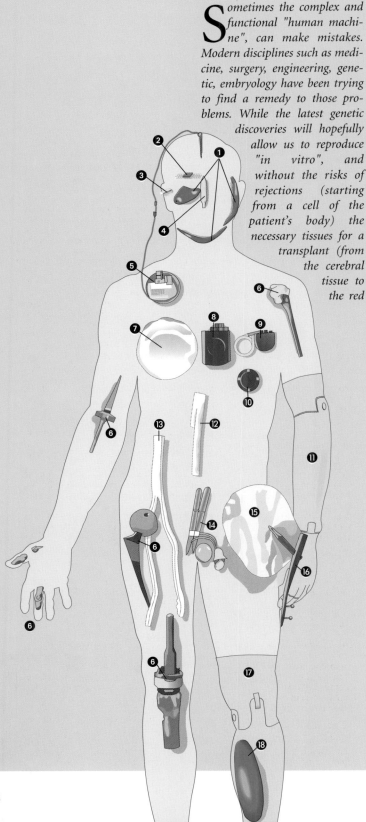

Sometimes the complex and functional "human machine", can make mistakes. Modern disciplines such as medicine, surgery, engineering, genetic, embryology have been trying to find a remedy to those problems. While the latest genetic discoveries will hopefully allow us to reproduce "in vitro", and without the risks of rejections (starting from a cell of the patient's body) the necessary tissues for a transplant (from the cerebral tissue to the red blood cells, from the hepatic tissue to that of the pancreas), the attempt to substitute the damaged pieces of the "human machine" have been involving for years other fields such as medicine, engineering and biotechnology. In the recent years we have made much progress in both the development and the use of synthetic materials better fitted for different surgical needs, and in the miniaturization of instruments to assist defective organs, and in the creation of new reconstructive and transplant surgical techniques. For example, we must not forget the recent development of many bioartificial organs: skin, cartilage, bones, semi synthetic ligaments and tendons are already a reality (for those that can afford it) in the US, but also in other advanced institutes.

Genetic engineering is still a big question mark: among thousands of ethical, moral and scientific controversies, research moves on. In the future we will probably have animals cloned with the same characteristics of humans, to be used for xenotransplants; we are already working on the human embryonal tissues, whose cells are still undifferentiated and therefore perfect for the reconstruction of organs perfectly compatible to those of the patients'…

For now, the most popular surgical interventions remain the "traditional" ones: transplants from compatible donors, use of mechanical instruments (such as pacemaker or the artificial articulations), prosthesis that can partially substitute the damaged pieces of our body.

◄ **Substitutive pieces**
Here are some popular elements surgically used to substitute organs or missing parts of the body:

❶ different materials prosthesis substitute certain bone structures of the face (zygoma, mandible, maxillary, etc.)

❷ electronic prosthesis used to increase vision

❸ electronic prosthesis to increase hearing (hearing aids)

❹ electronic prosthesis for the reconstruction the nasal septum

❺ electronic equipment for the regulation of motor impulses, in the cure of Parkinson's disease

❻ artificial joints: shoulder, elbow, hand, hip, knee

❼ mammary prosthesis

❽ cardiac pump

❾ pacemaker

❿ titanium cardiac stimulator

⓫ prosthesis of an arm, with electronic hand

⓬ artificial aorta

⓭ artificial veins and arteries

⓮ erectile penile prosthesis

⓯ artificial anus

⓰ articulated femoral prosthesis

⓱ prosthesis of the lower limb, with foot and knee joint

⓲ calf prosthesis

GLOSSARY

Å = **Angstrom**: it is the ten-billionth part of the meter (equal to 1/10.000.000 of 1 mm); 1 Å 10.000 times smaller than 1 mm.

Actin: a protein made up of two protein chains covered by a double helix. By joining with the filaments of myosin, it allows the muscular cell to contract.

Adenomere: functional unit of a gland. It is made up exclusively of cells that produce substances secreted by the gland.

Adolescence: period in which puberty takes place, along with many other somatic, emotional and intellectual changes that mark the passage from infancy to adulthood.

Agglutination: a reaction that causes the union of large numbers of corpuscles, which have the tendency to sediment (red blood cells or bacteria) and that are usually floating inside a fluid. Usually, agglutination is caused by the presence of antibodies.

Ameboid, movement: cellular movement that takes place through the movement of fluid material inside the cell (cytoplasm) and the estroflexion of pseudopodes. In other words, one side of the cell stretches toward the desired direction to then transfer the entire cellular content.

AMP: the initials of the adenosine-5-monophosphoric acid, a molecule that is able to bind to other phosphate molecules, through high-energy bonds, transforming therefore into ATP. It plays an essential role in the biosynthetic and muscular contraction processes.

Amino acid: organic compound whose molecule is made up of a carboxyl group (acid, COO_) and of an aminic group (basic). The amino acids make up proteins, binding with each other into long chains (polymers).

Aromatic: it contains one or more cyclical structures made up primarily of carbon atoms.

Terminal: it is the amino acid found at the end of a protein. Either the basic group or the acidic group is free.

Anastomosis: connection between blood vessels and nervous trunks, or the same canals of two hollow organs.

Renvier Ring: an area on the surface of a myelin neuron, where the myelin coating is interrupted due to the absence of Schwann cells.

Aponeurosis: tendon shaped as a flattened layer.

Asclepium: ancient Greek divinity, patron of medicine. The entire ancient world worshipped him as a healing divinity (in Rome he became Esculapio). He was characterized by the symbol of a snake rapped around a stick. The system used inside the sanctuary dedicated to him was quite mysterious, but we know that here, surgical practices and medical applications took place, along with other disciplines such as the interpretations of dreams, the purification of patients through bathing, fasting and propitiatory rituals.

Atrophy: shrinking, degeneration or reduction of

the capacities of an organ or of a part of the body.

Arteriosclerotic, formation: a formation found inside arteries affected by arteriosclerosis; a degenerative pathology characterized by a thickening of the internal layer of the walls of the arterial vessels, which causes the shrinking of the vasal lumen and a reduction of the blood flow. These are plaques (atheromes), made up of fats coming from the blood circulation that deposit inside the vessels.

ATP: the initial for adenosine -5-triphophate, an energy filed molecule that originates from the intercellular degradation of "combustibles" (fats, sugars….) stored as phosphoric bond. The AMP binds to the phosphate molecules through this bond. When required by the metabolic processes (syntheses of the macromolecules, transport of substances through the membrane, muscular contraction), it is able to release energy again.

Target organ: in the case of hormones, it is the organ that is selectively stimulated by a specific hormone; in the case of therapies or examinations that use radioisotopes or antibodies, it is the organ selectively involved in the specific method utilized.

Catabolism: a set of chemical processes which characterizes a metabolism; it releases chemical energy transforming larger molecules, rich in energy, into smaller ones.

Schwann cell: it develops by wrapping itself around the axon of a neuron, becoming one of the elements of the myelinic "cuff".

Cyst: membranous sac, containing fluid or other material (fat, hair, etc…) that can grow on the surface or deep inside any anatomical structure. It can be congenital or it can grow: usually the latter type develops due to the presence of foreign bodies (splinters, unabsorbed chemical substances), or due to the introduction of foreign substances inside a tissue that reacts to their presence, incorporating them into a connective membrane.

Cytoplasm: fluid part of the cell (gel) found between the plasmatic membrane and the nucleus. Inside the cytoplasm there are numerous cellular organelles, where the main metabolic activities of the cell are divided into compartments. It is made up primarily of water (80%), of proteins at different stages of aggregation, of nucleic acids, sugars and of aggregations of ions.

Blood coagulation: process that occurs when a blood vessel is damaged resulting in a blood spill; it creates some sort of tissue, more or less compact, made up of a fibrine reticulum. The erythrocytes deposit inside its web, until they obstruct the wound, by blocking the hemorrhage. In order for coagulation to take place, platelets (or thrombocytes) are indispensable.

Collagen: fibrous protein found inside the connective tissues, which makes up 30% of all the proteins in the body; it is made up primarily of glycine and of praline, and it creates triple chains, spirally

wrapped, which bind with each other and form fibrilles that are characterized by a limited elasticity and a strong resistance.

Complement: enzymatic system of the blood, made up of 9 essential factors, necessary to obtain the lyses of cells that have been recognized by the defense system of the body.

Composition: blood composition.

Chromosome: isolated or grouped filiform structure found inside cellular nuclei. Made up primarily of DNA and proteins, it contains the genes that determine the individual characteristics of each person. The number of chromosomes found in each cell of an organism is characteristic of that particular species. The somatic cells (of the body) contain double the number of chromosomes, compared to those of sexual cells (gametes).

Course: trend, development, and course of an illness or of a phenomenon. In anatomy, it also indicates the course of a vein or of a nerve.

Dentition: a change in the quaternary and tertiary structure of a protein, which causes the loss or the alteration of the specific chemical properties. It can be produced by heat, by Ph and by chemical reactions.

Dendrite: thin extension of the cellular body of a neuron or nerve cell.

Radiological Density (or opacity): quantity of X rays that a tissue undergoing radiography, can absorb. This quantity is high in bones, and very low in internal organs. In order to be able to highlight internal organs it is necessary to implement the use of some contrast.

Dermis: thicker and more internal layer of the cutis. It is made up of connective tissue rich in blood vessels, nerve endings, elastic fibers and smooth muscle fibers.

Differentiation: biological process that, starting from a cell without any specific characteristic or functions (non differentiated) causes the formation of a cell characterized by a particular structure and function (differentiated). This process occurs throughout one's life: it marks the passage of cells from the embryonal stage to the mature one, during both the fetal development and also during the continuous cellular regeneration of different tissues. According to most researchers, at the basis of this differentiation there is a complex game of activating and suppressing the genes found inside the DNA equal in all the cells of the same body. Such differentiation is also characterized by the extra cellular environment: growing factors, hormones,[230] and cellular interactions affect the process according to mechanisms which are still not quite clear.

Facilitated Diffusion: the transport of molecules through the plasmatic membrane of a live cell, thanks to a specific carrier (usually a protein). This is a process that doesn't require the expenditure of metabolic energy.

Dislocation or luxation: literally "removed from its

place", it indicates the shifting of a bone from its proper articular site.

Dissection: cut and removal of parts and organs of the human body.

Electrolyte: dissolved salt or ion that participates in one of the many series of chemical processes occurring inside the human body. Sodium and potassium are some of the most important ones.

Hematic flow: blood flow

Embryo: a living creature during its first developmental stage. In a human being, this name is given to an unborn, until it reaches the 2nd month of life.

Hemopoiesis (or hematopoiesis): literally the hematic production; it is the process of cellular differentiation, which produces blood cells from the staminal cells of the bone marrow, of the spleen, of the thymus and of the liver.

Hemorrhage: blood spill due to damage of the walls of the vessels.

Endometrium: mucous tunic that coats the uterine cavity and that, under the influence of the ovarian hormones, is subject to a cyclic hypertrophy and scaling. It adheres to the muscular tunic (myometrium) and is distinguished into two layers: a basal one that has a regenerative function, and a functional one. This is the layer that changes during the ovarian cycle, and is eliminated during the menstrual cycle.

Endothelium: tissue made up of flat cells connected to each other by a thin membrane. It covers the lumen of the blood vessels and of the lymphatic vessels.

Enzyme: a protein able to speed up a specific chemical reaction.

Epidermis: superficial layer of the cutis that covers and protects the body. It is made up of a series of cellular layers: one of the deepest ones is the germinative layer, directly connected to the dermis, from which new epidermal cells constantly originate. Progressively, they move away and are pushed toward the surface. The outer layer of the epidermis is made up of dead cells that constantly scale.

Phagocytosis: process during which the plasmatic membrane introflects and then closes around a solid particle, which is captured inside a vacuole. In turn, it fuses with a lysosome, a cellular organelle that contains enzymes, which are able to "digest" the particle.

Phagocytic activity: the ability of a cell to engulf other cells or foreign bodies (cellular fragments, microorganisms, organic particles such as pollen, etc..) and to destroy them.

Fetus: a developing unborn baby, from the 3rd month of pregnancy until birth.

Photoreceptor: cellular element able to react to different light intensities (literally, light receptor)

Ganglion: nodule made up of nervous cells.

Gel: the consistency of a semisolid body, made up of gelatinous substances (colloids) dissolved in a solvent. The most popular organic example is the albumen of the egg, which is filled with water.

Germinal: that which facilitates reproduction.

Gland: group of cells specialized in the production of one or more specific substances (secrete). There are two types of glands:

Endocrine: that which pours its secretion inside the blood. The secretion of an endocrine gland is called hormone.

Exocrine: …that which doesn't pour its secretion inside the blood flow, but outside of the body. Among these there are the salivary glands of the digestive system, the sudoriparous glands and the lachrymal glands.

Myelin sheath: a coating of Schwann cells, which surrounds the neuronal axon acting as an electrical insulator.

Hydrolysis: chemical reaction that breaks particular molecular bonds as they react with water.

Ion: atom or molecule with a charge (positive, if it has lost electrons, and negative, if it has acquired them) the positive charged ions are called cations, and those with a negative charge are called anions. Almost all of the substances that are dissolved in water are in an ionic state.

Hypertrophy: abnormal development of a tissue, of an organ or a part of it, which cause an increase in volume. It can be a normal process (endometrium), induced (as in the case of muscular development on body builders) or pathological.

Hyperventilation: increasing pulmonary ventilation causing deeper and more frequent respiration.

Peptic bond: chemical bond between the basic and the acidic extremities of amino acids molecules. The peptic bonds are the skeletons of proteins.

Ligament: fibrous connective tissue that binds two or more anatomical structures.

Lysis: literally "rupture"; it indicates the dissolution process of cells or microorganisms, produced by the attack to their plasmatic membrane by chemical physical or biological agents (antibodies and complement).

Wave length: the distance between two successive peaks or valleys of a wave. In the case of a radiation, the wavelength (1) indicates the distance between two consecutive peaks of an electromagnetic wave. In a vacuum, it is associated to the frequency (n), the physical measurement which indicates the number of repetitions of the periodical phenomenon (undulation) in a unit of time, from the formula l x v = c, where c indicates the speed of light (3x10 to the 8th power m/s).

Chemical mediator: a substance that "mediates" a physiological process through a specific chemical reaction. In particular, neurotransmitters facilitate the transmission of a nervous impulse by activating specific molecules of the membranes of postsynaptic cells.

Meiosis: cellular division that produces haploid gametes, which are reproductive cells whose number of chromosomes is halved, compared to the regular number of the somatic cells of the specie.

Plasmatic membrane (or cellular): lamellar tissue, which covers, connects or coats the cell and its corpuscular elements (organelles). It is semi permeable, and made up primarily of proteins and phospholipids. Many ions and water can go right through it, via osmosis. The biggest molecules can pass through, thanks to the phagocytic process. From a metabolic standpoint, some important molecules are actively transported through specific types of metabolisms (active transport, facilitated diffusion)

Menopause: interruption of the female ovarian cycle that takes place, on average, after the age of 40.

Mesentery: lamina of tissue outlined on both sides by the peritoneum. It supports the intestine and all the other organs inside the abdominal cavity.

Metabolism: a set of chemical transformations of energetic phenomena which takes place inside an organism or a cell (cellular metabolism), and that assures the conservation and the renewal of the live material that characterizes it.

Myelin: lipoid substance found inside the Schwann cells. Thanks to their characteristics, the coating of cells can act as an electrical insulator for the membrane of the myelin fibers.

Myofibril: bundle of contractile filaments (made of actin and myosin) that run along the major axis of the muscular fiber.

Myosin: protein made of two protein chains partially coated with a double helix and with a "head" that is able to bind with the actin molecules and change shape, depending on the concentration of Ca+. This mechanism allows for the creation of temporary actin-myosin bonds, at the basis of the muscular contraction.

Mitochondria: cellular organelle that has its own membrane and DNA located inside the cytoplasm of the nucleate cells and of the superior organisms. It plays an important role in cellular respiration, and in the synthesis of compounds filled with energy. During the cellular division, all the mitochondria of the mother cell reproduce through scission, and distribute inside the siblings cells: each baby receives the mitochondria only through the ovule of the mother.

Mitosis: cellular division that produces somatic cells, which contain the number of chromosomes characteristic of particular species.

Ìm, micrometer or micron: it is the millionth part of 1 meter (equal to 1/1000 of mm)

Molecule: union of two or more atoms of the same element or of different chemical elements (compounds) due to electromagnetic forces (Wan der Waals). Depending on the number of atoms that compose it, it can be very small (like the hydrogen molecule) or very big (like the DNA molecule).

Morphology: internal and external structural form

Mucosa: coating of the viscera and of the body cavities made up of epithelium.

Mucous: viscous fluid secreted by the exocrine glands located inside the mucosa. Its role is to lubricate, protect, and to keep this type of tissue clean.

Necrosis: a set of irreversible alterations of a cell or of a tissue that causes its death.

Neurotransmitter: a substance secreted by a nervous cell, which allows for the transmission of the nervous impulse to another cell (nervous, muscular, glandular).

Cellular nucleus: cellular element inside which we find the DNA, the molecule that contains the genetic set characteristic of a species.

Homeostasis: the process of maintaining the equilibrium of all the environmental conditions inside an organism.

Osmoreceptor: receptor that is able to detect the changes in osmotic pressure registered inside the blood, depending on the amount of water circulating. It causes a sensation of thirst.

Osmosis: the passage of a solvent through a semi permeable membrane separating two solutions that have different concentrations. It follows the concentration gradient, until it zeros it.

Otolith: tiny calcium crystal enclosed inside the internal sacs of the labyrinth. Under the pressure of gravity, it stimulates the ciliate receptors appointed to measure pressure. It is an essential part of the perception of the body posture in a three-dimensional space.

Para hormone: a substance that has some hormonal characteristics (but is not produced by endocrine glands). The histamine is a natural Para hormone; some polluting substances are artificial Para hormones.

Permeability: it is the characteristic of a solid body to be passed through by a fluid.

PH: it expresses the acidity or the basicity of a solution. A neutral solution, at 25° C and under atmospheric pressure (101325 Pa), has a pH value of 7. ▶[231] All the acidic solutions have pH values lower than 7 (and the lower the value, the more acidic the solution is), while basic solutions have values which are higher than 7. (The higher they are the more basic the solution is). We must remember that chemical reactions necessary for life are compatible only with minimal variations of the pH, and in order to neutralize the sudden changes in acidity, there are chemical systems both inside the cells and outside the organism. The pH scale ranges from 1 (very acidic) to 14 (very basic). As for plants, the limits are a little more than 4 and they go from 8,5 to 9.

Pigment: coloring substance.

Neuromuscular or motor plaque, or fuse: part of a

membrane of a muscular cell in contact with a motor nerve ending. The neurotransmitter released by the nerve, determines specific changes in this section of the muscular membrane, which cause the contraction or the relaxation of the fiber.

Placenta: an organ found in the fetus and in the mother, half of which has a fetal origin (coral villi, generated by the outer layer of embryonal cells, the corium, which branches like a felt) while the other half has a maternal origin (the modified endometrium, or decidua). At the end of pregnancy it reaches a diameter of about 25-35 cm and a thickness of about 3 cm. Its weight is about 500g (almost 1/6 of the weight of the fetus) and it covers a total surface of the villosity of about 10-14 m_. The spiral uterine arteries wet it, allowing for the inflow (through a capillary network of about 50 km) into a single cavity, of about 30 l/hr of maternal blood, which remains always separated from the fetal blood. The placenta has various functions: an endocrine one (it produces progesterone, which neutralizes during the entire pregnancy; pituitary oxytocin, along with all the hormones that guarantee a regular pregnancy and the appropriate development of the fetus), a respiratory one (it is the organ where the fetal blood is oxygenated), and a nutritional and excretory one (it is permeable to fat, sugar, protein, some vitamins, mineral salts and water found inside the blood, which the fetus absorbs through the villi. They are also permeable to waste products that go from the fetus go into the maternal blood). It also offers a certain degree of immune protection to the unborn: here, the maternal antibodies protect the baby from many infections, while the majority of the pathogen agents that could infect the mother (bacteria, viruses and protozoa) as well as many poisonous substances, are not able to pass through.

Plasma cell: a cell that produces antibodies. It develops from a lymphocyte, and an antigen inside the lymph nodes, the spleen and the bone marrow, activates it.

Plexus: a complex and branched intertwining of anatomical structures (nerves, blood vessels) that are in close contact with each other by anastomizing.

Pleura: a double membrane that covers the lungs; the external layer adheres to the thoracic cavity, while the internal one adheres to the lungs. The lubricated surfaces of the two membranes allow the lungs to move inside the thoracic cavity, without any friction.

Polyp: soft benign tumor, mostly pedunculated, that usually develops inside the mucosae or inside the cavities of the organism (bladder, intestine, larynx, nasal septum, uterus….) Rarely it develops on the cutis or inside the serosae.

Action potential: process of the nervous impulse, during which we have an inversion of the electrical charge of the neuronal membrane and its immediate reinstatement.

Membranous potential: a difference in the electrical potential between the outside and the inside of the cell. It is due to a different concentration of ions (positive and negative) between the inside and the outside, thanks to the active transportation and facilitated diffusion that take place inside

the plasmatic membrane.

Partial pressure: it is the pressure inside a mixture of gasses, at a constant temperature, exercised by each component; it is equal to the pressure that it would exercise if it were the only one to occupy the volume.

Proenzyme or zymogene: it is an enzymatic precursor. It's a protein molecule that, if modified, transforms into an active enzyme.

Proprioceptor: receptor of internal stimuli (special position of the body, muscular movements, movements of various organs, pain…)

Protein: from the Greek protos = first, it a very complex organic substance, extremely important from a biological stand point. A protein is a chain of amino acids with a variety of functions both inside the cell (structural, enzymatic, functional, contractile…) and in transmitting intercellular stimuli (for example…hormones). The construction of these biological polymers is genetically determined.

Prosthesis: from the Greek pro thesis = put in front. This term can indicate either the substitution of an organ, of a part of an organ, of a body segment with artificial instruments, but also the structures utilized for this specific purpose. In a more general sense, we can say that prosthesis is a particular tool or device utilized to improve the functionality of a part of the body, without having to completely substitute it.

Pseudopod: cytoplasmatic extroflexion of a cell (needed for phagocitosis or for movement).

Puberty: the developmental stage in which sexual maturity is reached and procreation is now possible. It is an anatomical and physiological stage.

Radiation: a type of propagation of electromagnetic energy. It is the rapid propagation (at the speed of light) of variations of the electric and of the magnetic field. It is characterized by wavelength and by frequency.

Radioisotope: radioactive isotope. In the clinical analysis, it indicates a chemical radioactive element, whose characteristics are similar to those of other elements which usually enter inside those metabolic processes we want to follow or monitor.

X rays: a radiation whose wavelength falls between 10^{-} and 10^{-9}.

Receptor: an organ or a cell that is able to perceive a stimulus arriving from the outside or from the inside of the body.

Amniotic sac: embryonal annex full of amniotic fluid, which protects the fetus and the embryo during its development.

Vitelline sac: embryonal annex rich in nutrients. It supports the embryo during the first few weeks of its development.

Sarcomere: contractile unit that divides the muscle. It is made of two types of proteins: the actin and the myosin.

Sciaman: magician or priest able to contact the dead, to cure and heal in a miraculous way.

Serosa, serous membrane: a double-layered smooth and shiny membrane that coats some cavities of the organism: the pleura, the peritoneum, and the pericardium.

Synapses: point of connection between a nerve ending and another excitable cell. They divide into a chemical synapse, which allows the action potential to pass from one cell to the next, thanks

to the mediation of some substances and of particular processes of the cellular membranes, and into a electrical synapse (the connecting point between nerve endings) allowing the action potential to go from one to the next, without the mediation of neurotransmitters.

Excitation threshold: minimal stimulus able to evoke a nervous response; in other words, able to trigger an action potential.

Somatic: of the body, bodily

Hydrophilic substance: substance that has an affinity with water.

Intercellular substance: a fluid that occupies the inner cell, connecting all of its elements (nucleus, mitochondria, ribosome, etc.) Generally speaking, it indicates a cytoplasm.

Sperm: seminal fluid, that contains the secretion processes of the testicle, of the deferent duct, of the seminal vesicles, of the prostate, of the Cowper's glands and of other minor glands of the male reproductive system.

Staminal: a cell from which an entire line of different cellular groups derives.

Stroma: connective network of a tissue, made up of fibers and cells.

Surfactant: a substance that, once dissolved inside a fluid, reduces its superficial tension. The surfactants found inside the lungs can also prevent the internal surfaces of the alveoli they have come in contact with, from sealing together.

Tendon: the ending part of a muscle made of fibrous, non-elastic connective tissue; it connects a muscle to a bone, to another muscle or to the dermis.

Nerve ending: distal part of a nervous dendrite. It can act as a sensitive receptor, as an element of contact and communication between nervous fibers, and also as an element of transmission of the nerve impulses.

Active transport: the movement of substances through the plasmatic membrane of a live cell, that often occurs in the opposite direction to that of a concentration gradient. This process requires a lot of energy and also the mediation of some specific proteins.

Ultraviolet (ultraviolet light): a set of electromagnetic radiations, whose length ranges between $400 \cdot 10^{-9}$ m.

Volemia: the total amount of circulating blood fluid.

Xenograft: the transplant of an animal organ (and not human) inside a human being.

Zygote: totipotent cell that derives from the fusion of the female and male gametes, from which the embryo originates.

ANALYTICAL INDEX

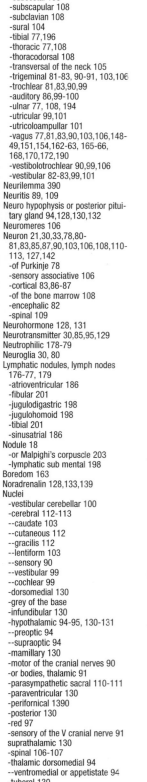